AF572349

WASTE DISPOSAL EFFECTS

ON

GROUND WATER

A Comprehensive Survey of the Occurrence and Control of Ground-Water Contamination Resulting from Waste Disposal Practices

Edited by

David W. Miller
Geraghty & Miller, Inc.
Syosset, New York

PREMIER PRESS

Berkeley, California

This book is a photographic reproduction of *The Report to Congress, Waste Disposal Practices and Their Effects on Ground Water* issued in January 1977 by the Office of Water Supply and the Office of Solid Waste Management Programs, U.S. Environmental Protection Agency.

Publication of the report in book form has been undertaken to make this valuable material available in a conventional format for general and reference use.

Library of Congress Catalog Card Number: 78-65680
ISBN: 0-912722-01-0

Printed in the United States of America

CONTENTS

CONTENTS (Continued)

SECTION I

FINDINGS

Ground water is a high quality, low cost, readily available source of drinking water.

- Half of the population of the United States is served by ground water.
- In many areas, ground water is the only high quality, economic source available.
- The use of ground water is increasing at a rate of 25 percent per decade.

Waste disposal practices have affected the safety and availability of ground water, but the overall usefulness has not been diminished on a national basis.

- Current data indicate that there are at least 17 million waste disposal facilities emplacing over 1,700 billion gal. (6.5 billion cu m) of contaminated liquid into the ground each year. Of these, 16.6 million are domestic septic tanks emplacing about 800 billion gal. (3 billion cu m) of effluent.
- Ground water has been contaminated on a local basis in all parts of the nation and on a regional basis in some heavily populated and industrialized areas, precluding the development of water wells. Serious local economic problems have occurred because of the loss of ground-water supplies.
- Degree of contamination ranges from a slight degradation of natural quality to the presence of toxic concentrations of such substances as heavy metals, organic compounds, and radioactive materials.
- More waste, some of which may be hazardous to health, will be going to the land because of increased regulation against, and the rising costs of, disposal of potential contaminants to the air, ocean, rivers, and lakes.
- Removing the source of contamination does not clean up the aquifer once contaminated. The contamination of an aquifer can rule out its usefulness as a drinking water source for decades and possibly centuries.

Almost every known instance of ground-water contamination has been discovered only after a drinking-water source has been affected.

- Few state or local agencies systematically collect data on contamination incidents, water supply wells affected, and drinking-water supplies condemned as unsafe.

- Effective monitoring of potential sources of ground-water contamination is almost non-existent.
- Typical water-well monitoring programs traditionally have not been directed toward protecting public health because water analyses normally do not include complete coverage of such significant parameters as heavy metals, organic chemicals, and viruses.
- There are potentially millions of sources of contamination and isolated bodies of ground-water contamination nationwide.
- While detailed national inventories of all potential sources of ground-water contamination have not been carried out, EPA and some states have begun some inventories and assessments of some waste disposal sources.

Waste disposal practices of principal concern are those related to industrial and urban activities.

- For every waste-disposal facility documented as a source of contamination, there may be thousands more sited, designed, and operated in a similar manner.
- The opportunity for severe contamination of ground water is greatest from industrial waste-water impoundments and sites for land disposal of solid wastes.
- Septic tanks and cesspools discharge large volumes of effluent directly to the subsurface. In many cases, the degree of treatment is not adequate to protect ground-water supplies.
- Contamination resulting from the collection, treatment, and disposal of municipal waste water exists but the magnitude is unknown.
- Because there is a known potential for contamination from the land spreading of industrial and municipal sludges, there is concern about the expected increase in sludge generation over the next decade.
- There have been far fewer reports of contamination of potable ground-water supplies by the several hundred industrial and municipal wells injecting into saline aquifers than from thousands of shallow wells used to dispose of sewage, runoff, and irrigation return flow to aquifers containing potable water.

Other waste-disposal practices, whose distribution is dependent upon geology, climate, and topography, have also contaminated ground water.

- Contamination from oil and gas field activities is caused primarily by improperly plugged and abandoned wells and, to a lesser degree, poorly designed and constructed oper-

ating production and disposal wells.
- Although specific case histories of ground-water contamination related to the disposal of mine wastes do exist, adequate documentation of the problem is unavailable.
- Ground-water contamination from the disposal of animal feedlot wastes is a relatively new environmental problem, and few cases of ground-water contamination have been reported.

Existing technology cannot guarantee that soil attenuation alone will be sufficient to prevent ground-water contamination from a waste disposal source.

- Proper site selection as well as proper operation and maintenance of facilities, is the principal technique available for minimizing ground-water contamination problems.
- Such technology as advanced treatment and physical containment play a major preventive role where economics dictate that sites be located in areas of critical ground-water use.
- Land disposal of wastes is not environmentally feasible in many areas and such alternatives as waste transport, resource recovery, ocean disposal, and surface-water or air discharge should be investigated and may be more environmentally acceptable.
- Federal demonstration grants and technical assistance are provided to assist the development of new technology and facilitate the application of existing technology.

Existing Federal and state programs address many of the sources of potential contamination, but they do not provide comprehensive protection of ground water.

- Existing Federal programs administered by EPA which address ground water are (1) the Federal Water Pollution Control Act Amendments of 1972; (2) the Safe Drinking Water Act of 1974; and to a lesser degree (3) the Solid Waste Disposal Act of 1965; and (4) the National Environmental Policy Act of 1969.
- The FWPCAA provide for a statewide and areawide waste treatment management planning function which may include identifying and controlling pollution from mine runoff, the disposal of residual waste, and the disposal of pollutants on land or in subsurface excavations.
- FWPCAA also include (1) a program to issue permits for point sources of water pollution, including some wells; (2) best practicable treatment standards for municipal sewage effluent disposal which must address ground-water protection; (3) guidelines for land spreading of munici-

pal sludges; and (4) municipal waste treatment facilities planning for areas where septic systems pose potential adverse ground-water impacts.

- FWPCAA do not address the discharge of contaminants to ground water from surface impoundments, land disposal of solid wastes, septic systems, or most wells.
- The SDWA provides for a Federal/state cooperative effort to prevent endangerment of underground drinking water sources from industrial and municipal waste disposal wells, oil-field brine disposal wells and secondary recovery wells, and engineering wells. At present, surface impoundments are not included in this program, but some types of impoundments may be included at a later time.
- SDWA also provides that EPA may review any commitment of Federal financial assistance in an area designated as having a sole source aquifer.
- SDWA cannot be used to regulate land disposal of solid wastes, land application of sludges and effluents, or septic systems except under the emergency powers provisions of the Act.
- The Solid Waste Disposal Act contains no specific reference to ground water, however, guidelines developed under the Act provide for ground-water protection from pollution activities and surface drainage. There are also site development guidelines which consider the impact on ground water. These guidelines are only mandatory for Federal agencies.
- The NEPA requires Federal agencies to prepare environmental impact statements on major actions. Ground-water protection is a significant need for writing an EIS.
- While site selection is an important parameter in preventing ground-water contamination, there are no direct Federal controls in this area. States are encouraged to develop site selection programs within the context of their land-use planning and control authorities.
- Most state laws give broad authority to protect all waters of the state, including ground water. Such language, plus deficiencies in budget and staffing, force state and local agencies to act on cases of contamination only after the fact.
- States are beginning to develop programs which encourage prevention of contamination from some waste disposal sources.
- Because clean-up of contaminated ground water is rarely economically or technically feasible, action by the states has been directed toward condemning the affected water supply.
- Legal action is seldom taken against a specific source of contamination because individuals, private organizations, and public agencies seldom have the resources required to

prove a specific source as the source of contamination.

A national strategy of ground-water protection will require a better understanding of the environmental, legal, technical, and economic complexities of dealing with the resource.

- Better coordination of existing regulatory programs and a better understanding of the impact of all regulatory actions on ground water is necessary. Regulatory programs need to reflect the close relationship between land, ground water and surface water.
- Inventories of ground-water contamination cases have shown that other contaminant sources including spills, salt-water intrusion, and highway deicing, have a significant impact on ground water. Many of these sources are not included within the scope of Federal/state ground-water protection programs, but may be addressed on a case-by-case basis.
- The most effective means for protecting ground water is to control and monitor the potential source of contamination and not the aquifer or point of withdrawal.
- New potential sources of contamination should be evaluated on a case-by-case basis.
- Existing potential contamination sources should be reviewed in order to develop control strategies that are instituted in accordance with local priorities.
- Increasing Federal regulation of surface-water and air discharge and ocean disposal may result in land disposal practices (particularly of sludge) which could contaminate ground water.
- At the present time, there does not exist a comprehensive Federal program for sludge management. However, EPA is developing a comprehensive program to address this issue.

SECTION II

INTRODUCTION

On December 14, 1974, the Safe Drinking Water Act became law (PL 93-523). Under Sec. 1442(a)(4) of the Act, the Administrator of the U. S. Environmental Protection Agency (EPA) was directed to conduct a survey of "(A) disposal of waste (including residential waste) which may endanger underground water which supplies, or can reasonably be expected to supply, any public water systems, and (B) means of control of such waste disposal." This report describes the results of the investigation.

Waste disposal practices discussed in this report include only those activities which result in the actual collection and disposal of liquid, semi-solid, and solid wastes. Such materials include: (1) industrial waste water that is contained in surface impoundments (lagoons, ponds, pits, and basins); (2) municipal and industrial solid refuse and sludge that are disposed of on land; (3) sewage wastes from homes and industries that are discharged to septic tanks and cesspools; (4) municipal sewage and storm-water runoff that are collected, treated, and discharged to the land; (5) municipal and industrial sludge that is land spread; (6) brine from petroleum exploration and development that is injected into the ground or stored in evaporation pits; (7) solid and liquid wastes from mining operations that are disposed of in tailing piles, lagoons, or discharged to land; (8) domestic, industrial, agricultural, and municipal waste water that is disposed of in wells; and (9) animal feedlot waste that is disposed of on land and in lagoons. The sources of potential contaminants and their various routes to the ground-water system are shown on Figure 1. Table 1 lists the waste disposal practices discussed in this report and their relative impact on the ground-water environment.

The first few sections of the report describe the use and occurrence of the ground-water resource along with the mechanisms of contamination. These are followed by a discussion of each of the major waste disposal practices. All of these latter sections are uniformly organized with (1) an explanation of the practice, (2) a listing of the characteristic potential contaminants, (3) an estimation of the extent of the ground-water contamination problem on a national basis, and (4) a review of the present prevention technology. The final portion of each section explores some of the typical institutional controls presently available to state agencies to prevent or correct ground-water contamination problems.

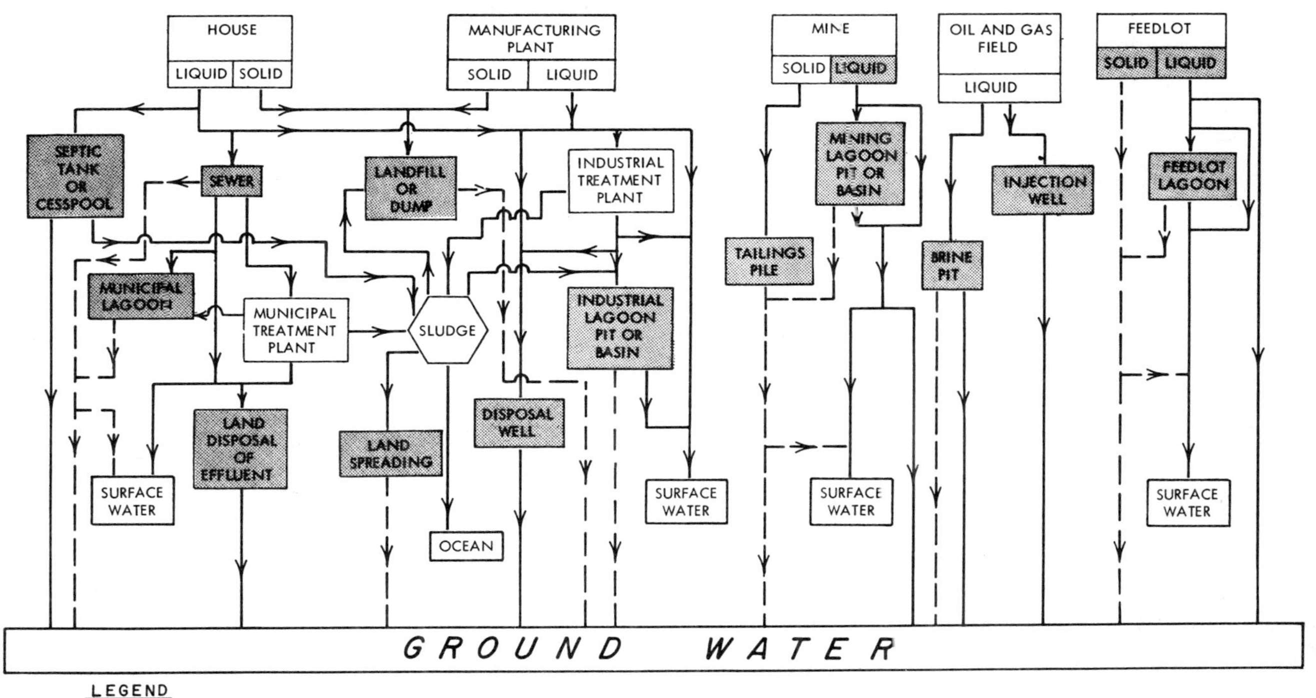

Figure 1. Waste disposal practices and the routes of contaminants from solid and liquid wastes.

Table 1. WASTE DISPOSAL PRACTICES AND THEIR RELATIVE IMPACT.

Waste disposal practice	Frequency reported	Principal contaminants reported	Typical hazard to health	Typical size of affected area	Relative national importance	Future national importance	Principal EPA regions affected [b)]
Industrial impoundments	I	Hm, P, A	I	y	I	I	II-V, VII, IX, X
Land disposal of solid waste		C, I, Hm, P					
Municipal	I		I	y	I	I	I-V, X
Industrial	III		I	y	I	I	– c)
Septic tanks and cesspools		B, N, C					
Domestic	I		II	x	II	II	I-V
Industrial	III		I	y	II	II	– c)
Municipal waste water		N, C, Hm					
Sewer systems	III		II	x	II	II	I-V, IX
Treatment lagoons	III		II	y	II	II	IV-VII, IX, X
Land spreading of sludge		N, Hm, P					
Municipal	III		II	y	III	I	I-V
Industrial	III		II	y	III	I	– c)
Petroleum exploration and development		S, C, O					
Wells	II		III	x	II	II	III-IX
Pits	I		III	x	II	III	III-IX
Mine waste		Sa, I, Hm [a)]					
Coal	III		III	y	II	I	III-V, VII, VIII
Other	III		II	y	II	II	IV-VI, VIII-X
Disposal and injection wells		Hm, A, T, M, N					
Agricultural, urban runoff, cooling water, and sewage	III		II	y	II	III	II, IV, IX, X
Industrial injection	III		I	y	III	III	III-IX
Animal feeding operations		N					
Cattle	III		III	z	III	III	VI-IX
Other	III		III	z	III	III	III-VIII

Table 1 (continued). WASTE DISPOSAL PRACTICES AND THEIR RELATIVE IMPACT.

EXPLANATION

I - high	A - Acids	N - Nitrate	x - small area, but can be regional due to high density of individual sources	a) dependent upon local mineralogy
II - moderate	B - Bacteria	O - Oil	y - can affect adjacent properties	b) for EPA regions, see figure attached
III - low	C - Chloride	P - Phenols	z - contained on one property	c) insufficient data for regional evaluation
	Hm - Heavy metals	S - Sodium		
	I - Iron	Sa - Sulfuric acid		
	M - MBAS	T - Temperature		

Frequency reported - based on the relative number of case histories described in published and unpublished sources.

Principal contaminants reported - based only on normally analyzed constituents. Unfortunately, for economic reasons, incomplete analyses are performed in most ground-water contamination cases. Therefore, the principal contaminants reported do not often include such hazardous substances as heavy metals, organic chemicals, or radioactive elements that may actually be present.

Typical hazard to health - based primarily on the nature of the contaminants with secondary consideration given to the typical volumes expected.

Relative national importance - based on the typical health hazard of the contaminants, the typical size of the area affected, and the distribution of the waste disposal practice across the U.S. A waste disposal practice may be a serious problem in certain areas; but if the number of such areas is relatively small, then the practice would not be given a high national rating. A very widespread practice which does not create serious problems even where sources of contamination are concentrated, would also be given a low rating with regard to national importance.

Future national importance - based on predicting what the relative national importance of each waste disposal practice will be in approximately ten years. The ratings take into account past and present technological trends in the treatment and disposal of wastes and evaluations of the impact of new and proposed regulations.

Principal regions affected - regions listed are the Environmental Protection Agency's Regions I through X as shown on the attached figure. Designations are based on present relative impact, taking into account the degree to which the waste disposal practice is carried out in the various regions and factors of geology and hydrology, where pertinent.

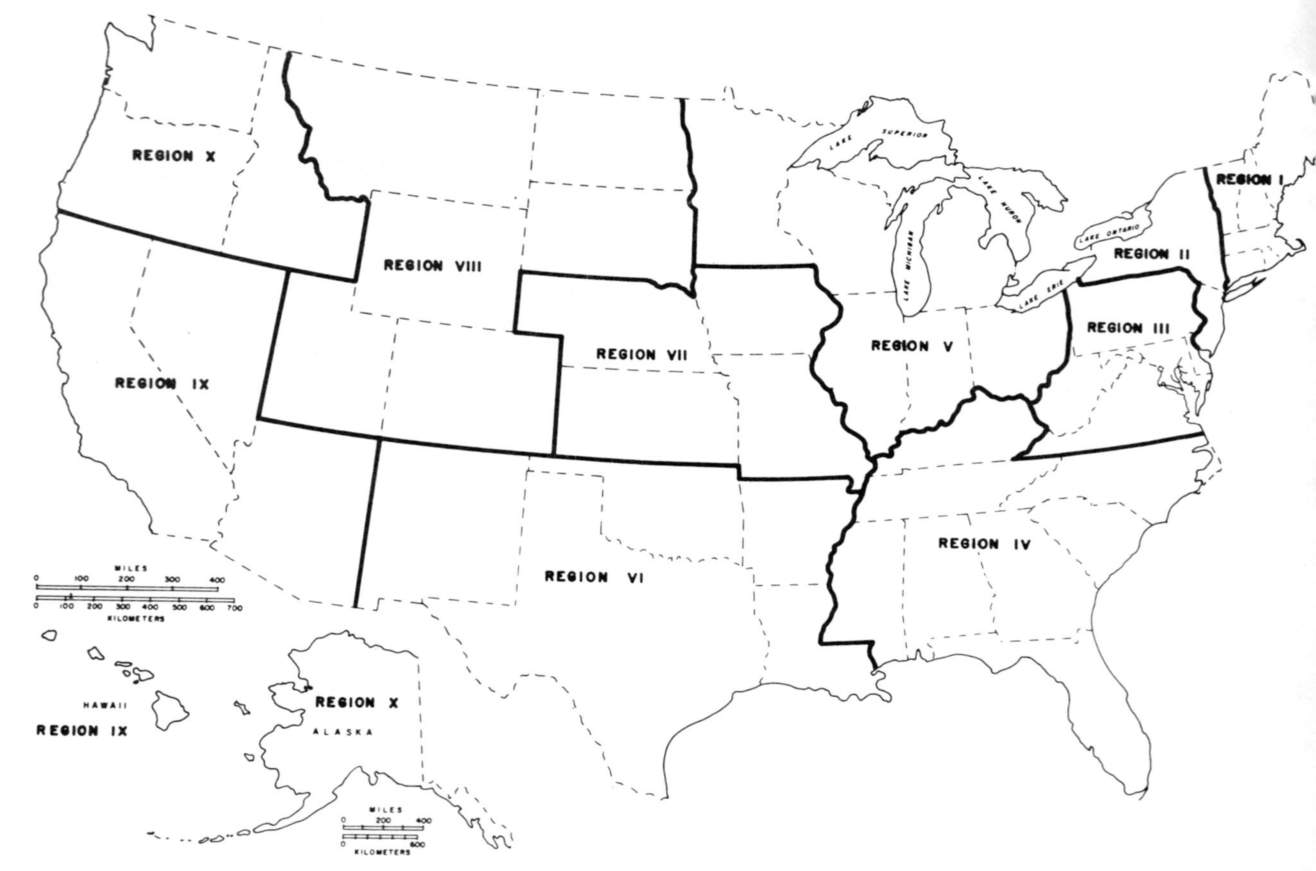
REGION X
REGION VIII
REGION VII
REGION V
REGION I
REGION II
REGION III
LAKE SUPERIOR
LAKE MICHIGAN
LAKE HURON
LAKE ONTARIO
LAKE ERIE
REGION IX
REGION VI
REGION IV
MILES
0 100 200 300 400
0 100 200 300 400 500 600 700
KILOMETERS
HAWAII
REGION IX
REGION X
ALASKA
MILES
0 200 400
0 600
KILOMETERS

In the next section of the report, there is a discussion of the importance of non-waste disposal practices as they affect ground-water quality. The final two sections of the report define the present status of Federal legislation that applies to ground-water quality protection and the various regulatory alternatives and strategies available to state and local agencies. Estimating the economic impact of technological or institutional controls was not one of the objectives of this survey.

The report is based on an evaluation and analysis of available data, a major portion of which has not been published. About 40 technicians in the ground-water and pollution-control fields contributed directly to this effort. Many more were contacted and provided the researchers with essential information. In addition, a working group consisting of representatives of various offices of EPA, plus personnel of state environmental agencies, periodically reviewed the report and contributed significantly to its content. This publication represents a compilation of key data taken from a larger draft Report to Congress on file with EPA.

Ground-water contamination is the degradation of the natural quality of ground water as a result of man's activities. The term "contaminant" is defined in the Safe Drinking Water Act as "any physical, chemical, biological or radiological substance or matter in water." In this report, only those contaminants which result from waste disposal activities are considered in detail.

In order to appreciate the magnitude and severity of ground-water contamination, the hydrologic system itself, mechanisms of ground-water contamination, and environmental hazards must be understood. Figure 2 illustrates these concepts.

The contamination process begins with sources of contaminants; the waste disposal practices. The type of contaminant, of course, depends on the source and can range from hazardous organic chemicals in landfill leachates to high concentrations of salt in oil-field brines. Either deliberately (septic tanks) or unintentionally (industrial wastewater impoundments), contaminants can leak, percolate, be discharged to, or injected into water-supply aquifers.

As the contaminant travels through the soil and into the ground-water system, it can be modified by various attenuation processes. These processes are very complex and not all are completely effective. In fact, once in an aquifer, certain toxic substances, such as some heavy metals, are

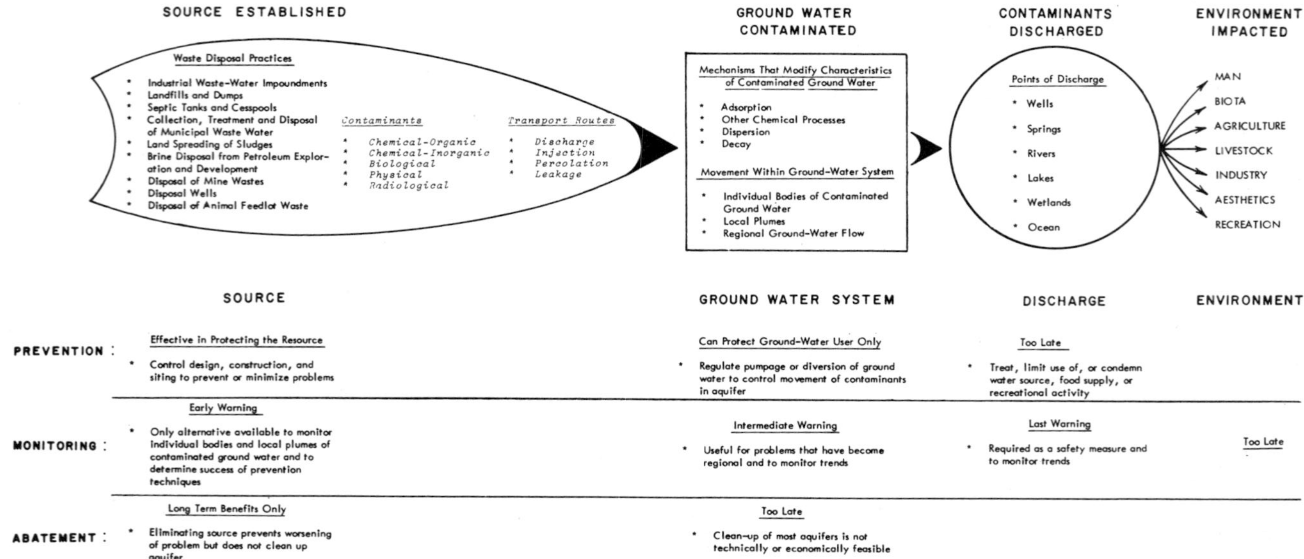

Figure 2. The hydrologic system controlling ground-water contamination and its constraints on methodologies for prevention, monitoring, and abatement.

highly mobile. Attenuation in an aquifer is extremely slow as is the movement of ground water (typically less than 2 ft/day or 0.6 m/day). Therefore, contaminants within the ground-water system do not mix readily with native water and move as: (1) individual bodies or slugs (e.g., caused by intermittent filling of and seepage from waste-water impoundments); (2) local plumes (e.g., caused by continual flow of leachate from beneath a landfill toward a pumping well); and (3) masses of degraded water (e.g., caused by a large number of septic tanks discharging nitrate-enriched water which travels with the regional ground-water flow pattern).

Although ground water travels through an aquifer slowly, it is in constant motion and must eventually discharge to the surface because all aquifer systems are being recharged to some degree. In humid areas, discharge of contaminants is relatively quick for shallow water-table aquifers and slow for deep artesian aquifers. In arid regions, recharge and discharge are so slow that some aquifers can actually be considered sinks similar to the ocean. Points of discharge include wells and springs used for water supply, and surface-water bodies such as rivers and lakes. In fact, the base flow of most streams is supported by ground-water discharge, and the quality of the surface water during low flow periods is dependent upon ground-water quality. The usefulness to man and his environment of both surface water and ground water is severely limited if ground-water quality is degraded.

The way the ground-water system works controls the methodologies available to prevent, monitor, and abate instances of contamination. Prevention must be directed toward the source, where proper design, construction and siting can help protect the resource or at least minimize problems. If the aquifer becomes contaminated, then the resource has already been degraded, and efforts must be shifted toward preventing the ground-water user from being damaged. Controls involve such actions as regulating pumpage patterns in order to contain or isolate the contaminant. When the contaminant reaches the point of discharge, it is too late except for such expensive alternatives as treatment or condemnation of a water supply.

Again, in monitoring, the most effective place to devote the greatest effort is at the source, where observation of water quality degradation allows enough time for minimizing the problem and for establishing a warning procedure. After contamination has affected enough of the aquifer, monitoring no longer becomes a protective measure but simply informs the regulator or the user of long-term changes. Also, random placement of monitoring wells on a regional basis can provide misleading information, because important plumes and individual bodies of contaminated ground water are overlooked. Monitoring of discharge points serves as a safety precaution and helps define trends. For this reason, it cannot be eliminated from the monitoring program.

The principal abatement procedure for surface-water problems is to eliminate or correct the source of contamination. Because streams are subject to the cleansing action of turbulent flow and the purifying effects of air, light, and biological organisms, they can recover quickly. The opposite is true for ground water. Removal of the source prevents the problem from becoming worse but does not lead to a cleansing of the aquifer. In addition, clean-up procedures such as removal of the contaminant by means of pumping wells followed by treatment of the water is almost never economically or technically feasible. For example, pumping may require the use of an inordinate number of wells and a complex collection and treatment system, which is only temporary and difficult to support with either private or public funds. Although containment of contaminants within a selected portion of an aquifer has been achieved to various degrees in certain instances, complete removal is rarely attempted and has not been successful.

SECTION III

IMPORTANCE OF THE GROUND-WATER RESOURCE

SUMMARY

At least one half of the population of the United States depends upon ground water as a source of drinking water. Of the total population, 29 percent use ground water delivered by community systems and another 19 percent have their own domestic wells. In addition, millions of Americans drink ground water from wells serving industrial plants, office buildings, restaurants, gas stations, recreational areas, and schools. Practically none of the domestic wells in the nation are subject to routine or even initial evaluation of water quality. Few of the several hundred thousand small water systems supplying industrial establishments, schools, etc., are monitored.

INTRODUCTION

There are many reasons why ground water has become a major source of drinking water in the United States. It is more widely available and accessible than surface water. A domestic or farm well can be successfully constructed almost anywhere; over one third of the nation is underlain by ground-water reservoirs generally capable of yielding at least 75,000 gpd (285 cu m/day) to an individual well; and there are large areas where hundreds or even thousands of gpm can be obtained from wells or springs. Ground water is a relatively reliable source, not subject to the rapid and sometimes great fluctuations in availability characteristic of a surface-water supply. Over much of the United States, the present utilization of ground water is small compared to the total supply potentially available.

Ground water is generally more mineralized than surface water, but its quality is more uniform at a specific locality from year to year. Likewise, the temperature of ground water is usually constant throughout the year. Because of the efficient filtering capacity of the unsaturated zone, ground water is normally free of most organisms and suspended solids and requires little or no treatment. As compared to surface water, the evaporation losses of ground water are minimal except in areas where it occurs at shallow depth.

The development of ground-water reserves commonly has little negative impact on the surface environment, and the costs of such development are, except for the required hydrogeologic investigations, relatively low.

USE OF GROUND WATER FOR DRINKING PURPOSES (1970)

The actual importance of the ground-water resource may best be illustrated by an examination of water-use data. Drinking water is supplied by three different types of systems:

1. Public or municipal water wells
2. Individual domestic wells
3. Self-supplied industrial or commercial wells

The specific data listed below in this section have been compiled from published and unpublished records of the U. S. Geological Survey, the U. S. Water Resources Council, and the U. S. Bureau of the Census.

Dependence on Ground Water as a Source of Drinking Water

Almost one half of the United States population (48 percent) depends on ground water for drinking (Figure 3). Of the total population, 29 percent obtains ground water through public supplies and 19 percent through individual domestic wells. The variation among states, as illustrated in Table 2, ranges from a 92-percent dependence in New Mexico to a 30-percent dependence in Maryland and Pennsylvania, still considerable.

The rural population dependent upon ground water is much higher (94 percent) than the population served by public supplies (37 percent). This is because it is almost always simpler and cheaper to install a domestic well than to pipe and treat water from the nearest surface-water body. Also, unless the surface source were a large lake, river, or reservoir, the supply would probably not be as reliable. The relationship of ground-water to surface-water use is shown in a somewhat different manner on Figure 4. The shaded states are those in which over one half of the total population relies on ground water as a source of drinking water. As can be seen, this is the case in nearly two thirds of the states.

Withdrawal Rates of Ground Water for Domestic Use

The total withdrawal of ground water for domestic purposes in 1970 was 9.4 bgd (35.6 million cu m/day). [2] As shown on Figure 5, of all water withdrawn for drinking water purposes, 45 percent was ground water. The per capita ground-water usage varies from state to state and from the urban to rural environment (Table 3). Per capita withdrawal from public supplies is considerably greater -- 114 gpd (431 litres/day)

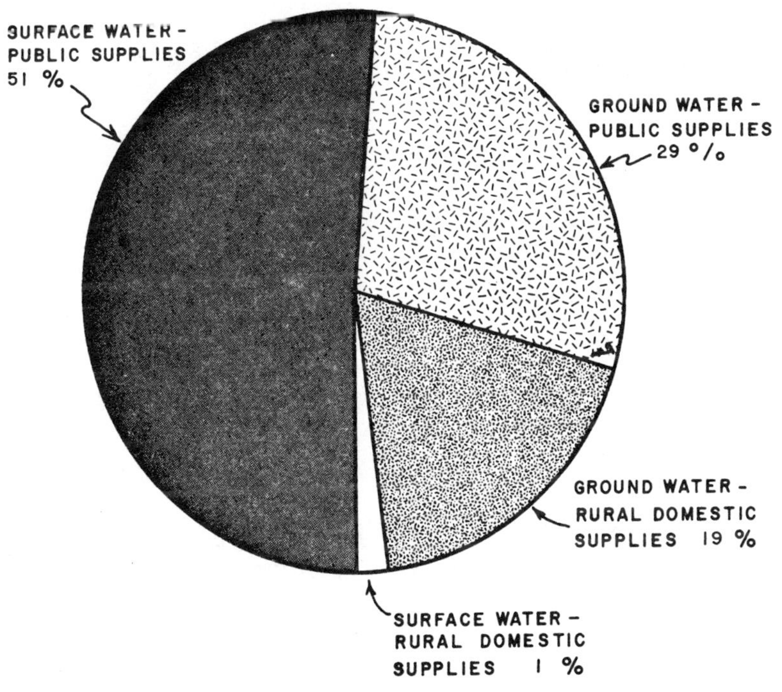

Figure 3. Population served by source and supply, 1970.[1)]

Table 2. DEPENDENCE OF THE UNITED STATES POPULATION ON GROUND WATER AS A SOURCE OF DRINKING WATER. 1)

State	Total Population (Thousands)	% Of Total Population Relying On Ground Water	Population Served By Ground Water From Public Supplies (Thousands)	Population Served By Ground Water From Rural Supplies (Thousands)	% Of Public Supply Population Relying On Ground Water	% Of Rural Population Relying On Ground Water
Alabama	3,444	59	884	1,139	38	100
Alaska	302	63	62	130	49	74
Arizona	1,772	71	989	274	66	100
Arkansas	1,923	67	605	681	49	100
California	19,953	46	8,000	1,164	43	93
Colorado	2,207	23	306	197	16	87
Connecticut	3,032	37	590	531	24	98
Delaware	548	65	217	138	53	100
Florida	6,789	91	4,819	1,379	89	100
Georgia	4,590	70	825	2,376	38	98
Hawaii	770	87	662	11	95	14
Idaho	713	88	407	218	87	90
Illinois	11,114	38	3,880	358	36	82
Indiana	5,194	58	1,497	1,494	43	87
Iowa	2,825	82	1,524	794	75	100
Kansas	2,249	62	817	565	50	93
Kentucky	3,219	39	304	936	14	88
Louisiana	3,643	62	1,291	927	48	100
Maine	994	37	153	219	20	91
Maryland	3,922	30	368	803	12	100
Massachusetts	5,689	31	1,464	286	27	100
Michigan	8,875	38	1,365	2,006	20	100
Minnesota	3,805	67	1,498	1,057	55	100
Mississippi	2,217	90	1,177	827	85	100
Missouri	4,677	31	942	529	24	74
Montana	694	47	152	174	30	93
Nebraska	1,484	86	908	365	81	100
Nevada	489	64	265	46	60	97
New Hampshire	738	61	261	191	48	98
New Jersey	7,168	53	3,032	746	47	100
New Mexico	1,016	92	645	293	91	96
New York	18,191	32	4,152	1,734	25	100
North Carolina	5,082	60	660	2,374	25	99

Table 2 (Continued). DEPENDENCE OF THE UNITED STATES POPULATION ON GROUND WATER AS A SOURCE OF DRINKING WATER. 1)

State	Total Population (Thousands)	% Of Total Population Relying On Ground Water	Population Served By Ground Water From Public Supplies (Thousands)	Population Served By Ground Water From Rural Supplies (Thousands)	% Of Public Supply Population Relying On Ground Water	% Of Rural Population Relying On Ground Water
North Dakota	618	66	189	222	48	99
Ohio	10,652	40	2,475	1,754	29	80
Oklahoma	2,559	40	551	475	28	85
Oregon	2,091	56	355	821	30	92
Pennsylvania	11,794	30	1,351	2,144	14	100
Rhode Island	950	33	213	103	25	100
South Carolina	2,591	61	221	1,359	18	100
South Dakota	666	79	284	240	69	94
Tennessee	3,924	51	1,194	805	38	100
Texas	11,197	58	4,584	1,949	50	100
Utah	1,059	58	502	113	53	99
Vermont	445	56	100	150	35	96
Virginia	4,648	34	477	1,109	14	98
Washington	3,409	44	1,072	433	38	78
West Virginia	1,744	53	380	551	32	97
Wisconsin	4,418	64	1,536	1,311	49	100
Wyoming	332	61	130	73	52	89
District of Columbia	757	0	0	0	0	0
Puerto Rico	2,712	26	396	302	17	80
United States Total	205,897	48	60,600	38,568	37	94%

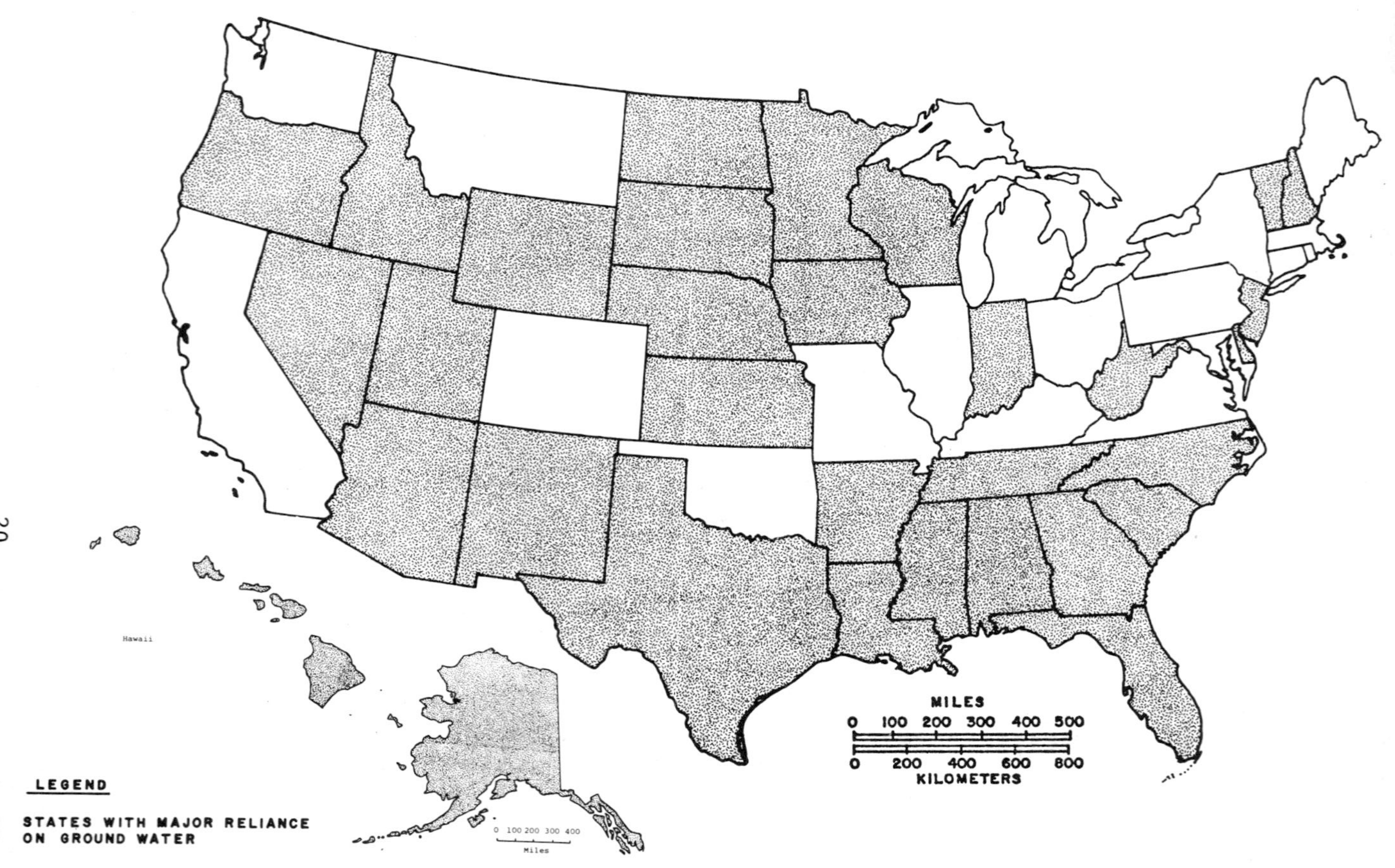
Hawaii
MILES
0 100 200 300 400 500
0 200 400 600 800
KILOMETERS
0 100 200 300 400
Miles
LEGEND
STATES WITH MAJOR RELIANCE
ON GROUND WATER

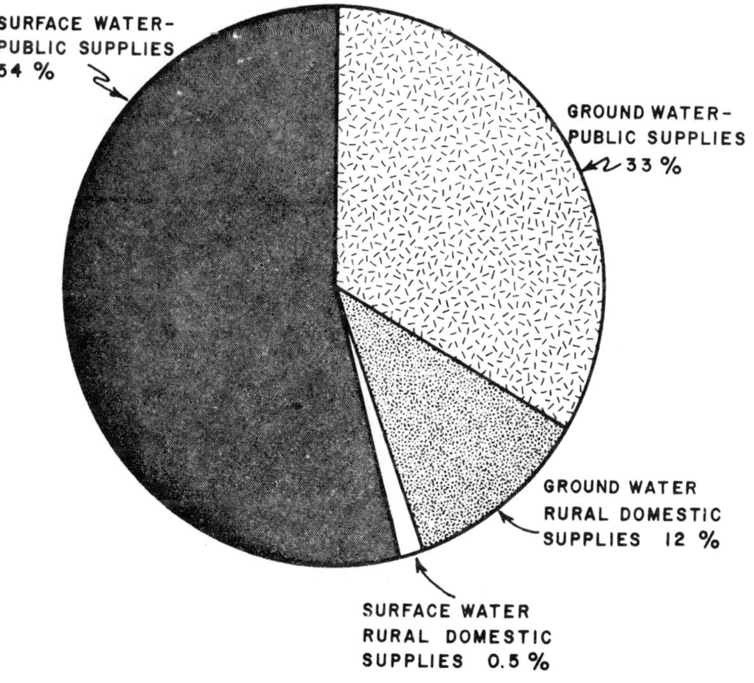

Figure 5. Water withdrawn for drinking water by source and supply, 1970. [1)]

Table 3. WITHDRAWAL RATES OF GROUND WATER FOR DOMESTIC USE (1970). [1)]

State	Total Water Withdrawn From All Sources For Domestic Use (MGD)	Ground Water Withdrawn By Public Supply Systems For Domestic Use (MGD)	Per Capita Ground-Water Withdrawals From Public Supplies For Domestic Use (GPD)	Ground Water Withdrawn From Rural Supplies For Domestic Use (MGD)	Per Capita Ground-Water Withdrawals From Rural Supplies For Domestic Use (GPD)
Alabama	268.5	78.9	89.2	62.9	55.2
Alaska	58.8	21.5	347.9	4.5	34.7
Arizona	286.5	174.7	176.6	21.9	79.9
Arkansas	151.3	49.9	82.4	48.9	71.8
California	2,882.8	1,179.9	147.5	116.3	99.9
Colorado	310.0	46.3	151.2	9.3	47.2
Connecticut	226.2	44.3	75.1	38.4	72.3
Delaware	54.8	23.2	106.8	11.0	79.7
Florida	882.3	638.8	132.6	165.0	119.7
Georgia	309.3	89.9	108.9	71.6	30.1
Hawaii	116.6	105.2	158.9	0.9	82.9
Idaho	130.0	91.4	224.6	21.9	100.5
Illinois	1,168.6	581.9	150.0	14.2	39.6
Indiana	435.1	149.7	100.0	75.7	50.7
Iowa	236.0	141.7	93.0	47.3	59.5
Kansas	245.6	96.6	118.2	48.0	84.9
Kentucky	156.4	14.3	97.1	48.4	51.7
Louisiana	392.8	119.7	92.7	66.9	68.4
Maine	83.2	14.5	95.1	10.6	48.3
Maryland	384.6	39.9	108.5	46.1	57.4
Massachusetts	465.7	118.5	80.9	28.4	99.3
Michigan	671.7	102.3	75.0	156.5	78.0
Minnesota	314.4	112.9	75.4	107.1	101.4
Mississippi	147.0	103.6	88.1	24.6	29.7
Missouri	290.1	59.5	63.2	29.3	55.4
Montana	88.8	23.9	156.5	8.8	50.7
Nebraska	163.8	115.3	127.0	21.7	59.5
Nevada	101.2	57.2	215.6	5.8	126.3
New Hampshire	59.5	23.1	88.4	11.2	58.8
New Jersey	860.2	294.4	97.1	79.5	106.6
New Mexico	121.2	94.4	146.4	16.3	55.6
New York	1,975.0	326.1	78.5	116.0	66.9
North Carolina	449.2	59.4	90.0	111.7	47.0

Table 3 (Continued). WITHDRAWAL RATES OF GROUND WATER FOR DOMESTIC USE (1970).[1]

State	Total Water Withdrawn From All Sources For Domestic Use (MGD)	Ground Water Withdrawn By Public Supply Systems For Domestic Use (MGD)	Per Capita Ground-Water Withdrawals From Public Supplies For Domestic Use (GPD)	Ground Water Withdrawn From Rural Supplies For Domestic Use (MGD)	Per Capita Ground-Water Withdrawals From Rural Supplies For Domestic Use (GPD)
North Dakota	64.8	23.0	121.8	16.6	74.9
Ohio	1,096.7	288.7	116.7	87.8	50.0
Oklahoma	202.5	48.0	87.1	23.9	50.3
Oregon	308.7	40.3	113.5	158.7	193.4
Pennsylvania	977.3	121.8	90.2	107.3	50.0
Rhode Island	45.9	10.4	48.8	4.6	44.7
South Carolina	225.2	32.0	145.0	46.5	34.2
South Dakota	56.9	28.3	100.0	14.9	62.1
Tennessee	297.9	98.9	82.9	39.4	49.0
Texas	1,061.3	478.9	104.5	94.9	48.7
Utah	275.8	134.2	267.6	22.7	200.9
Vermont	40.2	10.2	102.1	10.3	68.6
Virginia	300.4	30.7	64.3	72.8	65.7
Washington	491.6	163.7	152.8	43.3	99.9
West Virginia	123.2	34.2	89.9	17.1	31.1
Wisconsin	325.3	124.4	81.0	73.6	56.1
Wyoming	43.0	19.0	146.3	5.7	78.1
District of Columbia	110.0	0.0	0.0	0.0	0.0
Puerto Rico	131.2	21.3	53.8	4.6	15.0
United States	20,873.9	6,900.9	113.9	2,491.4	64.6

Note: Gallons times $3,785 \times 10^{-3}$ gives cubic meters.

-- than from rural supplies -- 65 gpd (246 litres/day). This may be partially explained by greater use of water-consuming devices by the urban dweller.

Of interest, and often not appreciated, is the fact that little of the so-called drinking water is actually consumed, and most is returned to ground waters through septic tanks and cesspools or to surface waters through sewage treatment plants. According to a 1964 study by the U. S. Geological Survey, the percentage breakdown of domestic water use is as follows: 1)

Flushing toilets	-41 percent
Washing and bathing	-37 percent
Kitchen use	- 6 percent
Drinking water	- 5 percent
Washing clothes	- 4 percent
General household cleansing	- 3 percent
Watering garden	- 3 percent
Washing car	- 1 percent

The recorded withdrawal rates may also include substantial unaccounted losses within the supply and distribution system due to leakage. Figure 6 is a state-by-state illustration of the comparative importance of ground water to surface water as a source of drinking water. The figures presented are a composite of the public-supply and rural withdrawals for domestic use.

Public Water Supplies

The total quantity of water withdrawn for public supplies in 1970 was estimated as 27 bgd (102 million cu m/day). Included in this quantity was water for domestic use; water for commercial and industrial use; water lost in the distribution systems; and water supplied for carrying out public services such as firefighting, street washing, and water for municipal parks and swimming pools. Of the total, ground water supplied approximately 34 percent, and was the major source of water to over 59 million persons served by public supplies. Water utilities supplied by surface-water sources, although furnishing almost twice as much water, are relatively few in number compared to the number of utilities using ground-water sources.

Figure 7 illustrates the breakdown by source and state for the total public supply for all uses.

Because of economic factors (including convenient access), many industrial and commercial establishments use public sup-

Figure 6. Total drinking water withdrawn, public supply and rural, mgd. 1)

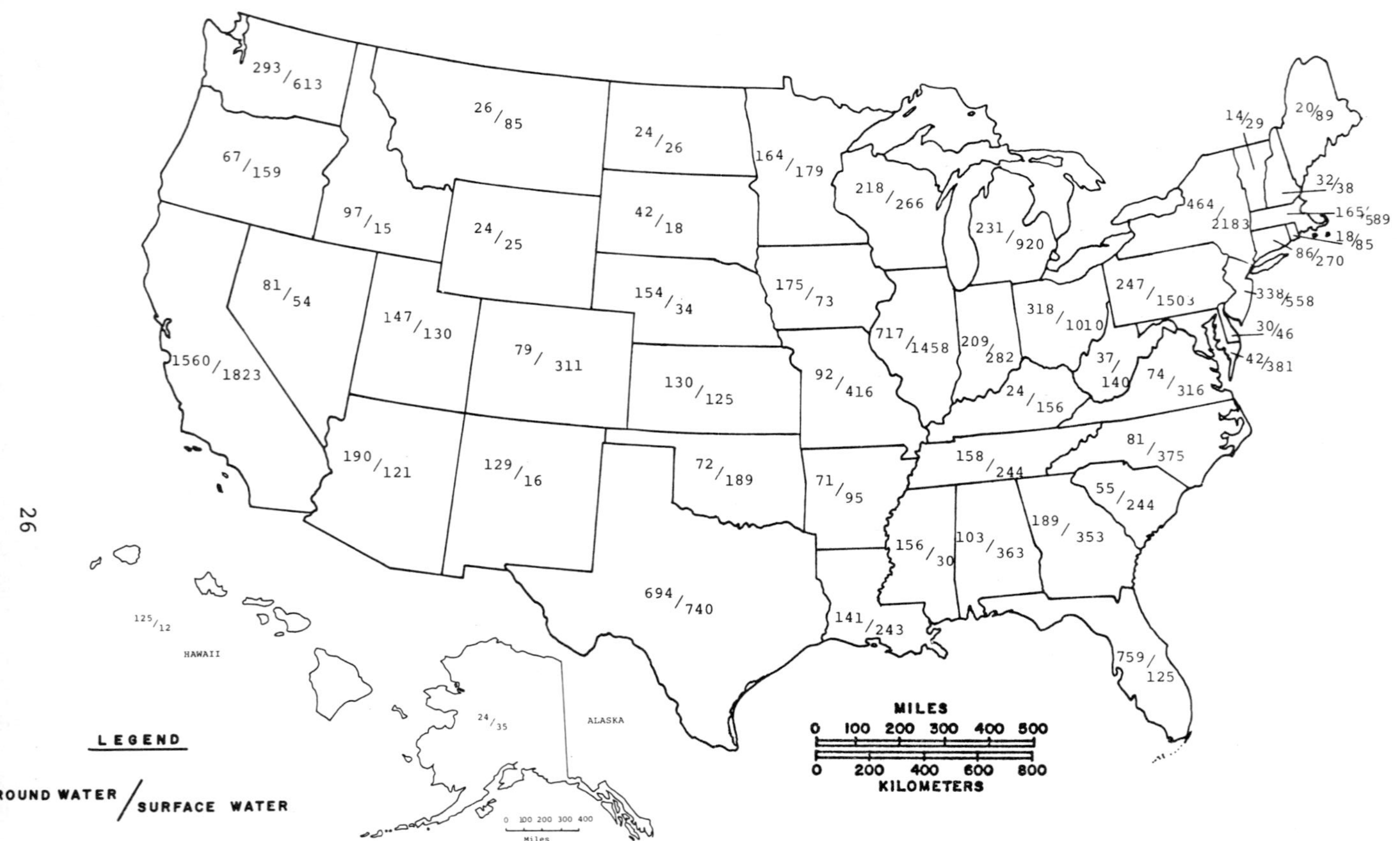
293/613
67/159
1560/1823
97/15
81/54
26/85
24/25
147/130
190/121
79/311
129/16
24/26
42/18
154/34
130/125
72/189
694/740
164/179
175/73
92/416
71/95
141/243
218/266
717/1458
231/920
209/282
24/156
318/1010
37/140
74/316
247/1503
464/2183
14/29
20/89
32/38
165/589
18/85
86/270
338/558
30/46
42/381
158/244
156/30
103/363
189/353
81/375
55/244
759/125
125/12
HAWAII
24/35
ALASKA
0 100 200 300 400
Miles
MILES
0 100 200 300 400 500
0 200 400 600 800
KILOMETERS
LEGEND
GROUND WATER / SURFACE WATER

plies, especially where the volume of water they require is small and the quality of the water must be high. Commerce and industry received approximately one third of the public supply withdrawals in 1970.

Individual Domestic Wells

In 1970, 41 million persons relied on their own supply of water and withdrew an average of 2.6 bgd (9.8 million cu m/day) for domestic use. Of this amount, ground water supplied 96 percent.

The per capita rate for domestic well use is about 65 gpd (246 litres/day). This represents a quantity intermediate between estimated low withdrawal rates in homes without running water and estimated high withdrawal rates in suburban homes that have running water and are equipped with modern high water-requirement appliances. Figure 8 illustrates the density of domestic wells by county.

Self-Supplied Industrial and Commercial Wells

Many of the large water-dependent industries of the country have installed their own ground-water supply systems for processing rather than depend on the more expensive purchase of water from public systems. In most of these installations, the drinking water for the employees is provided by the same wells. Examples of other small drinking-water systems, practically all supplied by well water when not served by a local public utility, include schools, restaurants and motels, highway rest stops, recreational areas, mobile home parks, and shopping centers. One requirement of the Safe Drinking Water Act of 1974 is that states provide water quality monitoring for drinking water systems serving 15 connections or more, or 25 people or more. It is estimated that there are some 200,000 of these small systems. [2)]

Specific information on water consumption and the actual number of people served by such systems is not readily available. However, for a few states, the number of small drinking-water systems not presently monitored has been estimated. For example, in Indiana, 10,000 additional water sources will require surveillance; [4)] in New Jersey the number is 7,000; [5)] and in Washington, 3,455. [6)] In Maine 2,555 systems will have to be brought under surveillance (a breakdown by type of system is presented in Table 4).

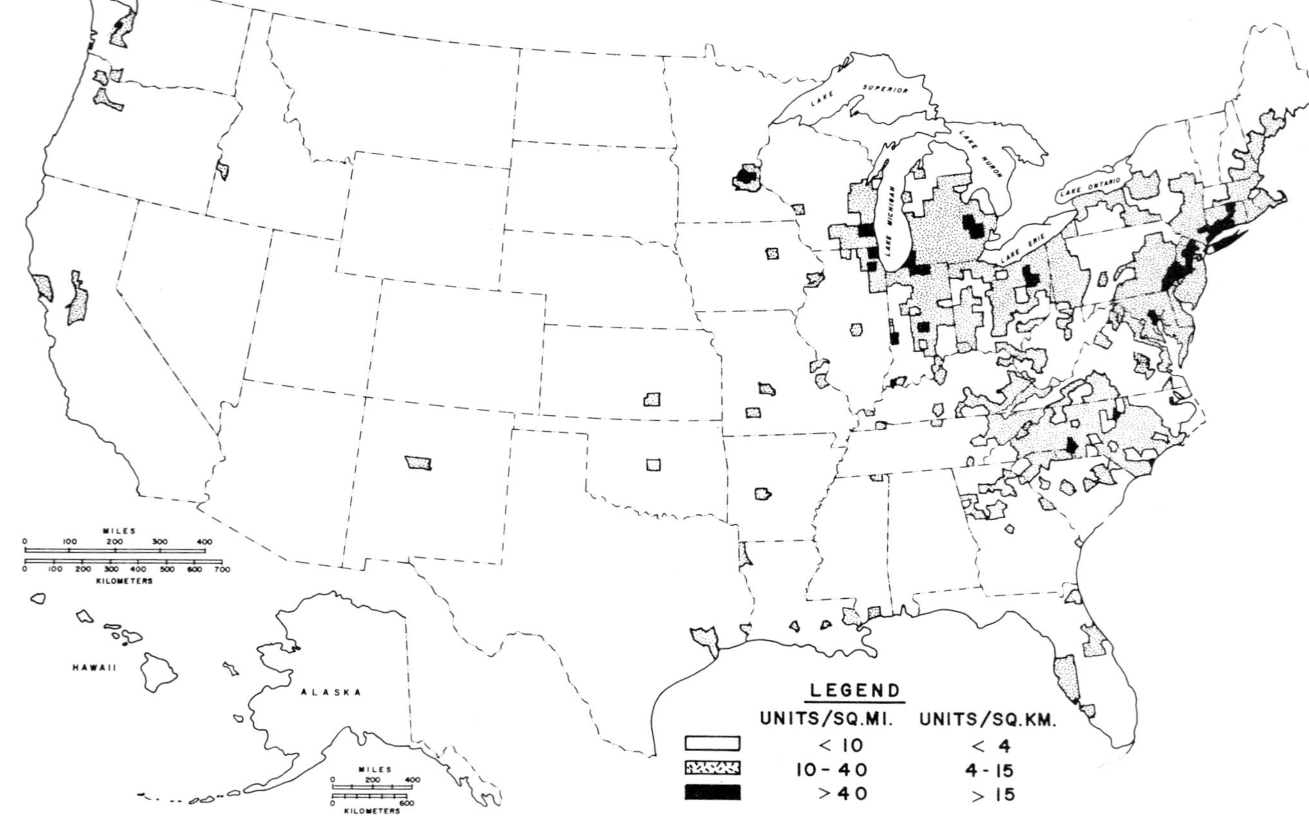

Figure 8. - Density of housing units using on site domestic water supply systems (by county)

Table 4. TYPES OF SMALL WATER SYSTEMS IN MAINE NOT PRESENTLY MONITORED, BUT REQUIRING SURVEILLANCE UNDER THE SAFE DRINKING WATER ACT. [7]

Small year-round systems serving groups of homes	30
Systems serving seasonally used groups of homes	50
Schools using well water	400
Industrial systems	50
Restaurants and motels	1,772
Highway Dept. rest centers	45
Dept. of Parks systems	30
Mobile home parks with 7 or more trailers	178
Total	2,555

DEGREE OF TREATMENT NORMALLY PROVIDED TO DRINKING WATER

Municipal Water Purification Works

The most common classes of municipal purification works as applied to ground-water sources are:

1. Iron and manganese removal plants -- reduce or eliminate excessive amounts of iron and manganese.

2. Softening plants -- remove excessive amounts of scale-forming, soap-consuming ingredients, chiefly calcium and magnesium ions.

Aeration in conjunction with filtration is the most common unit operation employed for the removal of dissolved iron and manganese from raw ground water. This operation also facilitates taste, odor, and corrosion control by removing dissolved volatiles, such as hydrogen sulfide.

One of the most important chemical precipitation techniques employed by municipal water treatment works is lime-soda softening. Calcium is precipitated as a carbonate ($CaCO_3$), and magnesium is precipitated as a hydroxide ($Mg(OH)_2$). Water softened by the lime-soda process is generally supersaturated with $CaCO_3$ and $Mg(OH)_2$ which can be stabilized by aeration with carbon dioxide. Normally this is done before filtration. Secondary carbonation or recarbonation relieves supersaturation and reduces precipitation of $CaCO_3$ on filter sand and in pipelines.

Today, most water supplies are chlorinated to assure their disinfection. Lime or other chemicals are often added to reduce the corrosiveness of water to iron and other metals, thereby preserving water quality during distribution and ensuring longer life to metallic pipes in particular. Odor- or taste-producing substances are adsorbed onto activated carbon, or destroyed by high doses of chlorine or chlorine dioxide. Numerous other treatment methods serve special needs.

Household Water-Conditioning Equipment

Individual water-supply systems, both domestic and commercial-industrial, encounter virtually the same quality problems with raw ground-water sources as municipal water-supply systems. Economic factors, however, limit the degree to which the raw water may be treated. Purification must normally be limited to simple water conditioning units.

Chlorination or ultraviolet treatment are the most common disinfection techniques employed by self-supplied dwellings.

Roughly 90 percent of all ground-water problems due to chemical characteristics in individual water supplies relate to either hardness or dissolved iron. Both characteristics are normally treatable to tolerable limits by simple water-softening units which employ ion-exchange techniques. If iron is not effectively controlled, an iron filter may be required ahead of the softener.

Feed pumps, chemicals, and specialty filters are commercially available for handling unusual water problems such as color, taste, odor, and corrosion control.

OVERALL GROUND-WATER USE IN THE UNITED STATES

Total fresh ground-water withdrawal is estimated to have been 66.6 bgd (252 million cu m/day) in 1970. [2] This amounts to 21 percent of the total fresh-water withdrawal. Common practice segregates usage into the following six categories:

1. Public supplies (for domestic, commercial and industrial use).
2. Rural domestic
3. Livestock
4. Irrigation
5. Self-supplied thermoelectric power generation
6. Other self-supplied industrial use

The usage category most dependent on ground water is rural domestic, which withdraws 96 percent of its water from the ground. Livestock use is 61 percent ground water.

The other large users of ground water are public supplies and irrigation which rely on ground water to the extent of 34 and 36 percent, respectively. In terms of absolute quantity, irrigation accounts for 67 percent of total ground-water withdrawal. Public supplies are the second largest consumer of ground water. At the other end of the spectrum, ground-water use by thermoelectric utilities is negligible. Figure 9 illustrates the breakdown of ground-water withdrawal by use, and Table 5 further differentiates these data by state.

HISTORICAL AND PROJECTED TRENDS IN FRESH-WATER WITHDRAWAL RATES FOR THE UNITED STATES (1900-2020)

Trends in overall water use have been compiled from various

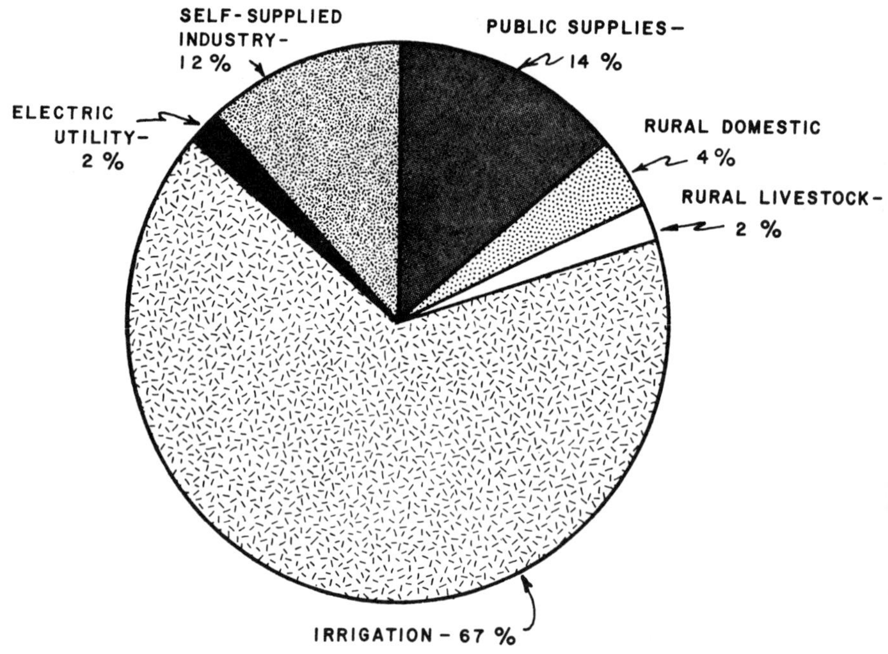

Figure 9. Total ground-water withdrawal, by use, 1970. 1)

Table 5. TOTAL FRESH GROUND-WATER WITHDRAWALS FOR ALL USES (1970). [1)]

State	Public Supplies (All Uses) (MGD)	Rural Domestic (MGD)	Rural Livestock (MGD)	Irrigation (MGD)	Self-Supplied	
					Electric Utility (MGD)	Others (MGD)
Alabama	100	63	13	5.4	26	93
Alaska	24	4.5	0	.4	1.4	8.0
Arizona	190	22	20	3,800	40	150
Arkansas	71	49	16	1,100	4.0	330
California	1,600	120	38	16,000	300	410
Colorado	79	9.3	20	1,900	30	55
Connecticut	86	38	2.1	.5	0	20
Delaware	30	11	1.6	2.2	.7	22
Florida	760	160	18	1,300	13	710
Georgia	190	72	31	6.5	6.7	330
Hawaii	120	.1	1.4	550	82	160
Idaho	96	22	10	2,100	0	340
Illinois	720	14	32	15	7.0	250
Indiana	210	76	29	13	1.0	140
Iowa	180	47	110	23	0	150
Kansas	130	48	31	2,800	38	120
Kentucky	24	48	3.7	.4	1.9	87
Louisiana	140	67	11	770	36	460
Maine	20	11	1.1	.2	1.0	3.0
Maryland	42	46	10	2.1	0	43
Massachusetts	170	28	1.3	13	0	140
Michigan	230	160	24	22	0	70
Minnesota	160	110	59	12	280	160
Mississippi	160	25	16	220	35	310
Missouri	92	29	28	70	6.0	200
Montana	26	8.8	17	63	0	34
Nebraska	150	22	80	2,700	270	100
Nevada	81	5.8	1.6	380	7.1	39
New Hampshire	32	11	.5	0	0	12
New Jersey	340	80	1.5	56	8.0	550
New Mexico	130	16	12	1,300	8.6	72
New York	460	120	24	14	130	140
North Carolina	80	110	43	50	1.0	140

Table 5 (Continued). TOTAL FRESH GROUND-WATER WITHDRAWALS FOR ALL USES (1970). [1]

State	Public Supplies (All Uses) (MGD)	Rural Domestic (MGD)	Rural Livestock (MGD)	Irrigation (MGD)	Self-supplied Industry: Electric Utility (MGD)	Self-supplied Industry: Others (MGD)
North Dakota	24	17	9.6	26	1.0	4.7
Ohio	320	88	24	9.0	49	390
Oklahoma	72	24	6.7	720	3.8	32
Oregon	67	160	2.6	630	0	110
Pennsylvania	250	110	14	.8	0	400
Rhode Island	18	4.6	.1	.4	0	15
South Carolina	55	46	4.0	8.9	.7	53
South Dakota	42	15	81	31	.7	14
Tennessee	160	39	5.4	1.3	0	88
Texas	[illegible]90	95	96	7,800	60	480
Utah	150	23	34	420	0	56
Vermont	14	10	5.6	0	1.0	12
Virginia	74	73	12	5.2	0	120
Washington	290	43	4.2	350	0	150
West Virginia	36	17	.9	0	0	25
Wisconsin	220	74	56	33	0	110
Wyoming	24	5.7	3.8	130	1.0	67
District of Columbia	0	0	0	0	0	.8
Puerto Rico	34	.4	1.3	67	.4	40
United States	9,400	2,500	1,100	45,000	1,400	8,000
	(14%)	(4%)	(2%)	(67%)	(2%)	(12%)

Total Ground-Water Withdrawal = 67,400 MGD

Note: Million gallons times 3,785 gives cubic meters

sources. Projections through the year 2020 have been added, using common population and economic growth assumptions. One must continually bear in mind, however, that these trends are merely indicative of the general direction of growth of water use. Future quantitative predictions must never be considered as established facts.

Table 6 lists water withdrawals by use, source, and demand for the years 1900-2020. Included, and used as a basis for many of the water projections, are trends and projections of total population, based upon U. S. Bureau of Census Series C projections. For all types of supplies, the withdrawal figures are subdivided into "all water" and "ground water" categories. The difference represents the withdrawal rate for surface water. All figures are in billion gallons per day. Figures 10 through 16 depict these trends.

It is estimated that between 1970 and 2020 the United States population will increase to 355 million persons -- 1.7 times the population in 1970. In addition to the expected increase in water demand from this increasing population, it is expected that the per capita withdrawal rate for all uses will increase from its 1970 figure of 1,571 gpd (5,946 litres/day) to over 2,100 gpd (almost 7,950 litres/day) in 2020. It is expected that a rise in our dependence on the products of water-using industries and utilities will create a greatly increased demand for water. Electric power plant cooling alone will account for over 50 percent of the total withdrawals.

By 2020, total fresh-water withdrawal is expected to rise to 760 bgd (2.9 billion cu m/day), of which ground water will provide almost one fifth or more than twice the amount of ground water withdrawn in 1970. The greatest increase in demand will be due to the expanding use of surface water for electric power plant cooling -- 300 bgd (1.1 billion cu m/day) more in 2020 than in 1970. The domestic use portion of the public supply water withdrawal will grow approximately at the same rate as the population, assuming that the nation's population growth will occur in areas served by public distribution systems.

It is projected that the commercial, public, and industrial use of public supply water will grow at an accelerated rate; almost twice that of the population. It is expected that the rural population, although growing, will do so at a much slower rate than the urban population. Rural domestic demand in 2020 is projected to be only 1.4 times that of 1970. Shifts in population, Federally-assisted rural community water developments, and expanding urban growth are all fac-

Table 6. HISTORICAL AND PROJECTED TRENDS IN FRESH WATER WITHDRAWAL RATES (UNITED STATES - 1970). 2,8,9,10,11)

Public Supplies (BGD)	1900[1]	1910[1]	1920[1]	1930[1]	1940[1]	1950[2]	1960[2]	1970[3]	1980[4]	1990[4]	2000[4]	2010[4]	2020[4]
All Water	3.0	4.7	6.0	8.0	10.1	14.1	21.5	27.2	33.6	42.1	50.7	62.5	74.3
Ground Water	1.0	1.5	1.8	2.3	2.8	3.8	5.7	9.4	11.5	14.2	17.1	20.8	24.4
Domestic													
All Water	2.4	3.8	4.8	6.3	7.7	10.3	15.3	18.5	21.0	25.5	30.1	33.2	36.3
Ground Water	0.8	1.2	3.3	1.8	2.1	2.7	3.9	6.9	7.8	9.4	11.1	12.3	13.4
Commercial, Public & Industrial													
All Water	0.6	0.9	1.2	1.7	2.4	3.8	6.2	8.7	12.6	16.6	20.6	29.3	38.0
Ground Water	0.2	0.3	0.3	0.5	0.7	1.1	1.8	2.5	3.7	4.8	6.0	8.5	11.0
Rural Supplies (BGD)													
All Water	2.0	2.2	2.4	2.9	3.1	3.6	4.0	4.5	5.1	5.8	6.5	7.4	8.3
Ground Water	1.6	1.8	1.9	2.4	2.6	2.9	3.3	3.6	4.0	4.5	5.0	5.6	6.2
Domestic													
All Water	1.4	1.5	1.6	1.9	2.0	2.2	2.3	2.6	2.7	2.9	3.1	3.4	3.6
Ground Water	1.3	1.4	1.5	1.8	1.9	2.1	2.2	2.5	2.6	2.8	3.0	3.3	3.5
Livestock													
All Water	0.6	0.7	0.8	1.0	1.1	1.4	1.7	1.9	2.4	2.9	3.4	4.0	4.7
Ground Water	0.3	0.4	0.4	0.6	0.7	0.8	1.1	1.1	1.4	1.7	2.0	2.3	2.7
Irrigation Supplies (BGD)													
All Water	20.2	39.0	55.9	60.2	71.0	100.0	110.0	129.9	145.9	152.8	159.8	165.4	171.0
Ground Water	2.2	5.3	8.2	9.1	11.2	19.8	33.0	45.0	55.4	62.6	71.9	81.0	90.6
Self-Supplied Electric Utility (BGD)													
All Water	5.0	6.5	9.2	18.4	23.2	45.9	98.7	120.0	141.5	204.3	267.2	343.4	419.6
Ground Water	0.1	0.1	0.1	0.2	0.2	0.6	1.2	1.4	1.7	2.5	3.2	4.1	5.0
Other Self-Supplied Industrial (BGD)													
All Water	10.0	14.0	18.0	21.0	27.0	30.4	36.2	38.8	46.3	53.9	64.1	78.0	91.8
Ground Water	2.4	3.0	3.8	4.2	5.8	6.5	7.2	8.0	9.3	10.8	12.8	15.6	18.4

Table 6 (Continued). HISTORICAL AND PROJECTED TRENDS IN FRESH WATER WITHDRAWAL RATES (UNITED STATES - 1970). [2,8,9,10,11]

	1900[1]	1910[1]	1920[1]	1930[1]	1940[1]	1950[2]	1960[2]	1970[3]	1980	1990	2000	2010	2020
Total U.S. Population (Millions) (Excluding Puerto Rico)	76.0[5]	92.0[5]	105.7[5]	122.8[5]	131.7[5]	150.7[5]	178.5[5]	203.2[5]	230.9[6]	266.2[6]	300.4[6]	338.0[6]	355.0[6]
Total Fresh Water Withdrawn (BGD)	40.2	66.4	91.5	110.5	134.4	194.0	270.4	320.4	372.4[7]	458.9[7]	548.3[7]	656.7[7]	765.0[7]
Total Fresh Ground Water Withdrawn (BGD)	7.3	11.7	15.8	18.2	22.6	33.6	50.4	67.4	81.9[7]	94.6[7]	110.0[7]	127.1[7]	144.6[7]
Total Per Capita Withdrawal (GPD) (All Fresh Sources)	529	722	866	900	1,036	1,287	1,344	1,577	1,613	1,724	1,825	1,943	2,155

[1] Withdrawal rates from U.S. Bureau of Domestic Commerce, 1900-1960 based principally on Committee Prints, "Water Resources Activities in the United States," for the Senate Committee on National Water Resources, U.S. Senate, 1960.

[2] Withdrawal rates from Source 1, and USGS Circular 676, "Estimated Use of Water in the United States in 1970."

[3] Withdrawal data from USGS Circular 676, "Estimated Use of Water in the United States in 1970."

[4] Water Resource Council, "The Nation's Water Resources," 1968 - Withdrawal data based on WRC projection modified where applicable to fit census projections.

[5] U.S. Bureau of Census, U.S. Census of Population: 1920 to 1970, Vol. 1.

[6] U.S. Bureau of Census, "Current Population Reports," Series P-25, Nos. 311, 483, and 493, Series C Projection. This projection assumes a slight improvement in mortality, an annual net immigration of 400,000 and a completed cohort fertility rate (average number of births/1,000 women upon completion of child bearing) of 2,800.

[7] Sum of individual withdrawal projections.

Note: Billion gallons times 3,785,000 equals cubic meters.

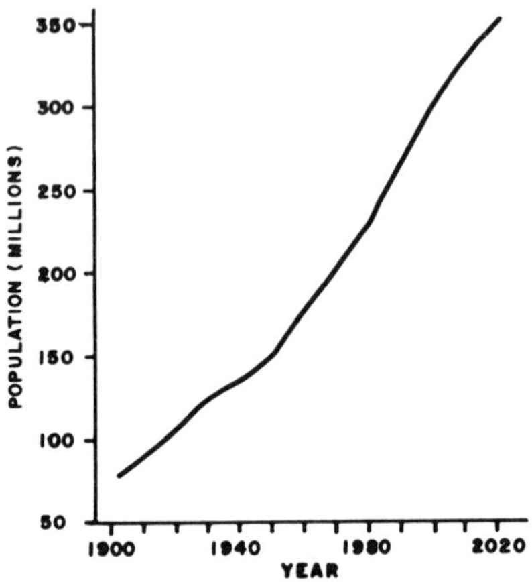

Figure 10. Historical and projected trends of total United States resident population.[8)]

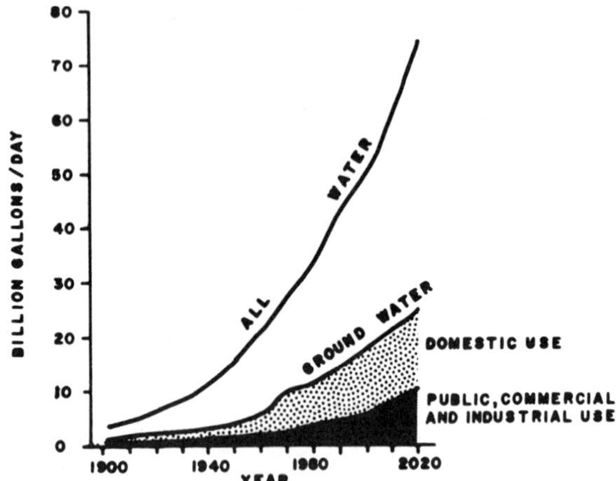

Figure 11. Historical and projected trends of fresh water withdrawal for public supply use.[8)]

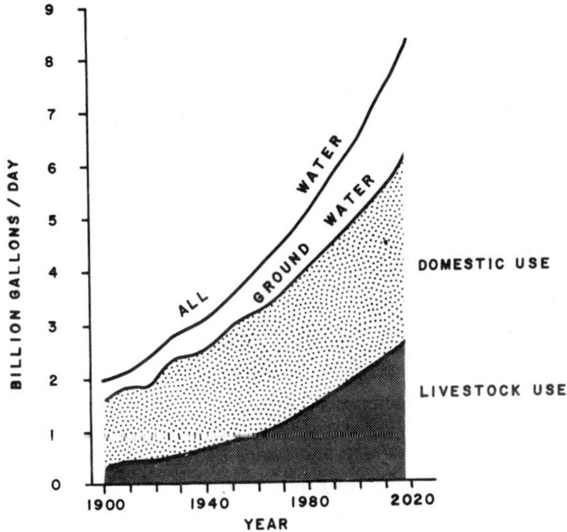

Figure 12. Historical and projected trends of fresh water withdrawal for rural supply use. [8]

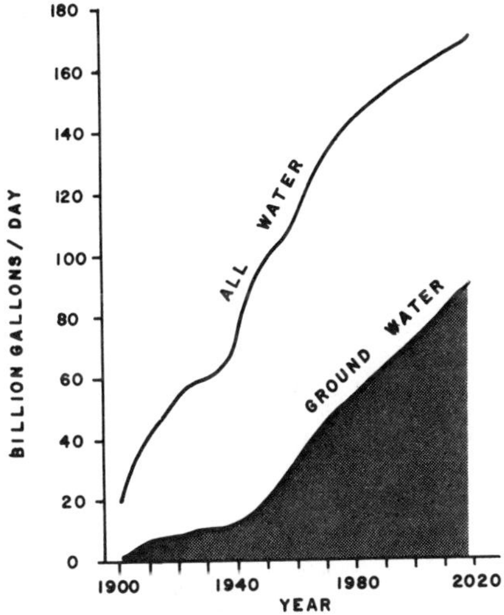

Figure 13. Historical and projected trends of fresh water withdrawal for irrigation use. [8]

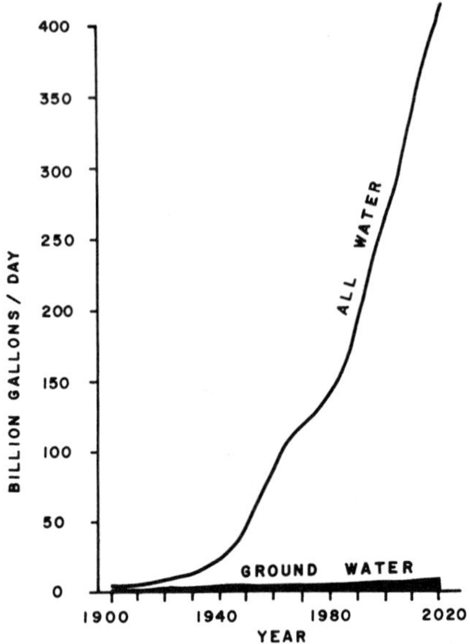

Figure 14. Historical and projected trends of fresh water withdrawal for electric utility use.[8)]

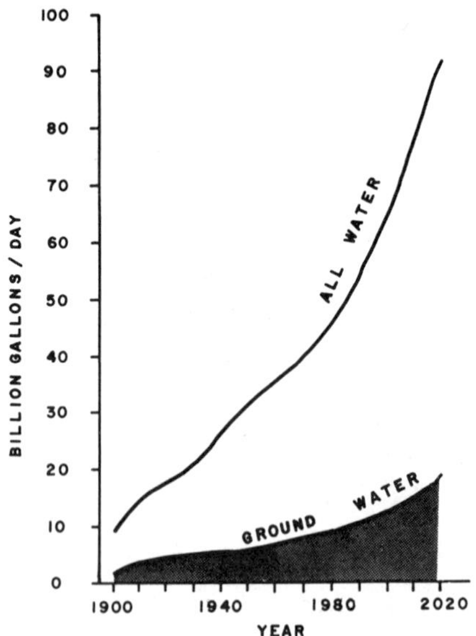

Figure 15. Historical and projected trends of fresh water withdrawal for other self-supplied industrial use.[8)]

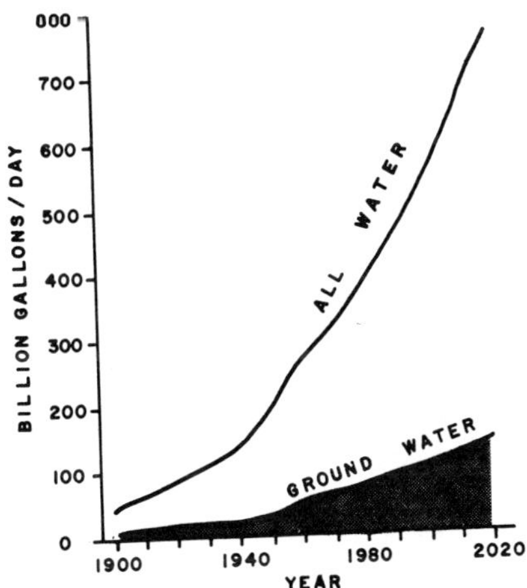

Figure 16. Historical and projected trends of total fresh water withdrawal.[8)]

tors that may decrease the reliance on individual water systems.

The use of water for irrigation will continue its upward climb of the past several decades. In addition, irrigation's reliance on ground water will increase from 35 percent in 1970 to over 50 percent in 2020. The long-term upward trend in irrigation is related to more efficient technology for storage, conveyance, and application; development of new water sources, particularly in the arid west; and socioeconomic factors including lowering of farm production costs and the benefits of a wider economic base in areas of heavy irrigation development.

Both the expanding population and the growth in real personal income will exert pressures on industry to increase its output. Considerable additional water will be required to satisfy this growth. The new water demand will be met partly by increased withdrawals, but in larger part by improved methods of water management including extensive recycling.

REFERENCES CITED

1. U. S. Geological Survey, Water Supply Division. 1970. Unpublished data.

2. Murray, C. R., and E. B. Reeves. 1972. Estimated use of water in the United States in 1970. U. S. Geological Survey Circular 676. 37 pp.

3. McDermott, J. H. 1974. Impact of the Safe Drinking Water Act. Page 3 in Proceedings national symposium on the state of America's drinking water. North Carolina Water Resources Research Institute, Raleigh, North Carolina.

4. Indiana State Board of Health. 1975. Personal communication.

5. New Jersey Department of Environmental Protection, Bureau of Potable Water. 1975. Personal communication.

6. Robischon, J. 1975. Optflo 1(7).

7. Maine Department of Health, Division of Health Engineering. 1975. Personal communication.

8. U. S. Bureau of Domestic Commerce. 1960. Water resources activities in the United States. National Water Resources Committee, Senate, United States Congress.

9. U. S. Water Resources Council. 1968. The nation's water resources - the first national assessment of the Water Resources Council. U. S. Government Printing Office, Washington, D. C. Parts 1-7.

10. U. S. Bureau of the Census. 1970. U. S. census of population: 1920-1970. U. S. Government Printing Office, Washington, D. C. Vol. 2.

11. U. S. Bureau of the Census. 1974. Current population reports. U. S. Government Printing Office, Washington, D. C. Series P-25: Nos. 311, 483, and 493.

SECTION IV

NATURE AND EXTENT OF THE RESOURCE

SUMMARY

At almost any location, ground water may be tapped to provide a supply sufficient for single-family domestic use, and more than one third of the nation is underlain by aquifers generally capable of yielding at least 100,000 gpd (380 cu m/day) to an individual well. In many regions, ground water is the only economic and high quality water source available. In others, ground water can be developed at a fraction of the cost of surface water.

Ground water in aquifers across the nation is generally suitable for human consumption with little or no treatment necessary, except for disinfection where large, piped water-supply systems are involved. Salinities tend to be higher in arid regions and areas where drainage is poor.

INTRODUCTION

As discussed in the previous section, ground water presently supplies almost one quarter of the nation's total water supply. It provides the dry-season flow (base flow) of streams that otherwise might cease flowing part of the year. Some of the nation's largest cities and most of the rural population depend on ground water as a source of drinking water.

It has been estimated that total ground water in storage in the United States greatly exceeds the combined volume of all the Great Lakes, and that the amount of useable ground water is 150 times the amount of water presently used. At almost any location, ground water may be tapped to provide a supply sufficient for single-family domestic use. However, the distribution of ground-water reservoirs (aquifers) capable of supplying communities, towns, and cities is more limited.

DEFINITION OF GROUND WATER

In the hydrologic cycle, water is continually evaporated from the oceans, moves through the atmosphere, and eventually returns to the ocean through one or more paths. Of the water that precipitates, a portion infiltrates into the ground under the influence of gravity. It moves first through an unsaturated zone known as the "zone of aeration." Passing downward, the water arrives at the zone of saturation where the voids between the rock particles are completely saturated. The water in the zone of saturation is called ground water. Figure 17 illustrates the relationships within the hydrologic system.

THE OCCURRENCE OF GROUND WATER

The ability of an aquifer to store and transmit water is a function of its porosity and permeability. Porosity reflects the volume of void space (pores) in a rock, and is an index of how much ground water can be stored in the saturated material. Porosity is usually expressed as a percent of the bulk volume of the material. Permeability is an index of how much ground water can be transmitted through a rock. The coefficient of permeability is expressed as the rate of flow of water (gallons per day) that will flow through a one-foot square area per unit of time under a hydraulic gradient of one, at a temperature of 60°F (16°C).

An index closely related to permeability is transmissivity. Transmissivity is simply permeability multiplied by aquifer thickness; it is indicative of the water-transmitting capacity of the entire aquifer thickness. Where the saturated rock is sufficiently permeable to store and transmit significant quantities of water, the rock is called an aquifer. Aquifers are defined by the ability to store and transmit water and not by rock type directly.

Major Types of Aquifers

The two major types of aquifers are: unconfined or water-table aquifers, and confined or artesian aquifers. Less permeable zones are called aquitards or confining layers. Figure 17 illustrates the major aquifer types.

Unconfined Aquifers -

When an aquifer is unconfined, the water is under atmospheric pressure. The upper surface of the aquifer is known as the water table and is free to rise and fall with changes in volume of stored water.

Under nonpumping conditions, the water level in a well and the adjacent water table are at the same elevation. The water table is responsive to changes in the amount of stored water, and fluctuates seasonally in response to variations in the rate of natural recharge. In the humid eastern states, for example, the water-table elevation is normally highest in spring and lowest in autumn.

The principal source of natural recharge to a water-table

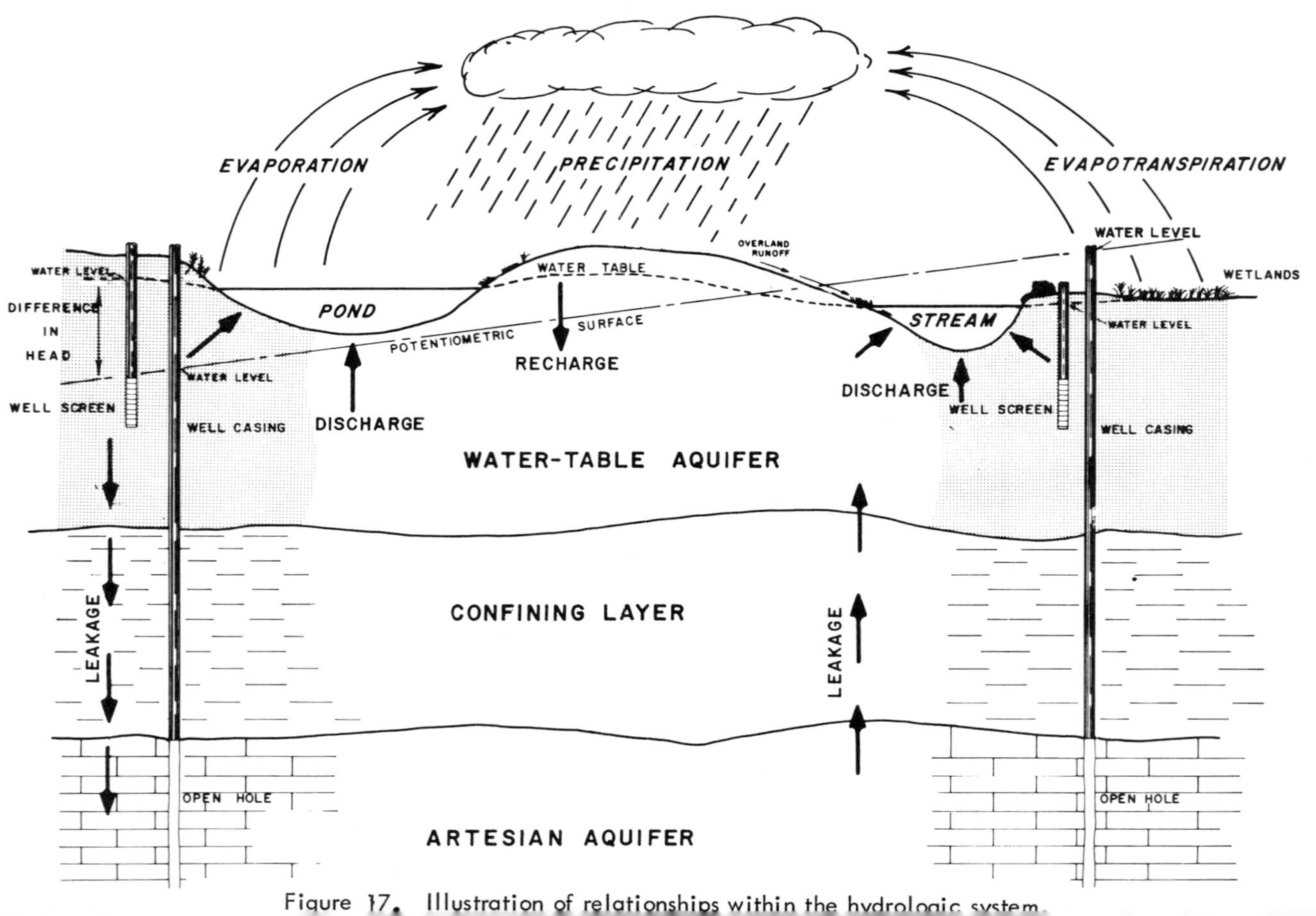

Figure 17. Illustration of relationships within the hydrologic system.

aquifer is precipitation. In arid regions, because precipitation is infrequent, intermittent surface streams carrying runoff from other parts of the region may provide significant recharge. Perennial through-flowing streams of more humid regions can be areas of recharge to or discharge from water-table aquifers.

A variant type of water-table aquifer is a perched aquifer. Occurring within the zone of aeration are beds of relatively low permeability, but of limited areal extent. Precipitation moving downward cannot pass easily through these beds, so a thin zone of saturation is created above the bed, forming a perched water body. Although perched aquifers are sometimes tapped by wells, they are usually not sufficiently thick or extensive to provide a significant supply of water. They do, however, restrict and control recharge to the underlying aquifer.

Confined Aquifers -

Confined or artesian aquifers are bounded below by geologic formations of relatively low permeability. In addition, an artesian aquifer is separated from the zone of aeration above or from shallow aquifers by geologic formations of low permeability. The aquifer is completely saturated with water, and the upper surface is defined and fixed by the lower limit of the overlying confining unit. Under nonpumping conditions, when a well is constructed and open only to an artesian aquifer, the water level in the well stands above the top of the aquifer at a height dependent upon the pressure in the confined aquifer (artesian pressure). Where sufficient pressure is encountered, the water level may stand above the top of the well casing, causing the well to flow. The hypothetical projection of the water levels is known as the potentiometric surface.

An artesian aquifer does not receive recharge everywhere uniformly. Most recharge is received in one or more general areas known as recharge areas. Rather than being sensitive to volumetric changes, the water levels in wells in artesian aquifers respond principally to changes in artesian pressure.

Rocks with identical characteristics may form an aquifer in one area, yet may act as a confining unit for a more permeable zone in another area. No confining unit is completely impermeable. Where an aquitard is sufficiently permeable to allow significant volumes of water to leak into or out of an aquifer, the aquifer is called semi-confined or leaky artesian. A water-table aquifer can overlie an artesian aquifer, separated by an aquitard. Two artesian aquifers can be separated by a confining unit.

Recharge and Discharge

Ground water is constantly moving from a point of recharge toward a point of discharge. If a particular region is a recharge area, the recharging water exerts a stress on the aquifer in the form of increased hydrostatic head. This head seeks release in areas of low head, which are designated discharge areas. Thus, movement of ground water is from regions of high hydrostatic head toward those of low hydrostatic head. In practice, recharge and discharge areas of an aquifer are indicated by relative water levels. Within an aquifer, areas of high water-level elevations indicate higher hydrostatic head and areas of lower water-level elevations indicate lower hydrostatic head, so ground water moves from areas of high water-level elevations toward areas of low water-level elevations. The hydraulic head difference divided by the distance along the flow path is known as the hydraulic gradient.

Head differences can be induced artificially by pumping wells. As water is withdrawn from a well, a hydraulic gradient is produced, which causes water to move toward the well. A cone-shaped depression in the water table or potentiometric surface is produced (Figure 18). As more water is extracted, the depth and radius of the cone increase, but at a decreasing rate, until the volume of water leaking into the aquifer exactly equals the withdrawal rate. At this point, the cone will stabilize (stop growing). Cones of depression from more than one well can overlap if leakage does not stabilize them first. In some cases, particularly in aquifers in arid western basins, the volume of water leaking to the cone (or cones) of depression never equals the total volume withdrawn. The cones continue to expand downward and laterally indefinitely. This activity is known as ground-water mining.

In addition to precipitation, a water-table aquifer can be recharged where it is hydraulically connected to a surface-water source, such as a stream or a pond. A water-table aquifer can receive leakage through semi-permeable confining beds of an underlying artesian aquifer. Artesian aquifers can receive recharge from confining beds or from precipitation and surface-water bodies in the outcrop area of the aquifer.

Recharge locations can be points, lines, or areas. Natural point recharge locations are infrequent; individual sinkholes in limestone terrane are an example. Artificial point

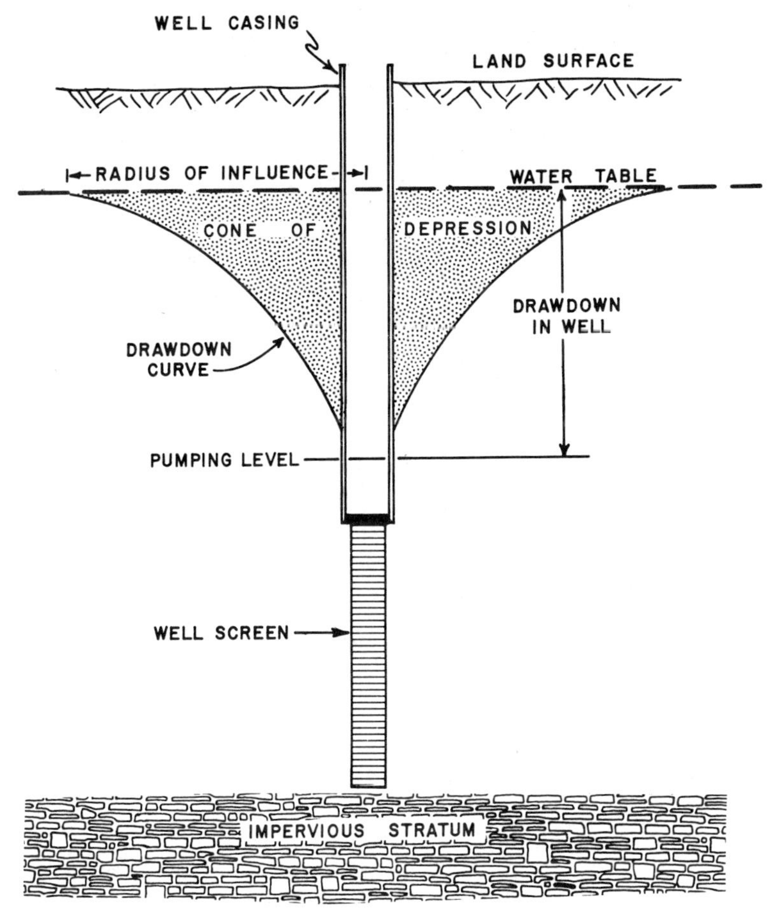

Figure 18. Cone of depression created by pumping in a water-table aquifer. [1)]

recharge locations are very common, and in fact are of major concern in a later section of this study. Examples include waste-disposal or recharge wells and individual septic tanks and cesspools. Natural line recharge is related to leakage from the beds of streams. This is a common situation in the western states where mountainous areas tend to capture precipitation, pass it to streams as runoff, and the streams carry it across valley-fill deposits where recharge to aquifers occurs. Natural line recharge also occurs along the edge of valley-fill deposits, the coarser layers of the fill receiving direct overland runoff from the adjacent mountains. Leaky sewage transmission pipes are an example of artificial line recharge. Most natural area recharge occurs across broad regions and is derived directly from precipitation. Artificial area recharge occurs where homes in subdivisions, as a group, each have septic tanks which recharge the aquifer. Reservoirs and large waste-water disposal ponds are also examples of artificial area recharge.

Discharge locations for aquifers can also be points, lines or areas. A spring is a natural point discharge location while a pumping well is an artificial point discharge location. Gaining streams can be line discharge areas. In this case, precipitation falling on adjacent upland areas infiltrates the water-table aquifer, and the ground water moves toward a nearby stream where it is discharged. Area discharge locations are swamps, ponds, lakes, and the sea. The volume of ground water naturally discharged to the ocean along the Atlantic coast is many times that discharged to wells, springs, and streams.

Climatic Effects

The amount of precipitation and the percent returned to the atmosphere (evapotranspiration) vary according to climatic conditions. Variations in the average precipitation in any region may create exceptional surpluses or deficits -- evidenced by floods or droughts -- during individual years. Figure 19 illustrates the average annual precipitation over the United States.

The processes which return water from the land surface to the atmosphere are evaporation and transpiration. The combined term evapotranspiration represents the amount of water lost to the atmosphere from the land surface. A distinction has been made between potential evapotranspiration and actual evapotranspiration in an area. Potential evapotranspiration represents the volume of water which would be lost from a completely vegetated area if there were no water deficiency at any time. On the other hand, actual evapotranspira-

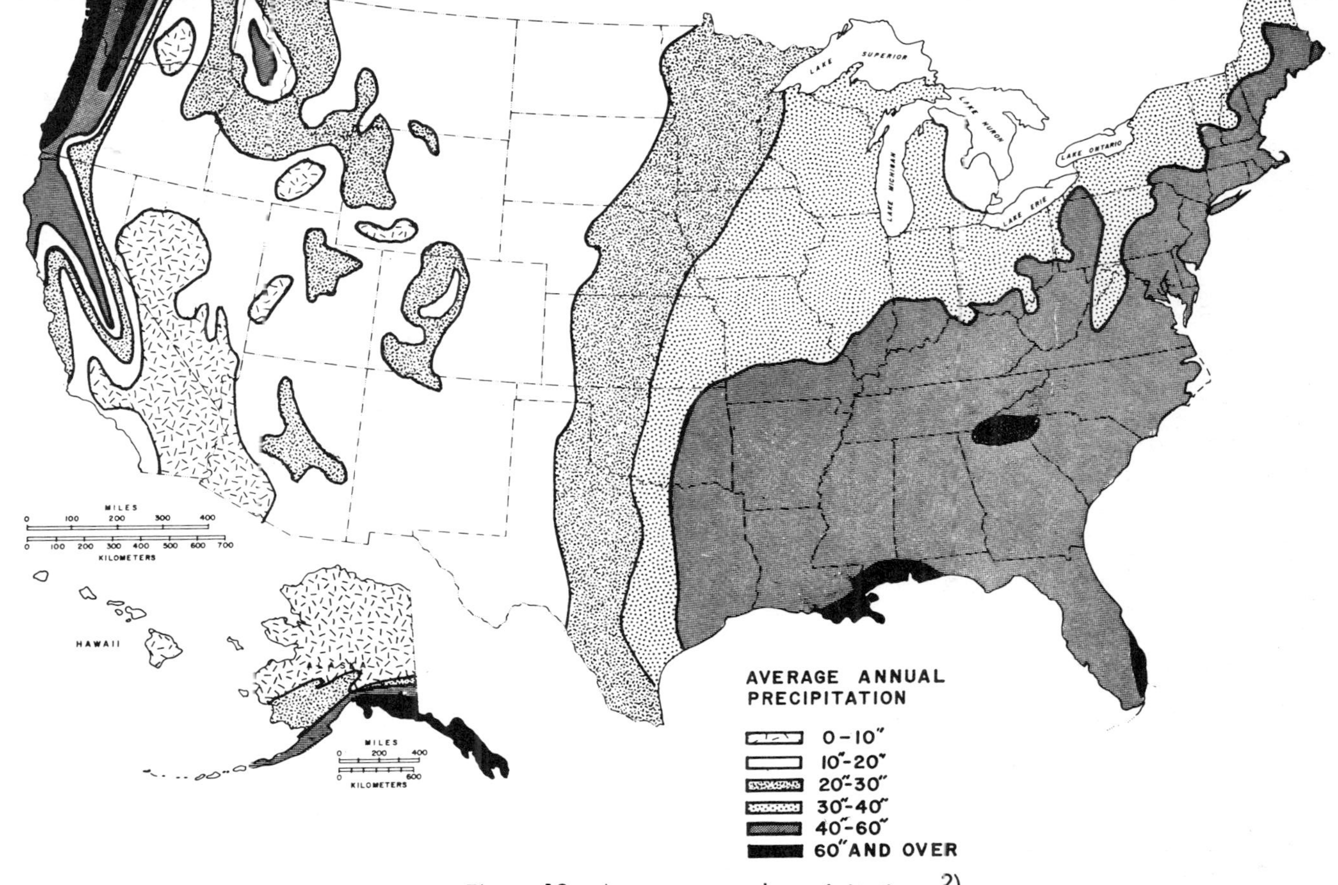

Figure 19. Average annual precipitation. 2)

tion is the real volume of water lost under prevailing conditions. Except in rain forests, potential and actual evapotranspiration are seldom equal.

Wherever evapotranspiration is greater than the precipitation, ground-water recharge by downward percolation through the zone of aeration is minimized. Such water-deficient areas exist, especially in the desert lowlands of the southwest, where annual precipitation is less than 10 in. (25 cm), and potential evapotranspiration is 4 to 20 times greater.

In areas of prevailing water deficiency in the western states, ground-water recharge may result from abundant precipitation during "wet" years or multi-year cycles, rainy seasons, or prolonged storm periods. In many valleys that are practically rainless throughout the warmer half of the year, winter precipitation provides the major portion of recharge to shallow ground waters. In other valleys, there may be evidence of ground-water recharge only during years of greater than average precipitation.

In the eastern states, annual precipitation exceeds evapotranspiration, creating surpluses which discharge to and form the base flow of perennial streams and springs. But the locale of the water surpluses may vary from season to season. In the winter, with minimum evapotranspiration, water may accumulate in the soil and percolate downward. In the growing season, vegetation depletes the soil moisture and, even with frequent rains, may leave nothing for ground-water recharge.

Temperature

Temperature is also an important factor in ground-water recharge. In northern latitudes and western mountainous regions, floods have resulted from rain falling upon accumulated snow during unseasonably warm periods in winter or early spring. The flood runoff is increased if the underlying soil is frozen, thus preventing infiltration. The persistence of extremely low temperatures may cause unusual conditions which acutely affect the existence and flow of ground water. Permafrost, or permanently frozen ground, is common in Alaska, and exists over 60 percent of the state. Within these regions, the soil from a few feet to several hundred feet below the surface is continuously frozen with the exception of a relatively thin, seasonally-thawed surface layer.

Permafrost zones act as confining beds, and both their composition and distribution have a significant influence on patterns and rates of ground-water flow. In a number of basins, the artesian pressure of water confined below permafrost causes wells drilled through the permafrost to flow. Ground-water discharge may be restricted to the lower, central part of many river valleys where the permafrost is discontinuous. In the region of continuous permafrost, unfrozen zones penetrate the permafrost only where salinity of the ground water prevents freezing, or where heat transfer from a body of surface water or from discharging subpermafrost water is sufficient to maintain the unfrozen conditions.

Water in the Unsaturated Zone

The unsaturated zone occupies a critical position in the hydrologic cycle. The relationship between the unsaturated and saturated zones is shown in Figure 20.

From land surface, the unsaturated zone receives water from precipitation to the limit of its infiltration capacity; the rest is left for surface storage, runoff, or evaporation. In most places, the upper part of the unsaturated zone is soil, which absorbs the infiltrating water, and retains much of it against the force of gravity until such time as the water is taken up by plant roots or otherwise returned to the atmosphere.

Some water, in excess of the retention capacity of the soil, percolates downward through the soil. In some places, the unsaturated zone is permeable enough to receive water rapidly and permit downward percolation with little retention. Under most of the land, the unsaturated zone extends below the soil and below plant roots to depths ranging from a few feet to hundreds of feet.

Topography

The amount of precipitation which recharges to ground water in any specific area depends, to some degree, upon topography. Rolling terrain, particularly when underlain by soils of low infiltration capacity, facilitates rapid runoff of precipitation to surface-water bodies. In valleys surrounded by mountains, the mountains tend to capture precipitation and direct it into the valleys, where it can recharge underlying aquifers. Spring runoff from snow melt in mountain areas is the principal source of recharge to many arid and semi-arid valleys in the western states.

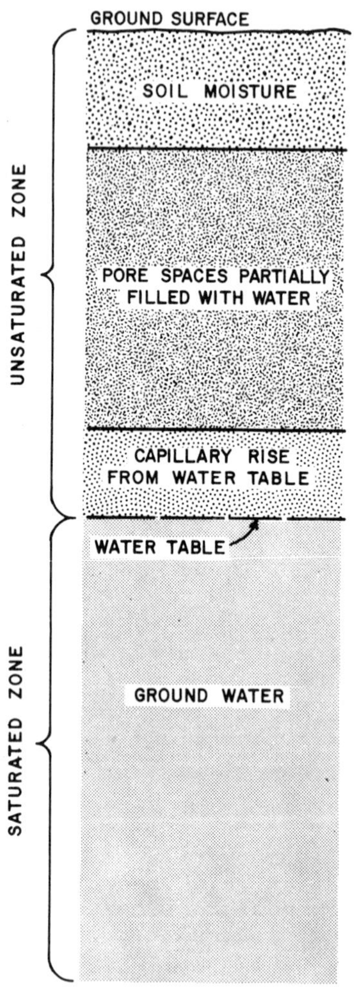

Figure 20. Relationship between unsaturated and saturated zones. [1]

ROCK TYPES COMPRISING AQUIFERS

The principal aquifers in the United States consist of saturated sedimentary, igneous, and metamorphic rocks. Among the sedimentary rocks are clastics, evaporites, and carbonates. Clastics may be subdivided into consolidated and unconsolidated rocks. Igneous rocks also are divided into two classes: plutonics and volcanics. Volcanics are subdivided into flows and pyroclastics. Metamorphic rocks are not subdivided. Figure 21 shows several types of interstices (openings or void spaces) found in aquifers.

Clastics

Clastic sedimentary rocks are composed of fragments of other rocks transported from their sources and deposited by water or glacial ice. Clastics include both unconsolidated and consolidated rocks. Unconsolidated deposits are relatively uncemented and loosely compacted. The degree of consolidation is determined by the degree of cementation and compaction. In unconsolidated clastics such as gravel and sand, ground water is stored and transmitted in the interconnected voids which occur between individual grains.

Water availability in unconsolidated rocks is greatly affected by sorting and grain size. Deposits which are well sorted have many particles of the same or similar size. This assures that very little of the available pore space will be occupied by grains which are either overly large or overly small.

A high degree of sorting alone, however, does not insure high ground-water availability. Water moving through rock has a tendency to cling to the rock by capillary and molecular attraction, forming a thin coating of water on the individual grains of rock. This water is unavailable to wells. Where grain sizes are small, as with silt and clay, even a well-sorted deposit will have a significant percentage of pore space occupied by retained water. Although the quantity of water in storage is great, that which is available is so small that clays and silts are normally considered to be confining beds. Well sorted sands and gravels, on the other hand, are considered aquifers.

As much as 30 percent of an unconsolidated rock may consist of pore space. Where sedimentary rocks are partially consolidated, a precipitated substance like silica or calcium carbonate occupies some of the space and cements some of the individual grains together. As a result, the intergranular space available for ground-water storage is decreased, and

ROCK TYPE *INTERSTICES*

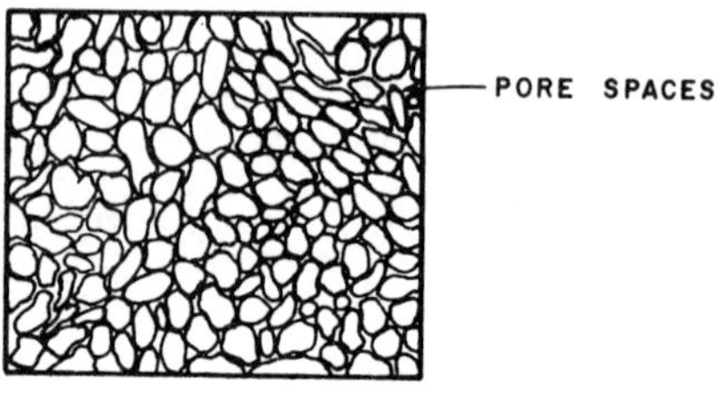

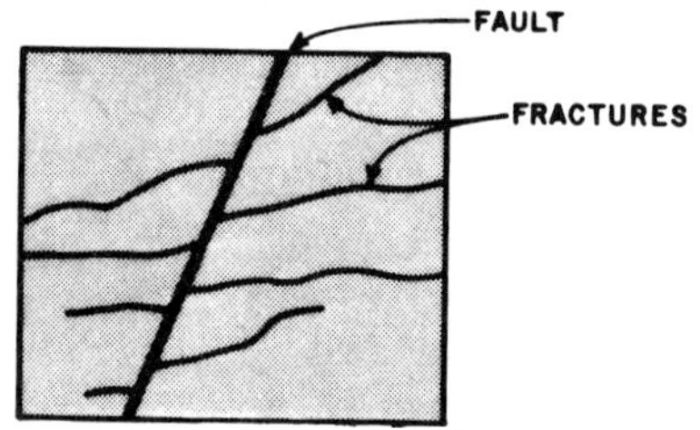

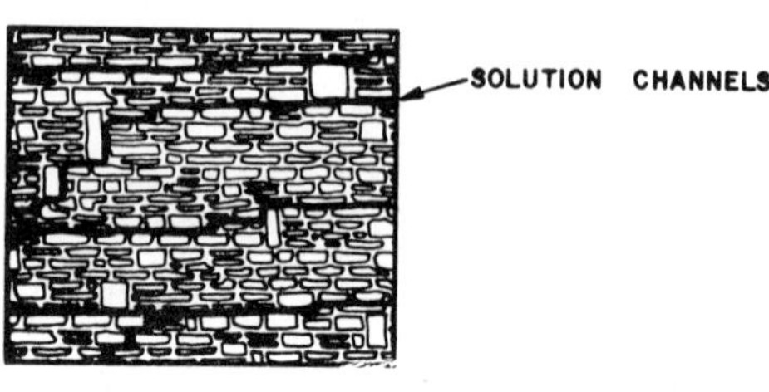

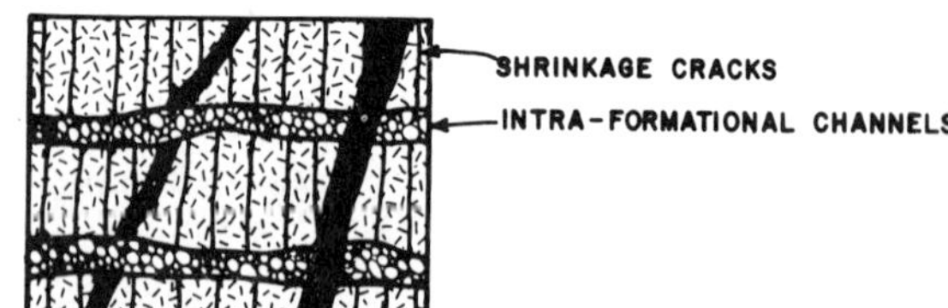

Figure 21. Rock texture in major aquifer types. 3)

the available interconnected spaces needed for ground-water transmittal are also decreased. Where completely consolidated, a major portion of the available pore space has been decreased by cementation and/or compaction.

In consolidated coarse-grained sedimentary rocks like sandstone and conglomerate, pore space is usually small, and ground water is stored and transmitted chiefly in fractures and joints, between layers, and along fault zones. In fine-grained, consolidated sedimentary rocks, pore space is virtually non-existent.

Evaporites

Sedimentary rocks which form by precipitation of dissolved minerals are called evaporites. When such deposits come into contact with fresh ground water, rapid dissolution occurs. Where other rocks are interbedded with evaporites, the voids produced by dissolution produce highly permeable aquifers. Unfortunately, ground water in these aquifers is normally so high in total dissolved solids that it is useless as a drinking-water source without costly treatment.

Carbonates

Rocks produced by secretions from organisms form a third class of sedimentary rocks known as carbonates. Shells and bones from aquatic animals collect on floors of seas, lakes, and streams. The matter is compacted and crystallized to form carbonate rock. Natural ground water, which is slightly acidic, can slowly dissolve carbonate rocks along joints and fractures. The resultant porosity and permeability may range from low values where the rock is slightly fractured to extremely high values where extensive fracturing and solution have taken place.

In some carbonate rocks, intergranular permeability is much more important than that attributable to fractures and solution openings. Some of the best aquifers of the southeastern coastal plain consist of soft coquinoid or bryozoan limestone or of slightly dolomitized limestone apparently owing most of its permeability to crystal-volume changes during dolomitization. In such aquifers, high permeability is so widespread that properly completed wells can obtain large yields almost everywhere.

Igneous

Igneous rocks form by crystallization of molten rock. Plutonic rocks cool and crystallize deep beneath the land surface, and volcanic rocks cool and crystallize on or near the surface. As with consolidated sedimentary rocks, ground water is available from fractures and joints, between layers, and along faults. A special class of volcanic rocks is known as pyroclastics -- unconsolidated to semi-consolidated deposits of fragmental material blown from volcanoes. However, pyroclastic deposits are rarely extensive. Ground-water availability in pyroclastics is variable but similar to that in semi-consolidated sedimentary rocks.

Metamorphics

Metamorphic rocks are recrystallized deposits of previously formed sedimentary and igneous rocks. No distinction is made here between metamorphic rocks of sedimentary origin and those of igneous origin. The distinction has little bearing from the standpoint of ground-water availability. Metamorphic rocks store and transmit water in a manner similar to plutonic rocks.

Other Factors

Other factors exert major influences on the availability of ground water from certain rock types. A principal factor is the variation in structure of geologic formations from one place to another. For example, in consolidated rock aquifers, a well drilled through a fault (a major break in the rocks) may be significantly more productive than a well in the same aquifer away from the fault. Faulting of the rocks produces more fractures along which ground water can move and be stored.

Weathering may also affect ground-water availability. In unconsolidated deposits, weathering of rock fragments may turn some of them to clay, thereby decreasing permeability. For consolidated rocks, weathering may enlarge or increase the number of joints and fractures, thereby improving the ability of the aquifer to yield water to wells.

PRINCIPAL AQUIFERS

Valley-Fill Aquifers

Valley-fill aquifers are composed of sand, gravel, and silt and generally lie along the course of present-day streams and rivers. Figure 22 indicates the locations of the major valley-fill aquifers. They are comprised of channel, floodplain, and terrace deposits, and are usually in direct hydraulic connection with surface streams. The deposits in each valley act as a single hydrologic unit, existing under

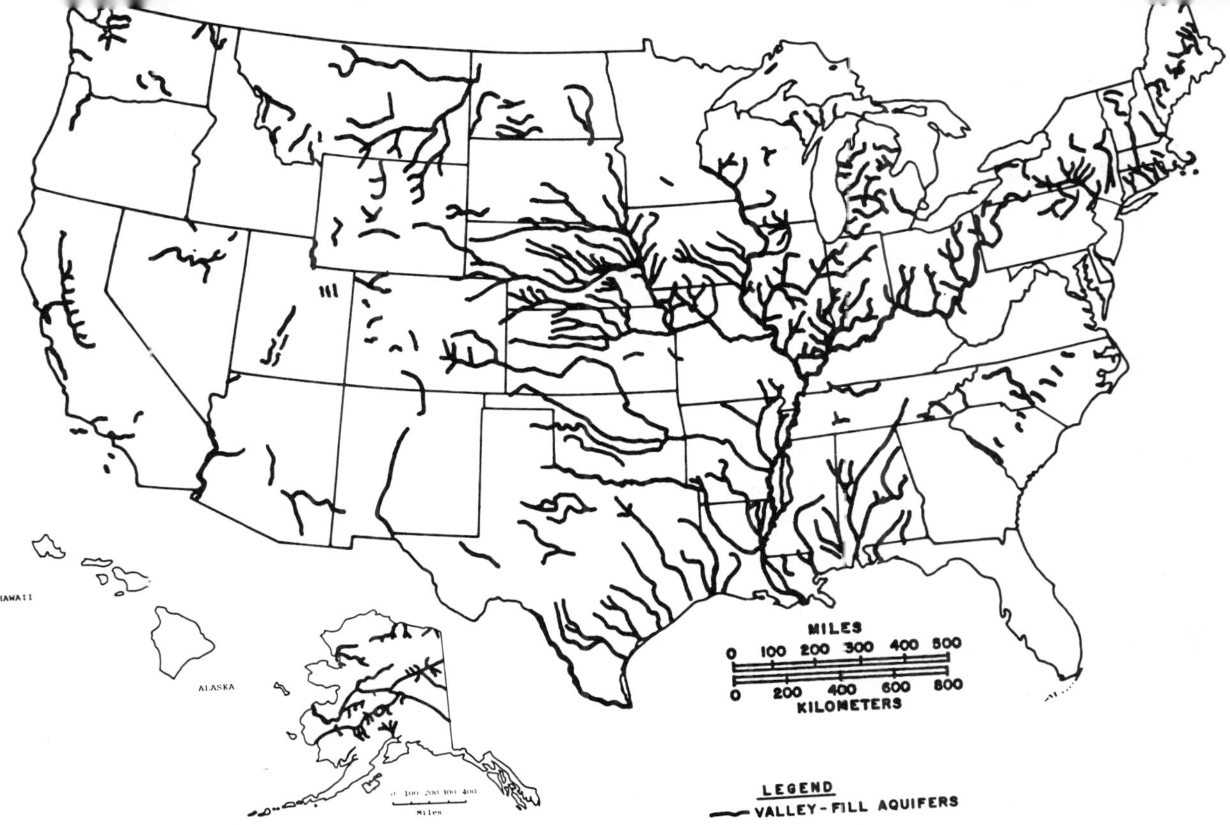

Figure 22. Valley-fill aquifers.

water-table or leaky artesian conditions. Where permeable valley-fill aquifers exist adjacent to perennial streams, large potential for ground-water development exists because of the opportunity to supplement natural recharge with infiltration of surface water.

The availability of potable water and the gentle topography of stream valleys has made them popular areas for urban development -- for example, the Ohio River valley and the Susquehanna River valley. Valley-fill aquifers are extremely susceptible to contamination from infiltration of poor quality surface water, or from wastes dumped on the land surface.

Because of the widespread distribution of valley-fill aquifers, no general statement can be made with regard to natural ground-water quality. However, the heavy use of these aquifers by industries and municipalities indicates the availability of generally good quality water.

Sands and Gravels of the Coastal Plain

Extensive deposits of clastics are deposited seaward of ancient uplands from which they were eroded. The principal water-bearing units are sands and gravels, which are interbedded with silts and clays, and occasionally marls and limestones. The sediments were deposited on plains only slightly above sea level or in the shallow near-shore marine environment. Land emergence has since raised these sediments above sea level. The most extensive coastal plain in the United States (the Atlantic-Gulf Coastal Plain) extends from Cape Cod, Massachusetts, to Texas (Figure 23).

Coastal plain sediments thicken seaward, and progressively younger geologic units outcrop in seaward direction. Although the outcrop areas of all units are under water-table conditions, the deeper sections are strictly artesian. Where some confining units are thin or moderately permeable, leaky artesian conditions may allow ground-water flow between artesian aquifers.

The outcrop areas of coastal plain aquifers receive recharge by direct precipitation and leakage from surface-water bodies. This recharge is transmitted downgradient within the aquifer to replenish the artesian portion. An additional source of recharge, where artesian conditions prevail, is inter-aquifer flow. Natural discharge areas for coastal plain aquifers are near the present shorelines.

The chief threat of contamination to coastal plain aquifers occurs where they exist under water-table conditions (in the

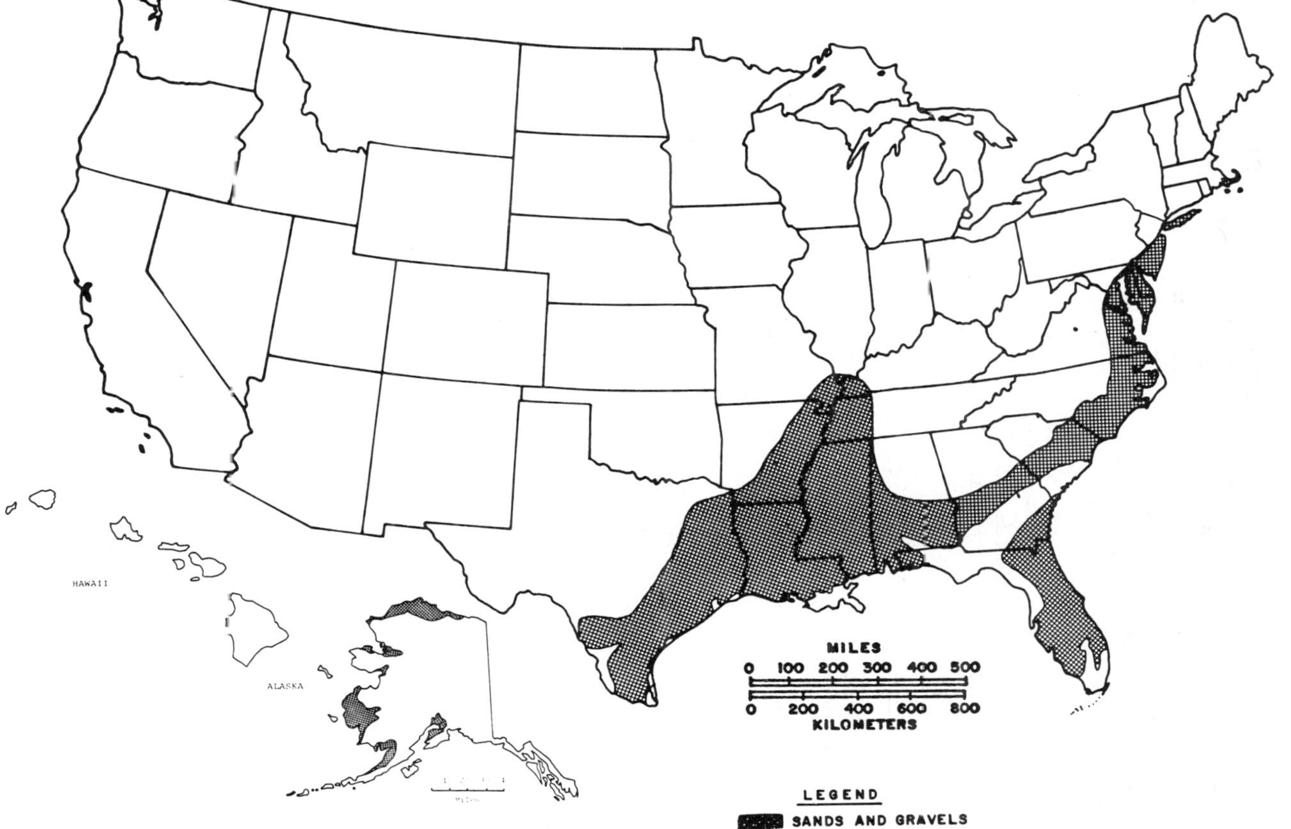

Figure 23. Sands and gravels of the coastal plain.

outcrop areas of the principal aquifers). Coastal plain aquifers are also vulnerable to inter-aquifer flow of contaminated water through leaky confining beds. Inter-aquifer flow has been a particular problem where an aquifer has been abandoned because of salt-water encroachment and is separated from an adjacent heavily pumped aquifer by leaky confining beds.

The natural water quality of coastal plain aquifers is generally good, particularly near outcrop areas. Problems do exist in some areas, however, principally related to low pH, high concentrations of iron, and the presence of connate saline water.

Sands and Gravels of the Intermontane Valleys

Mountain building periods in the western states have created intermontane valleys (Figure 24). These valleys have filled with sediment eroded from the adjacent mountains. The sediments include rock detritus, alluvial sand and gravel, and silts and clays. The permeable alluvia constitute excellent aquifers, and the valleys contain enormous quantities of water in storage. Because the sediments were transported by surface runoff from adjacent mountains, the aquifers are generally coarse grained toward the edges of the valleys and finer toward the centers. Occasionally, extensive clay and silt deposits are encountered within the geologic sequence.

Water-table and leaky artesian conditions prevail except where the extensive silts and clays overlie water-bearing zones (producing tightly confined aquifers). Because the intermontane valleys occur in generally water-deficient areas, little recharge is received by direct, downward percolation. The major source of recharge is runoff from adjacent mountains -- particularly from snow melt and spring rains -- which flows down mountain canyons and percolates into the coarse deposits at the edges of the valleys. In many intermontane valleys, pumping from wells far exceeds annual recharge, seriously depleting the resource.

Like other aquifers exposed at the land surface, those in intermontane valleys are susceptible to contamination. Continuous sources of contaminants (cesspools and septic tanks, leaky lagoons, mine drainage) and accidental spills are always a threat, particularly in the recharge areas. Also, in some closed basins, ground water is high in dissolved solids content as a result of continuous evaporation and accumulation of residual salts.

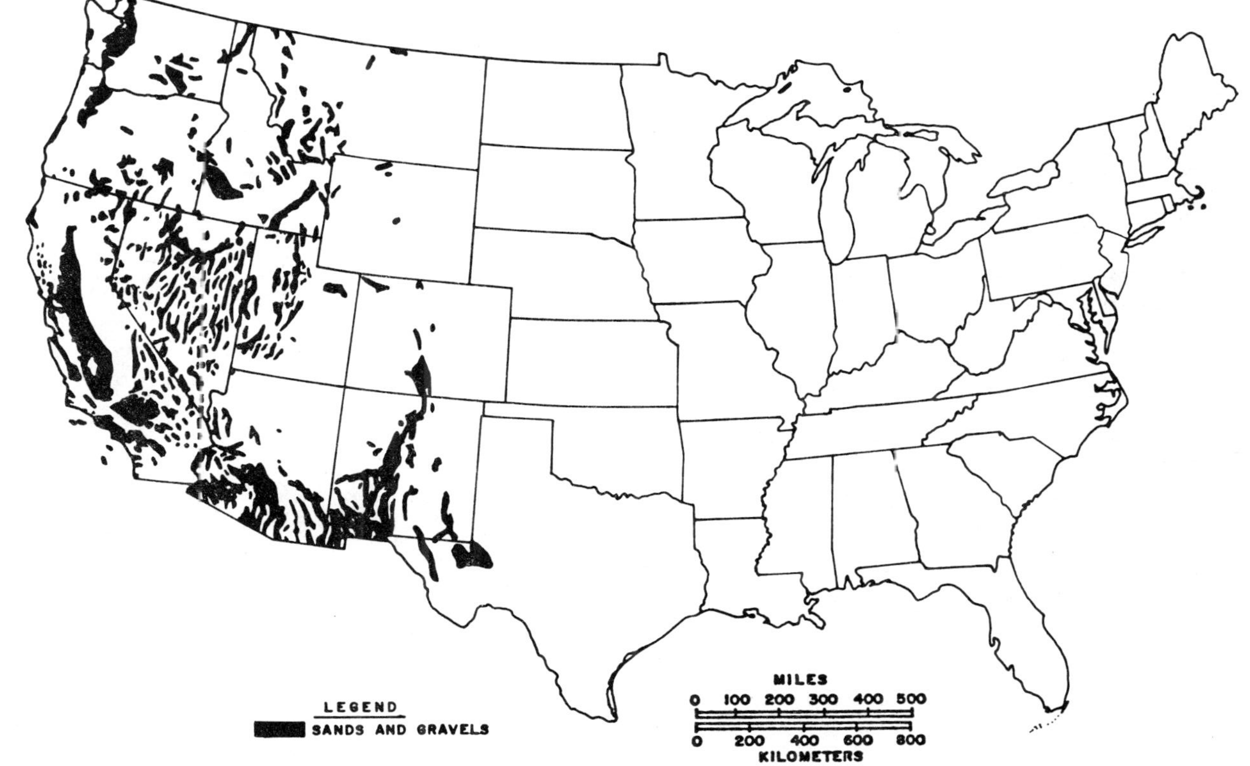

Figure 24. Sands and gravels of the intermontane valleys.

Alluvium of the High Plains

Alluvium, derived from the Rocky Mountains and laid down by eastward flowing streams, was deposited on a vast plain stretching from Wyoming to Texas (Figure 25). A large part of the original plain has been eroded by streams along its margin, but the remnants exist as the High Plains. The region is an important ground-water area because of the abundance of saturated sand and gravel, interbedded with silt and clay. The chief water-bearing unit is the Ogallala Formation.

Leaky artesian conditions prevail in the High Plains. Recharge is chiefly by downward leakage of direction precipitation through water-table beds. The High Plains lies completely within the water-deficient region of the United States, so recharge is variable. Generally, the southern part receives no recharge, and the northern part receives as much as 5 in. (12.7 cm) annually. Major streams can provide additional recharge where they have eroded into water-bearing zones.

Except in the Sand Hills region, the Ogallala aquifer appears to be well protected by overlying fine grained sediments from direct infiltration of contaminants. In some places, water levels are declining so rapidly from pumpage that the water table is falling at a rate greater than contaminants can percolate down to it. This pumpage also has resulted in the upward movement of saline water from deeper formations in some locales. Where streams have dissected the Ogallala, the near-stream portions of the aquifer are susceptible to contamination from infiltrating surface water.

The chemical quality of water in the High Plains is satisfactory for irrigation and generally meets the requirements for drinking water. Total hardness ranges from 200 to 600 ppm, although the average is less than 300 ppm. Excessive fluoride is a problem in some areas with local concentrations up to 5 ppm, and commonly exceeding 1.5 ppm. Silica may also be high, ranging up to 40 ppm. Total dissolved solids concentration averages less than 300 ppm, but may be as high as 1,000 ppm or greater. The quality of the water tends to deteriorate toward the southern end of the High Plains, and also tends to degrade with decreasing depth to the water table -- a condition produced by evapotranspiration, which has concentrated dissolved salts near the surface.

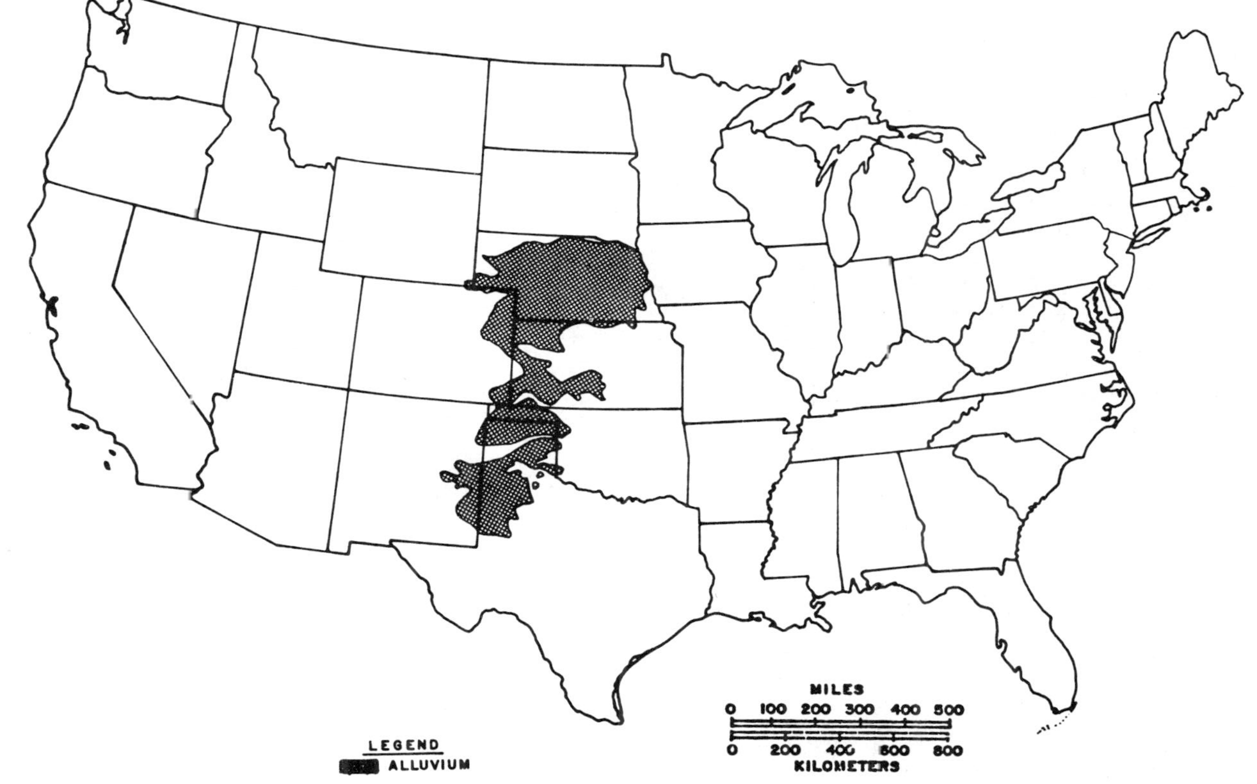

Figure 25. Alluvium of the High Plains.

Glacial Drift

During the Ice Age, glacial ice (continental glaciation) covered northern portions of the country (Figure 26). Glaciers also occupied major river valleys. Repeated glacial advances eroded the soil and bedrock and incorporated the material in the ice. When the ice melted, these particles were left behind or were carried across the land surface by melt water. Glacial drift is composed of all particles carried by the ice, regardless of size. Those deposits left in place (till) are unsorted; those sediments that were transported by water before deposition are generally better sorted. Where many small sediment-laden streams issued from the melting glaciers, broad extensive outwash deposits occurred, which are usually very productive water-table aquifers. Where fine grained glacial drift was deposited in standing bodies of water, like lakes in ice-dammed stream valleys, the resulting low permeability deposits constitute confining beds for underlying valley-fill aquifers.

The presence of low permeability till is much more common than that of more permeable water-borne deposits. Although till is tapped for small domestic supplies using dug wells, it more often acts as the confining bed for an underlying artesian sand and gravel aquifer.

Water-borne glacial drift is commonly a productive aquifer. Where it occurs in stream valleys, it can receive recharge from perennial streams and surface runoff from adjacent bedrock uplands.

Because glacial drift aquifers are commonly under water-table conditions, and highly permeable, they are vulnerable to downward percolation of contaminants. Precipitation is abundant over the glaciated region, and production of leachate from landfills, for example, in this water surplus area is of particular concern.

The natural quality of water in the glacial drift reflects the regional geology. In New England, where much of the bedrock is crystalline, the ground water in the glacial drift is normally of good quality and low in mineral content, although high concentrations of iron and manganese are not uncommon. In the Ohio River basin, the water in glacial-drift aquifers is generally hard and can be high in mineral content, especially calcium bicarbonate and sulfate. Similarly, in the Upper Mississippi River basin, much of the ground water has a total dissolved solids content ranging from 300 to 1,000 ppm, with hardness from 120 to 700 ppm. Still farther west, in the semi-arid Dakotas, ground water in the glacial

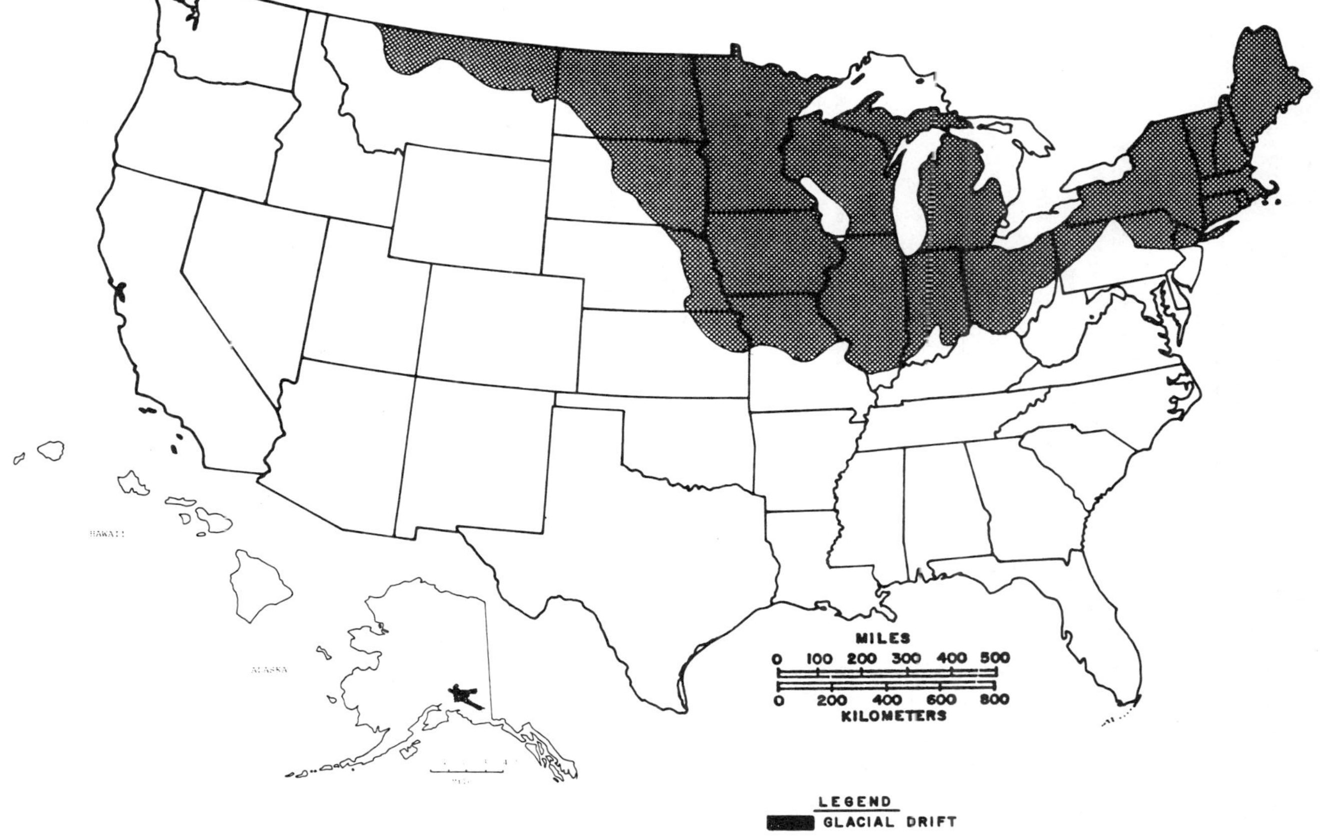

Figure 26. Glacial drift.

drift, although of better quality than in the underlying bedrock aquifers, may have a mineral content exceeding 1,000 ppm and may be locally brackish or even saline.

Basalt Aquifers

Basalt is volcanic rock which has flowed as lava across the land surface, or has intruded near the surface, and subsequently cooled. The basalt aquifers are thick, extensive sheets of rock piled in layer-cake fashion and interbedded with unconsolidated sediments. High capacity wells tap the natural openings between the basalt flows and depending upon their permeability, the interbedded sediments. Figure 27 shows the distribution of the principal basalt aquifers.

Ground water in the basalt aquifers occurs under artesian to leaky artesian conditions, produced by the varying permeabilities of individual beds in the aquifers. Recharge is almost exclusively by direct precipitation. Streams and rivers are incised deeply into the aquifers and serve to receive discharge from the aquifers rather than provide recharge to them.

In Hawaii, where basalt aquifers constitute most of the islands' ground-water reservoirs, water-bearing zones occur under both water-table and artesian conditions. Recharge is by both direct precipitation and stream flow. Hawaii's ground-water conditions are unusual because the basalt aquifer is sloping and is cross cut by many vertical dikes, most of which are impermeable and divide the aquifer into compartments. Individual compartments, if untapped, can fill and overflow to a compartment at a lower elevation through seeps and springs. A natural system of reservoirs is thus provided, which can be tapped by wells and tunnels for water supply.

In most basalt terranes, liquid contaminants can readily enter rock openings and move quite easily through the aquifer. In the northwestern states, the volume of ground water underflow is very large and there is an opportunity for some dilution.

The water in the basalt of the northwestern states typically has a total dissolved solids content in the range of 200 to 300 ppm. The best quality is found near recharge areas and in shallow aquifers. The waters are generally of a calcium-magnesium bicarbonate type, with total hardness ranging from 50 to 250 ppm. In addition, the water in these aquifers generally contains 40 to 80 ppm of silica, and locally, excessive iron. The ground water of Hawaii is of excellent qual-

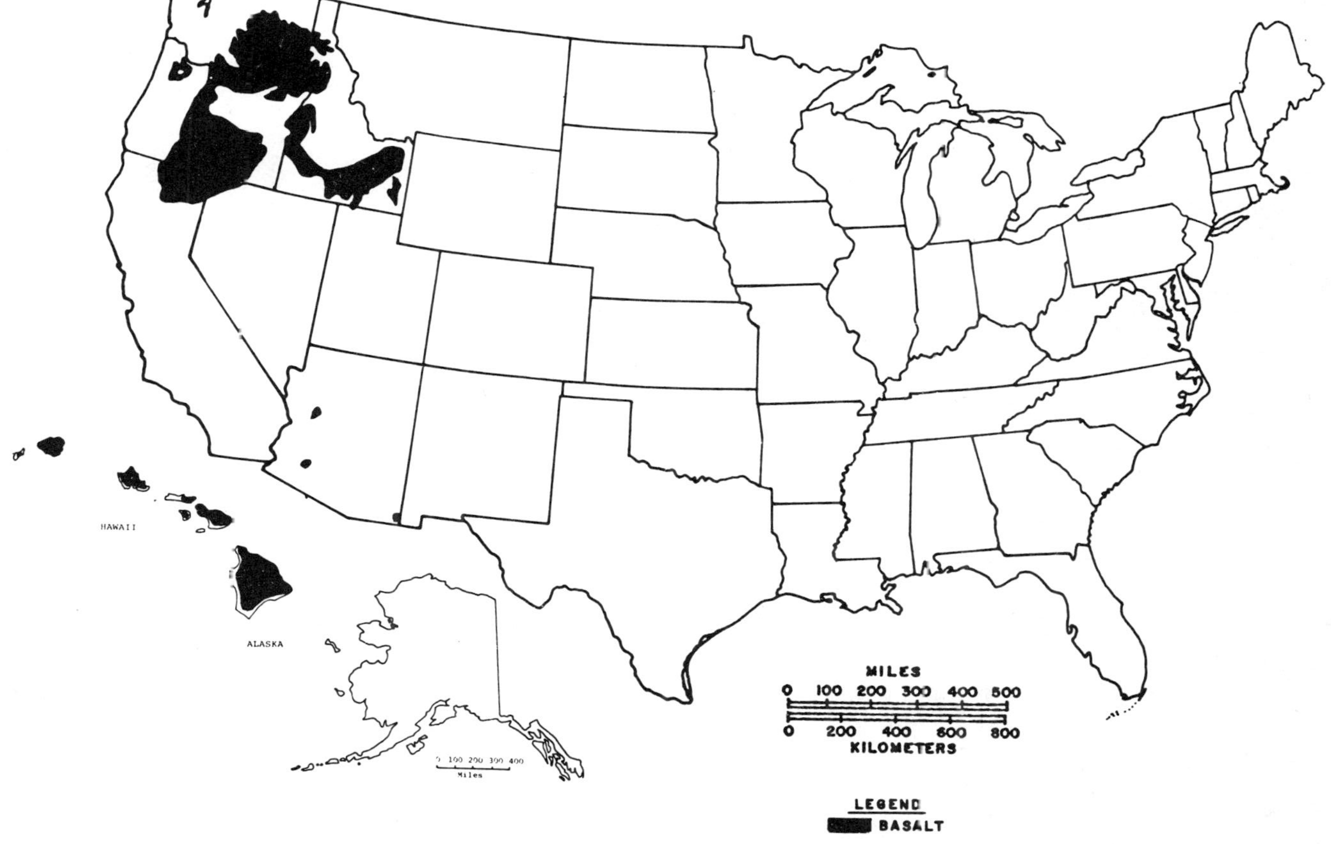

Figure 27. Basalt aquifers.

ity with total dissolved solids concentrations in the range of 100 to 300 ppm.

Carbonate Aquifers

Limestone and dolomite are relatively dense rocks composed of calcium and magnesium carbonate. However, in some places, ground water has partly dissolved the rock, increasing pore space and permeability. As a result, some carbonate rocks are among the world's most prolific aquifers.

Carbonate aquifers underlie large areas of the United States. The principal carbonate aquifers are shown in Figure 28. Carbonate aquifers may exist under either water-table or artesian conditions, but artesian conditions are most common, except where the aquifers outcrop at the surface. Recharge is by direct precipitation and leakage from surface-water bodies.

Karst topography is the ultimate development by erosion of a carbonate aquifer. In this situation, ground water has so dissolved the rock that extensive subterranean caverns and channels form. At the surface, karst topography is manifested by the lack of surface drainage, rivers that disappear underground and emerge at another location, and undrained surface depressions.

Especially where carbonates outcrop at the surface, the aquifers are highly vulnerable to contamination. The enlarged pore spaces in the rock provide easy movement for contaminants and very little treatment by natural filtration.

Water from carbonate aquifers is typically hard (high in calcium bicarbonate content), and high in dissolved solids. Other ions, sulfate for example, are present in excessive concentrations in some regions.

Sandstone Aquifers

Major sandstone aquifers occur in many states and constitute the principal water source of many urban and suburban areas (Figure 29). One important aquifer, the Dakota sandstone and its geologic equivalents, underlies all of the north-central and western Great Lakes states. Other productive sandstone aquifers are found in New Jersey, Pennsylvania, Connecticut, Alabama, Georgia, South Carolina, Oklahoma, and Texas. These aquifers usually do not contain just sandstone but are commonly interbedded with shales.

Because of the interbedded shales, which are of low permea-

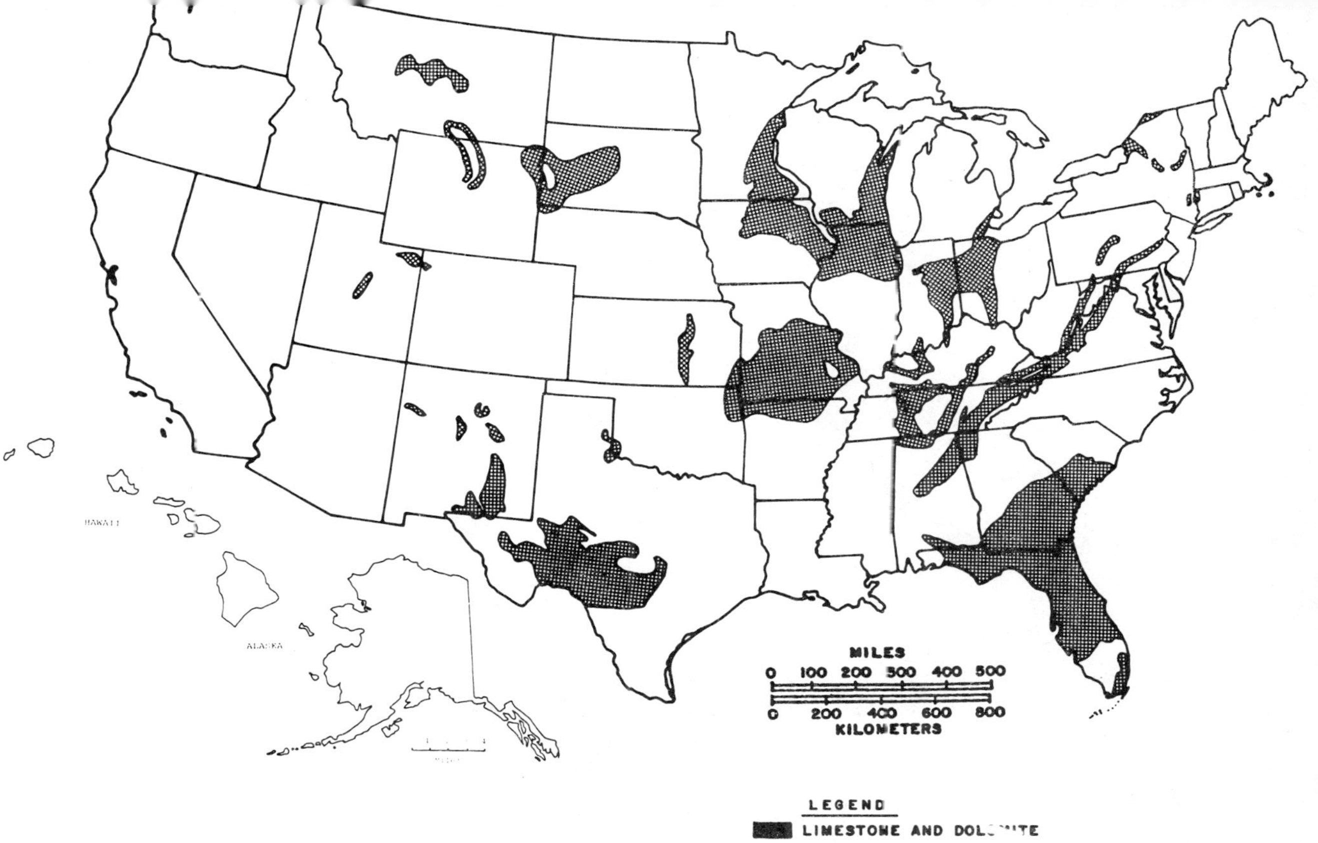

Figure 28. Carbonate aquifers.

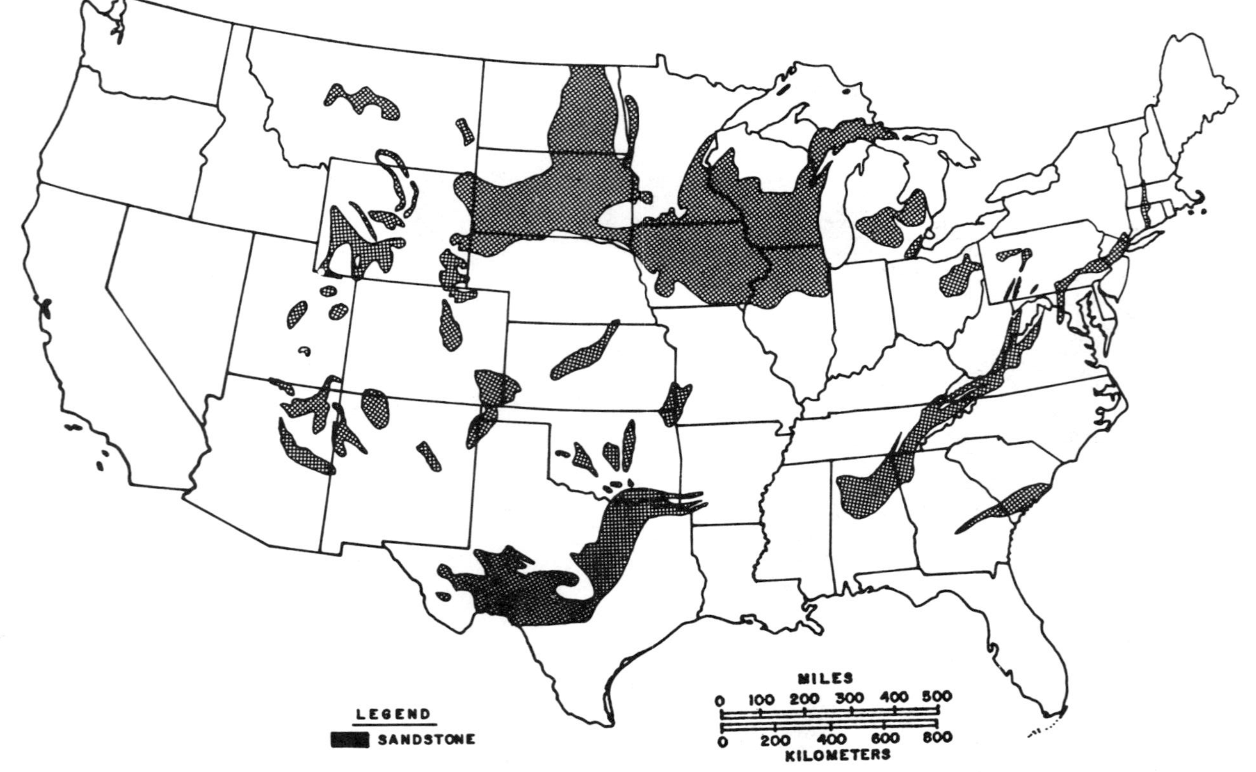

Figure 29. Sandstone aquifers.

bility, water in sandstone aquifers exists under artesian conditions except in the outcrop area. Artesian sandstone aquifers have been heavily pumped, particularly in the north-central states. Withdrawals have not been balanced by natural recharge and serious water-level declines have resulted. When the northern Great Plains were settled, wells tapping the Dakota sandstone flowed naturally; now, they must be pumped.

Where sandstones are exposed at the surface or underlie thin soils, contaminants can enter the aquifer directly. Under artesian conditions, sandstones are better protected, but in areas where significant overpumping of deeper zones has occurred, contamination does occur by inter-aquifer flow. Little natural treatment or filtration is provided by the more permeable sandstones.

Water quality in the outcrop and recharge areas of sandstone aquifers is typically good, but mineralization increases rapidly with depth and distance from recharge areas. High dissolved solids and excessive concentrations of iron and manganese occur in some regions, particularly where the interbedded shales comprise a considerable portion of the aquifer.

Crystalline Rocks (Igneous and Metamorphic)

In the unglaciated areas (the Appalachian Piedmont and California, for example) and on the inter-stream uplands in the glaciated region (New England and the Adirondack Mountains of New York), crystalline rock aquifers are tapped for small ground-water supplies (Figure 30). Individual well yields average 2 to 10 gpm (0.008 to 0.04 cu m/min) and rarely exceed 50 gpm (0.19 cu m/min).

Water from crystalline rock occurs principally in fractures and joints in the weathered zone; intergranular porosity is nil. Although wells in crystalline rock commonly are 165 to 330 ft (50 to 100 m) deep, most of the water is derived from the upper 100 ft (30 m). An exception is when a well penetrates a deep fault zone in crystalline rock.

Ground water in crystalline rock is generally under water-table conditions because the network of joints and fractures in the weathered zone extends to the surface. Recharge is by direct precipitation, but where a fracture connects to an adjacent surface-water body, considerably more recharge may take place.

Crystalline rock aquifers are highly susceptible to contamination. Overlying soils are commonly thin, and provide lit-

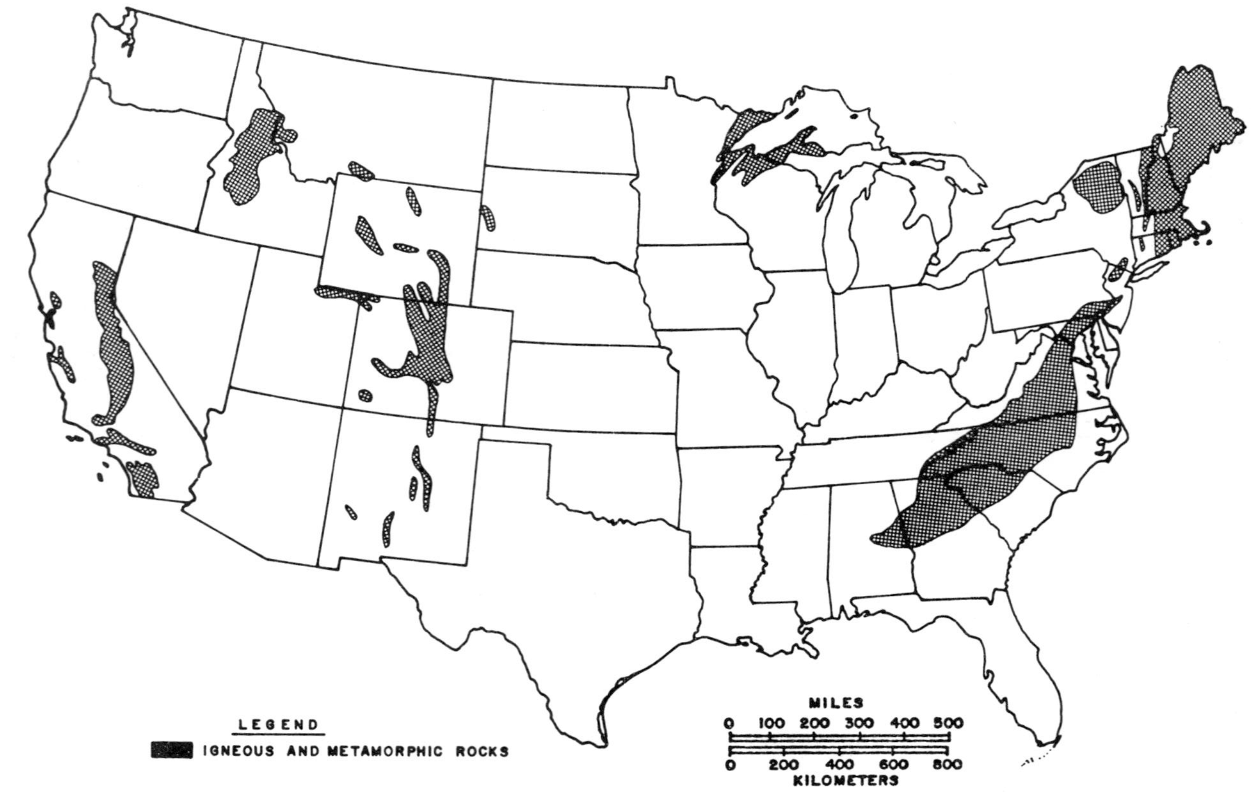

Figure 30. Crystalline rock aquifers.

tle retention of contaminants. Because ground-water movement through fractures can be rapid, little treatment takes place in the aquifer.

The water in crystalline rocks is generally soft and of excellent quality, reflecting the low solubilities of the major minerals in the rock. At isolated locations in some western states, however, highly mineralized water, often of high temperature, may discharge from springs fed by very deep fractures.

NATURAL CHEMICAL QUALITY OF GROUND WATER

All ground water contains chemical constituents in solution. The kinds and amounts of constituents depend upon the environment, movement, and source of the ground water. Typically, concentrations of dissolved constituents in ground water exceed those in surface waters. Salinity tends to be higher in arid regions and in areas where drainage is poor.

Chemical constituents originate primarily from solution of rock materials. Common chemical constituents of ground water include:

Cations	Anions	Undissociated
Calcium	Carbonate	Silica
Magnesium	Bicarbonate	
Sodium	Sulfate	
Potassium	Chloride	
	Nitrate	

Within a large body of ground water, the natural chemical composition or type of water tends to be relatively consistent, although the concentrations of individual minerals in solution may be variable from place to place. Time variations of ground-water quality under natural conditions are minor in comparison with surface-water quality changes. In a few isolated cases, significant concentrations of such hazardous constituents as arsenic and radioactive elements have been found to occur naturally. These instances are related to the unique aquifer materials of the area.

The chemical quality of ground water is often conveniently described for domestic and industrial use in terms of its hardness and salinity. Hardness is a measure of the calcium and magnesium content and is usually expressed as the equivalent amount of calcium carbonate. Figure 31 shows ranges of hardness in ground water in the United States.

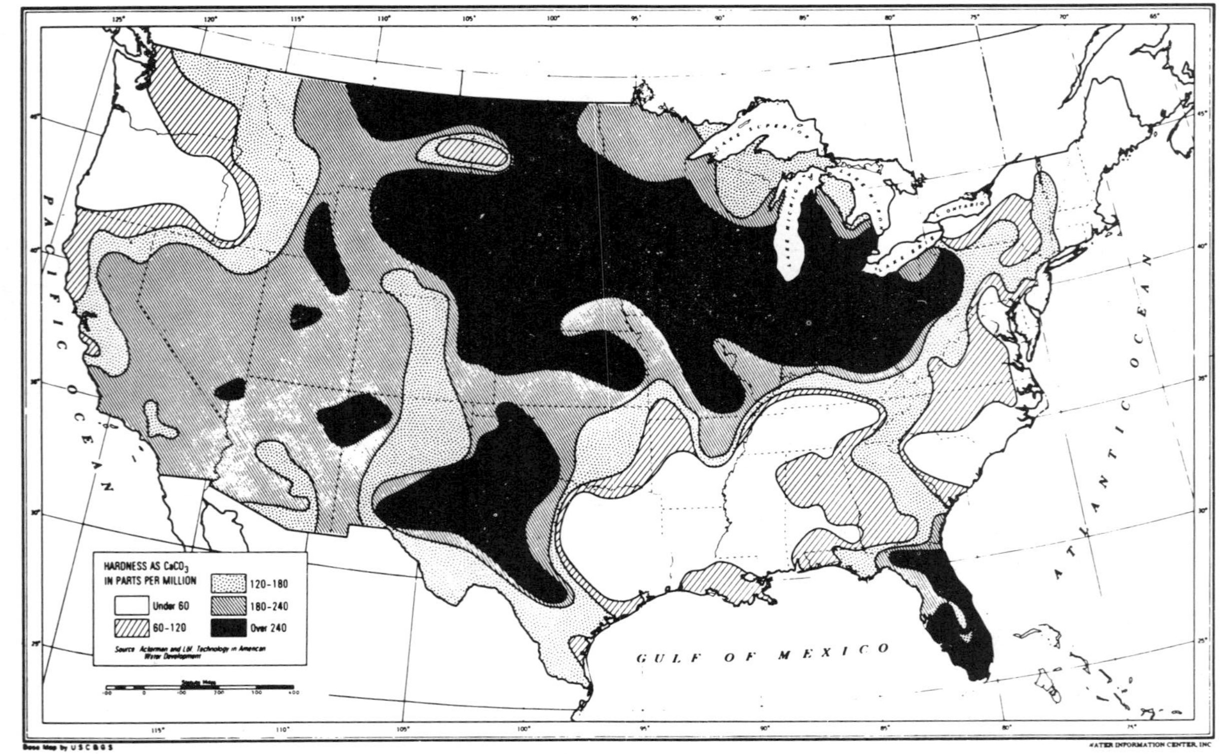

Figure 31. Hardness of ground water. 2)

The definition of saline water varies somewhat depending on the intended use of the water. The recommended limit on total dissolved solids for drinking water established by the U. S. Public Health Service is 500 ppm. However, where such water is not available, water containing 1,000 ppm or more is used for drinking water purposes. The Safe Drinking Water Act of 1974 considers waters containing up to 10,000 ppm as "potential" sources of drinking water.

Naturally occurring saline ground water may be classified by origin into four types, all of which generally exceed 10,000 ppm total dissolved solids content:

1. Connate water
2. Intruded sea water
3. Magmatic and geothermal water
4. Salt leaching and evapotranspiration products

Many sedimentary rock formations were originally deposited in a marine environment. Saline water may have remained trapped in the material throughout geologic history until the present. Such water is termed connate water. In many aquifers, infiltration and subsurface flow have flushed the connate water from the aquifers and replaced it entirely with fresh water. In other areas, for example in western Washington, such flushing is incomplete, and connate water is still present.

Sea-water intrusion has occurred in several coastal groundwater basins hydraulically connected to the sea where, as a result of pumping, the head of fresh water has been lowered relative to that of sea water. This lowering has resulted in landward movement of sea water in the aquifer.

Magmatic water is water derived from molten igneous rock or magma. It is also called juvenile water. Geothermal water is water of any origin, including precipitation, which has been heated by a geothermal source. These mineralized waters may issue forth as hot springs. The concentrations of some constituents, particularly sodium, potassium, calcium, and chloride, may be very high, ranging in some cases up to many tens of thousands of ppm. Various heavy metals are also commonly present. Most of these types of saline water occurrences are concentrated in the arid states, in areas of relatively recent volcanic and intrusive activity.

Evapotranspiration and accumulation of residual salts may produce relatively large bodies of shallow saline water. Precipitation, percolating through the unsaturated zone, continually dissolves various soluble salts and flushes them through to ground water. This action produces a large portion of the total dissolved solids found in subsurface water. Ground water also is changed in composition as it moves through an aquifer because of contact with the rock materials.

Vegetation tends to absorb relatively pure water through its root systems leaving behind dissolved salts. In areas where precipitation is insufficient to provide flushing to the water table (notably the arid regions of the west), salts may accumulate in the unsaturated zone. The natural accumulation of salts is greatest in the areas of lowest precipitation and areas where natural drainage is restricted. During times of unusually heavy precipitation, these minerals may be leached from the unsaturated zone and carried to ground water.

Most of the geologic formations containing fresh ground water are underlain by waters varying from brackish to highly saline. Approximately two thirds of the United States is underlain by aquifers containing at least 1,000 ppm of dissolved solids. These aquifers may occur very close to the surface or at depths of many thousands of feet. As a general rule, the salinity of ground water increases with depth. Figure 32 shows the distribution of depth to shallowest saline ground water, with saline water defined as that containing 1,000 ppm or more total dissolved solids. It should be noted that in much of the shaded area on the figure, ground water exceeds 10,000 ppm total dissolved solids. The blank areas on the map denote either a lack of information or geologic and hydrologic conditions that rule out the presence of saline ground water at depths within 1,000 ft (300 m) of land surface.

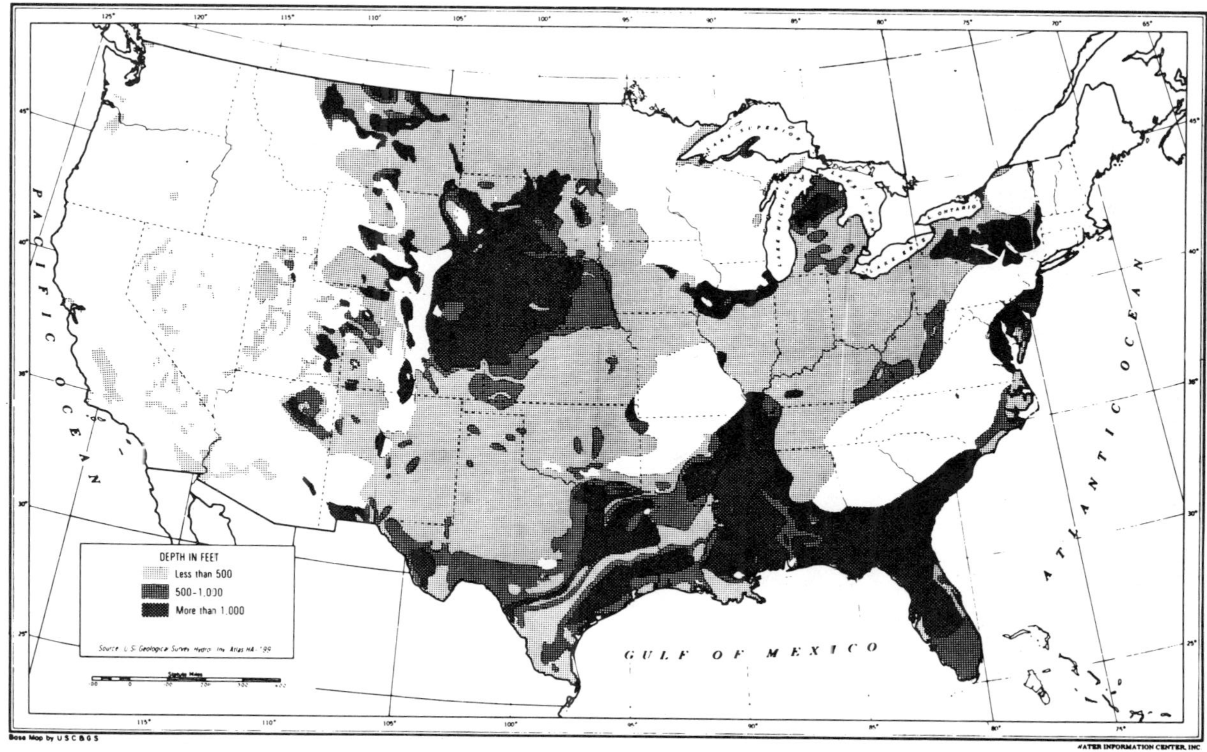

Figure 32. Depth to saline ground water. 2)

REFERENCES CITED

1. Edward E. Johnson, Inc. 1966. Ground water and wells. St. Paul, Minnesota.

2. Geraghty, J. J., and others. 1973. Water atlas of the United States. Water Information Center, Port Washington, New York.

3. Walton, W. C. 1970. Groundwater resource evaluation. McGraw-Hill Book Co., New York.

SECTION V

HOW GROUND WATER IS CONTAMINATED

SUMMARY

Innumerable waste materials and natural and man-made products, with the potential to contaminate ground water, are stored or disposed of on or beneath the land surface. Contaminants found in ground water cover the entire range of physical, inorganic chemical, organic chemical, bacteriological, and radioactive parameters.

Contaminants that have been introduced into ground water can move horizontally or vertically, depending on the comparative density and natural flow pattern of the water already contained in the aquifer. They tend to travel as a well-defined slug or plume but can be reduced in concentration with time and distance by such mechanisms as adsorption, ion exchange, dispersion, and decay. The rate of attenuation is a function of the type of contaminant and of the local hydrogeologic framework, but decades and even centuries are required for the process to be completely effective.

Under the right conditions and given enough time, contaminating fluids invading a body of natural ground water can move great distances, hidden from view and little changed in toxicity by the processes of attenuation. The eventual point of discharge of the contaminated ground-water body can be a well used as a drinking water source.

INTRODUCTION

The many and diverse activities of man produce innumerable waste materials and by-products; these are often deposited or stored on land surfaces where by percolation they may eventually be carried downward modifying the natural quality of the underlying ground water. Because of the large number of such locations, the sources and causes of ground-water contamination in the United States total in the millions. Fortunately, most are small sources whose contaminating effects are rapidly dissipated after they enter the ground. A few are widespread enough to affect large volumes of ground water.

The mechanisms of ground-water contamination are shown by illustrating the flow paths of contaminants for a variety of situations. The flow of ground water within underground formations affects the sizes and shapes of typical zones of contaminated ground water.

Ground-water contamination is the degradation of the natural quality of ground water as a result of man's activities. Contamination may impair the use of the water or may create hazards to public health through poisoning or the spread of disease. The term "contaminant" as defined in the Safe Drinking Water Act, means "any physical, chemical, biological, or radiological substance or matter in water."

Sources of contamination related to waste-disposal practices and described in detail in the following sections are:

1. Industrial Waste-Water Impoundments

2. Landfills and Dumps

3. Septic Tanks and Cesspools

4. Collection, Treatment, and Disposal of Municipal Waste Water

5. Land Spreading of Sludges

6. Brine Disposal from Petroleum Exploration and Development

7. Disposal of Mine Wastes

8. Disposal Wells

9. Disposal of Animal Feedlot Wastes

How contaminants from these waste disposal practices enter the hydrologic cycle via the ground-water system is illustrated in Figure 33.

MECHANISMS OF CONTAMINATION

If it were possible to see zones of ground-water contamination from an aerial vantage point, most would appear so small in relation to the total areas as to be termed scattered points of contamination. Areally extensive sources such as irrigation return flows and sea-water intrusion would be identified as non-point sources. A line source would result, for example, from recharge of sewage effluent in an ephemeral stream channel.

Shallow aquifers are normally the most important sources of ground water for water-supply purposes, but the upper por-

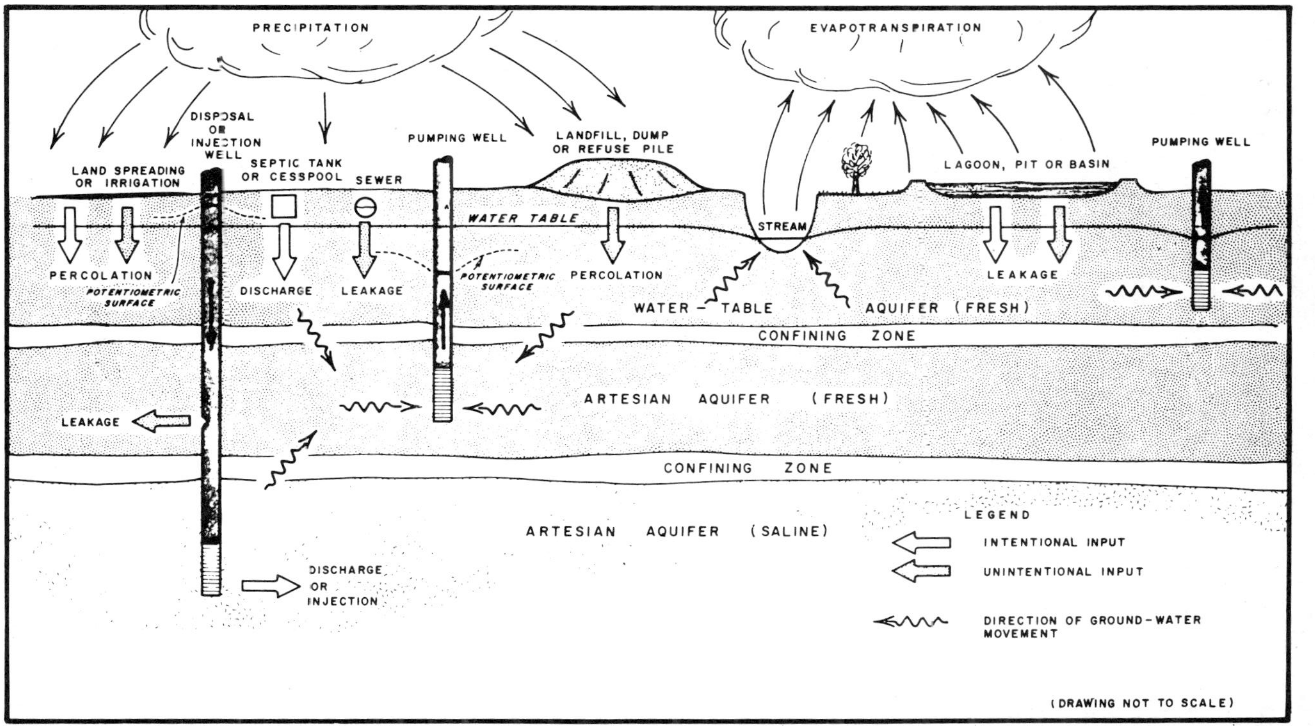

Figure 33. How waste disposal practices contaminate the ground-water system.

tions of these aquifers are also the most susceptible to contamination.

It should be recognized that the configuration of contamination entry into and movement within the underground is unique for each individual source of contamination. Furthermore, because there are many millions of ground-water contamination sources in the United States, it becomes apparent that the possibilities in terms of contaminant movement and distribution are virtually limitless. Notwithstanding this fact, typical flow patterns of ground-water contaminants for a variety of common situations can be described.

The diagrams on the following pages depict some of the frequently occurring contamination geometries. These emphasize vertical cross sections at sources of contamination; horizontal movement of contaminants thereafter is discussed later. Whatever the particular source of contamination may be, these diagrams indicate the hydraulic relationships for a given situation. Where the local hydrogeology is known, paths of probable contaminant movement can be defined. With estimates of permeability and hydraulic gradient available, rates of ground-water movement can be ascertained. Rates of contaminant movement are based on ground-water flow rates, chemical interactions with aquifer materials, and changes in water chemistry. Thus, contaminants travel at velocities equal to, greater than, or less than that of average ground-water flow.

Figure 34 illustrates the flow of contaminants from a surface source such as a disposal pit, lagoon, or basin. Note that the contaminated water flows downward to form a recharge mound at the water table and then moves laterally outward below the water table.

Figure 35 shows cross-sectional and plan views of ground-water contamination caused by a leaking sewer. The contaminant drains downward to the water table and then flows laterally thereafter to form a line source of contamination beneath the sewer.

Figure 36 indicates how contaminated water leached from a chemical or waste stockpile moves downward to the water table and thereafter laterally and vertically to a nearby pumping well.

Figure 37 indicates contaminant movement from a surface stream or lake to a nearby pumping well. The drawdown of the water table induces recharge of surface water to ground water. Because so many municipal water-supply wells are lo-

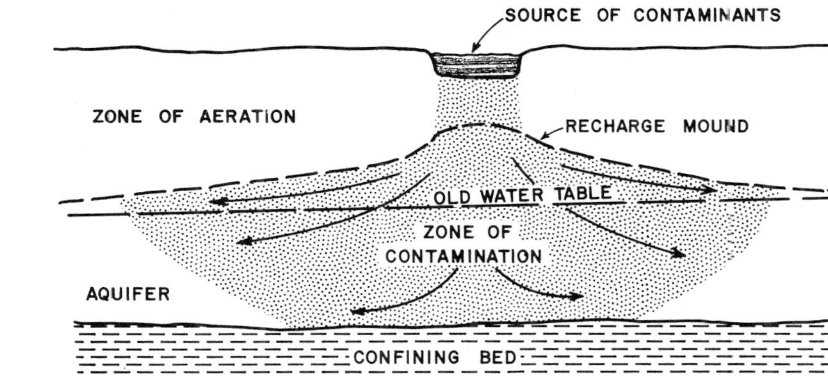

Figure 34. Diagram showing percolation of contaminants from a disposal pit to a water-table aquifer. after 2)

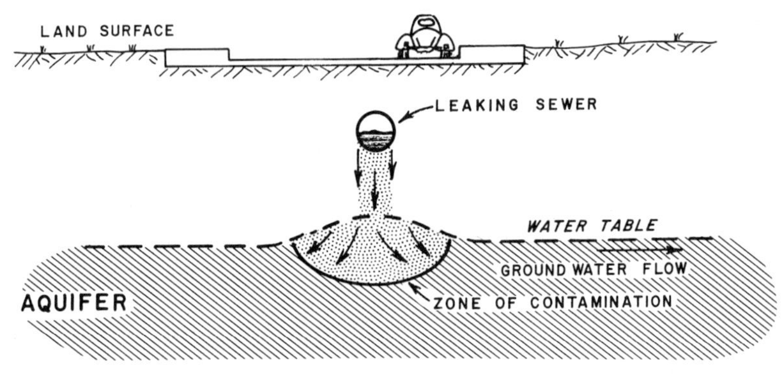

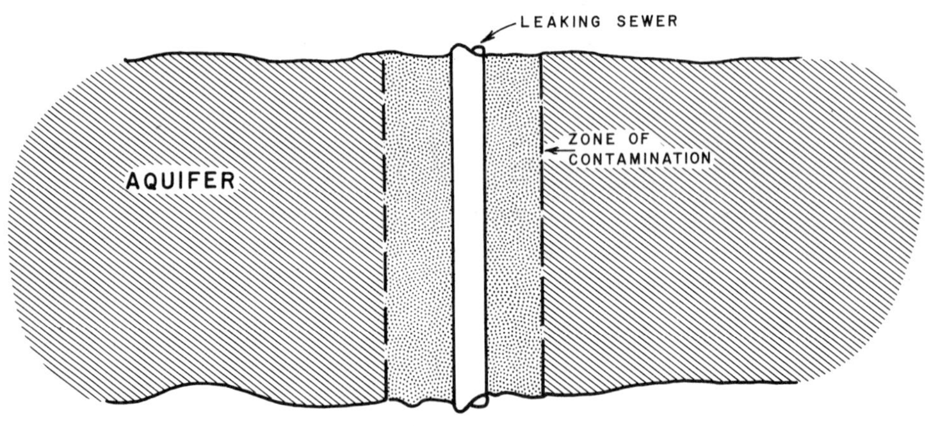

Figure 35. Illustration of a line source of ground-water contamination caused by a leaking sewer.

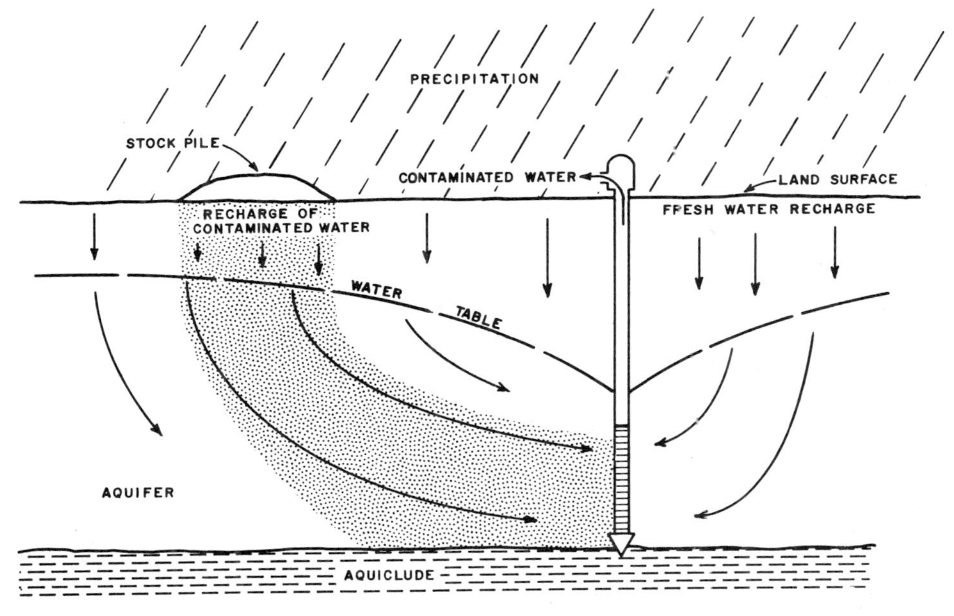

Figure 36. Diagram showing contamination of an aquifer by leaching of surface solids.[2)]

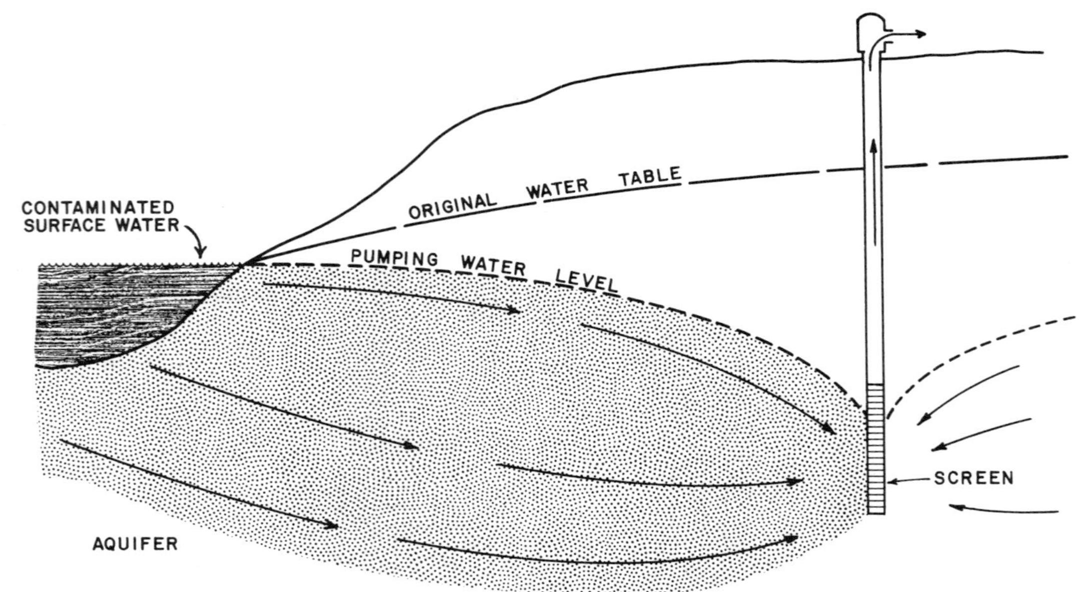

Figure 37. Diagram showing how contaminated water can be induced to flow from a surface stream to a well.[2)]

cated adjacent to rivers in order to insure continuous water supplies, this is an important ground-water contamination mechanism where rivers are polluted.

Figure 38 suggests how temporary flooding of a well can lead to ground-water contamination. Downward flow of polluted surface water occurs around the well casing if the well has been improperly sealed at ground surface.

Figure 39 indicates how contaminants introduced into a disposal well can be transported through the aquifer and lead to contamination of a nearby pumping well. Because a pumping well is a convergence point for ground water over an area, this collection mechanism increases the opportunity for obtaining contaminated water from a pumping well.

Figure 40 illustrates the reversal of underground flows due to pumpage from one aquifer and hence the possibility to degrade the ground-water quality by interaquifer flow. Under natural conditions shown in the upper diagram, the water table of Aquifer A is higher than the potentiometric surface of Aquifer B; therefore, ground water tends to move downward through the semi-permeable zone separating the two aquifers. In the lower diagram, however, pumping has interchanged the relative positions of the two water levels. As a result, the greater pressure in Aquifer B causes water to migrate upward into Aquifer A. If, as is often the case, the lower aquifer is more saline, this will cause the salt content of the upper aquifer to increase.

Figure 41 shows plan and profile views of a recharge pond overlying an unconfined aquifer with a sloping water table and with ground water flowing from left to right. Under these conditions contamination from the pond extends a short distance upstream and is stabilized. The bulk of the contaminants moves away from the pond in a downgradient direction within clearly defined boundaries. For given aquifer and recharge conditions, the lateral spread of the contamination as it moves downstream can be determined. Waste water from a disposal well penetrating an aquifer having the same conditions would move in a similar flow pattern.

Figure 42 suggests how underlying saline ground water can rise due to deepening of a stream channel with a resultant lowering of the water table. This intrusion of saline water occurs because of the reduced head of fresh water.

ATTENUATION OF CONTAMINATION

Contaminants in ground water tend to be removed or reduced

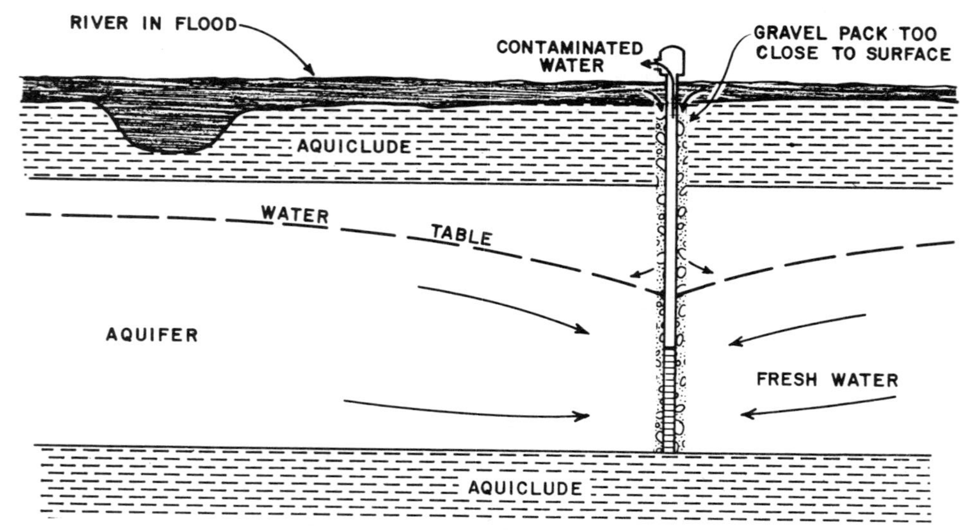

Figure 38. Diagram showing flood water entering a well through an improperly sealed gravel pack.[2)]

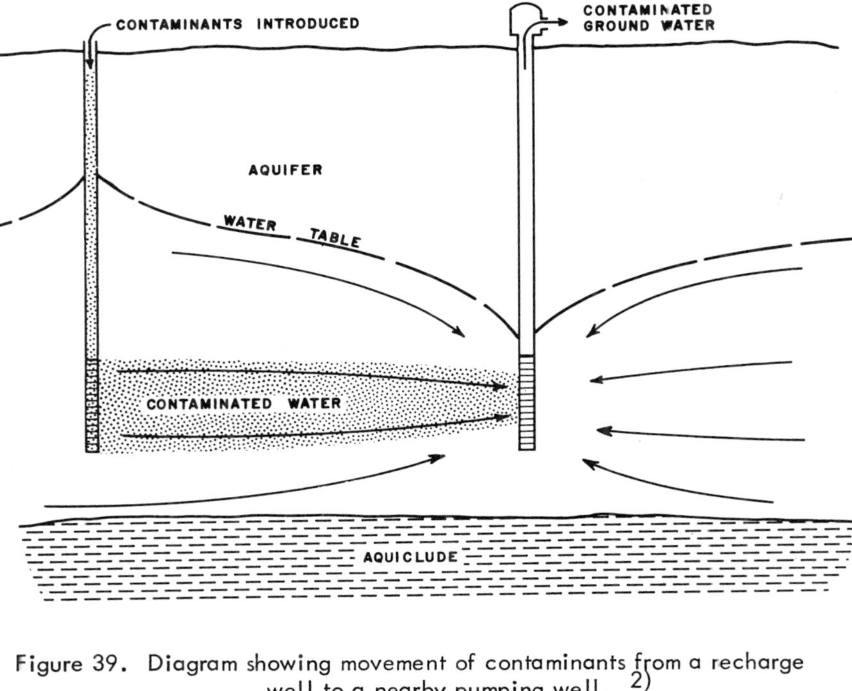

Figure 39. Diagram showing movement of contaminants from a recharge well to a nearby pumping well. [2)]

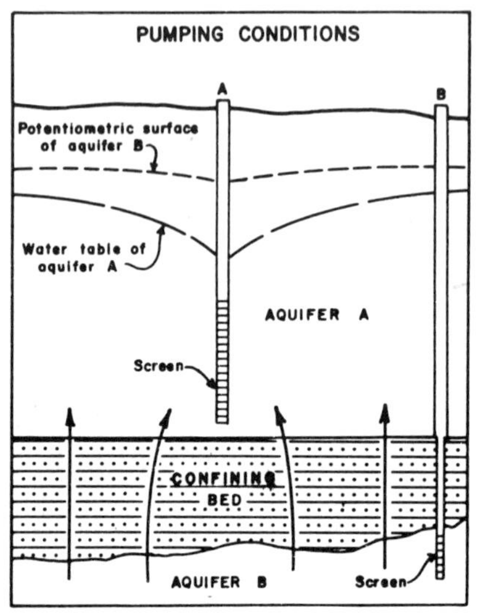

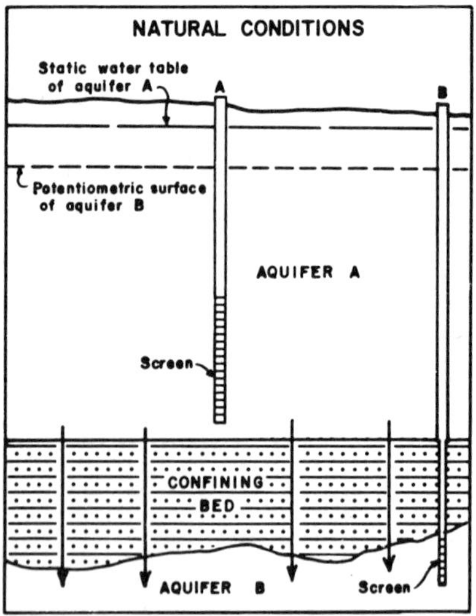

Figure 40. Diagrams showing reversal of aquifer leakage by pumping. 2)

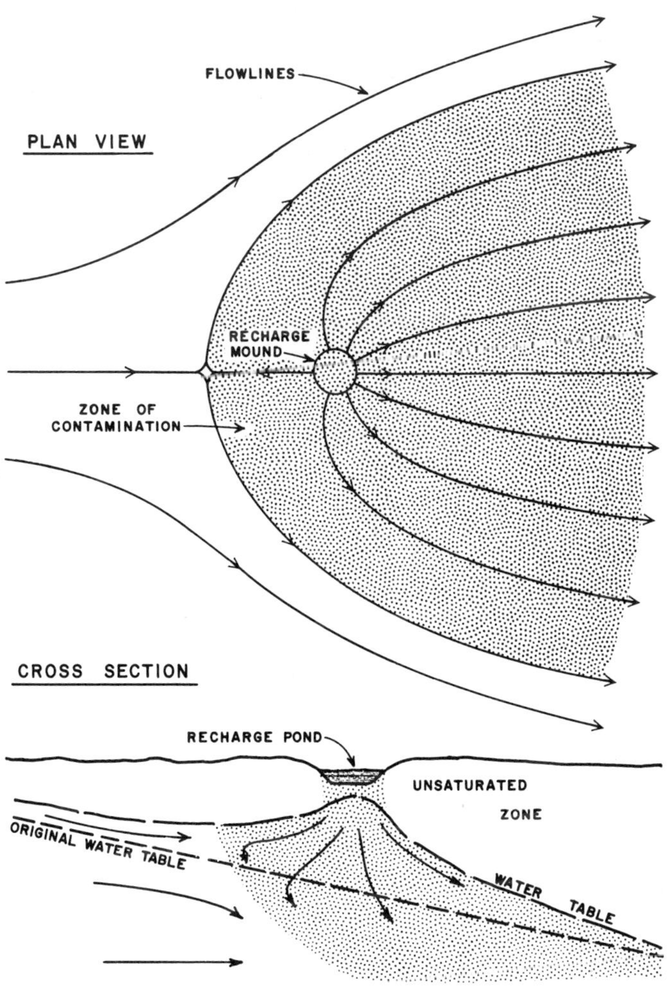

Figure 41. Diagrams showing lines of flow of contaminants from a recharge pond above a sloping water table. 2)

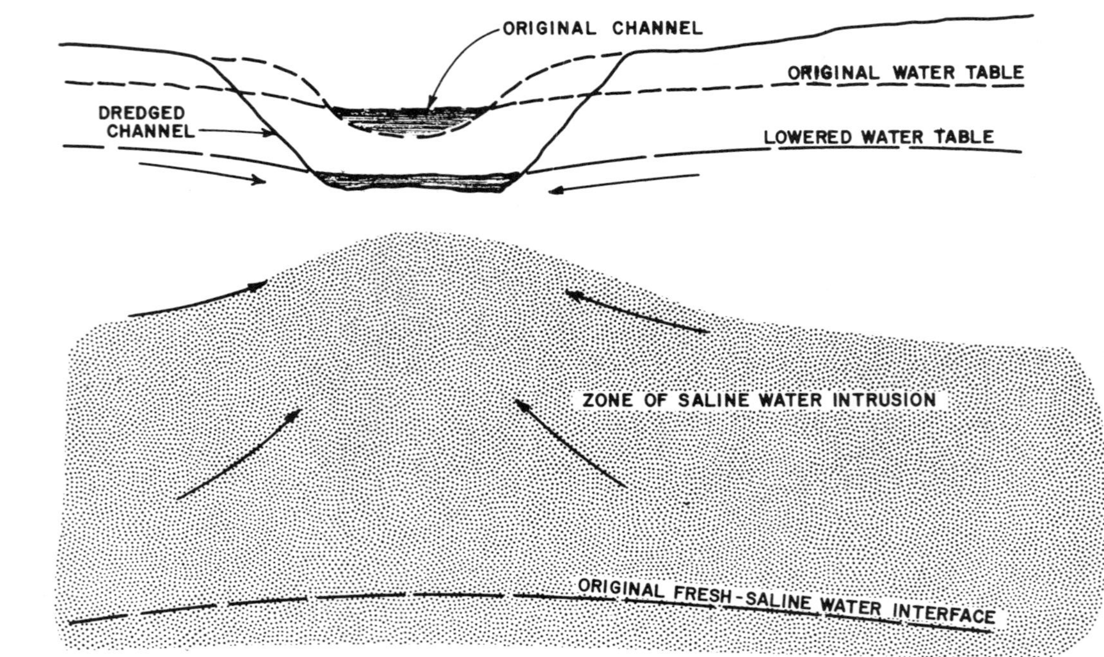

Figure 42. Diagram showing migration of saline water caused by lowering of water levels in a gaining stream. [2)]

in concentration with time and with distance traveled. Mechanisms involved include adsorption and other chemical processes, dispersion and dilution, and decay. The rate of attenuation is a function of the type of contaminant and of the local hydrogeologic framework. Predicting the degree to which contaminants will become attenuated is one of the most difficult -- but also one of the most important -- problems in the design of subsurface waste disposal systems.

Adsorption

Adsorption, in the context of this report, is the phenomenon whereby the surfaces of solids in contact with water are covered with a thin layer of molecules or ions taken up from the water and held tightly by physical or chemical forces. The more finely divided the solid, the greater the surface area per unit volume, which is one of the reasons that clays and silts have greater adsorptive capacities than do sands. When all potential adsorption sites on a surface become occupied, the process becomes one of ion exchange. This is the case through much, or all, of the subsurface-water system.

Percolating water has four options in passing through the unsaturated zone. It can move virtually unchanged, can show a net gain of solute, show a net loss of solute, or keep the same total ionic concentration with a net exchange of ions. Since few soils or sediments are chemically inactive, changes in transported solute are to be expected.

Clay minerals carry a net negative charge on their surfaces. The amount of charge and surface area depends on the mineral type. The negatively charged points on the clay surface hold cations (which carry a positive charge) by electrostatic and van der Waals forces. Usually the attraction is proportional to the positive charge on the cation.

A quantitative exchange is usually observed in which two monovalent ions replace a divalent ion, etc. Heavy metal ions, for example, having more than one unit charge, are attracted to the exchange sites and tend to displace hydrogen, sodium, and potassium ions which are already adsorbed. A net reduction of heavy metal concentrations can occur in this way if percolating water contacts clay in the unsaturated zone. The limit for fixation is the cation exchange capacity (CEC) of the sediment, which can range from nearly zero to probably not more than 60 milliequivalents per 100 grams. When the saturation point is reached at which cations have occupied the available sites, the percolate composition will remain stable. Solution concentrations, pH, and percolation rate affect the reactions quantitatively;

thus, no quantitative predictions can be made without specific operating parameters.

Many soils and sediments have coatings of hydrous oxides of manganese and iron which exert controls on the availability of metal ions, and heavy metals in particular. [3] In fact, the hydrous oxide coating frequently covers clay-mineral surfaces and becomes the truly effective sorptive surface. These coatings exist in amorphous or microcrystalline forms and in themselves exhibit a high specific surface area; up to 300 square meters per gram. The oxygen and hydroxyl groups of the hydrous oxides exert electrical charges which are pH dependent. Therefore, their capacity for sorption is pH dependent.

The dissolution and deposition of the coatings are also dependent upon the oxidation-reduction (redox) potential in the system. This parameter then becomes indirectly important in the adsorption or desorption of heavy metals. Sorption and desorption of metals further depends upon their concentrations in the percolate and upon which ones are present. As with clays, there is an order of selectivity in adsorption. It is quite possible, however, that some heavy metals may move into the ground-water system prior to the exhaustion of exchange capacity.

Dispersion

An understanding of the flow pattern of contaminants is of considerable importance to the understanding of dispersion, and indeed of the entire ground-water contamination picture. Figure 43 illustrates an idealized flow pattern. From this it is seen that the contaminated water moves to its discharge area by a definite route, and is not (as is often imagined) subject to dilution by the entire body of ground water lying between the disposal area and the area of discharge. There is, however, dilution caused by mechanical dispersion, which results from the complexity (on a microscopic level) of the paths followed by the fluid, and (on a macroscopic level) inhomogeneities within the aquifer. Because of this, the contaminated fluid invades the natural ground water to some extent and is concurrently invaded by the latter. Molecular diffusion also takes place, but this is relatively unimportant except when the flow rate of ground water is very low, or the concentration of the contaminant is very high. The latter is associated with high density percolates, which will also distort the idealized pattern by tending to sink to the bottom of the aquifer.

Two related parameters are commonly used in dispersion stud-

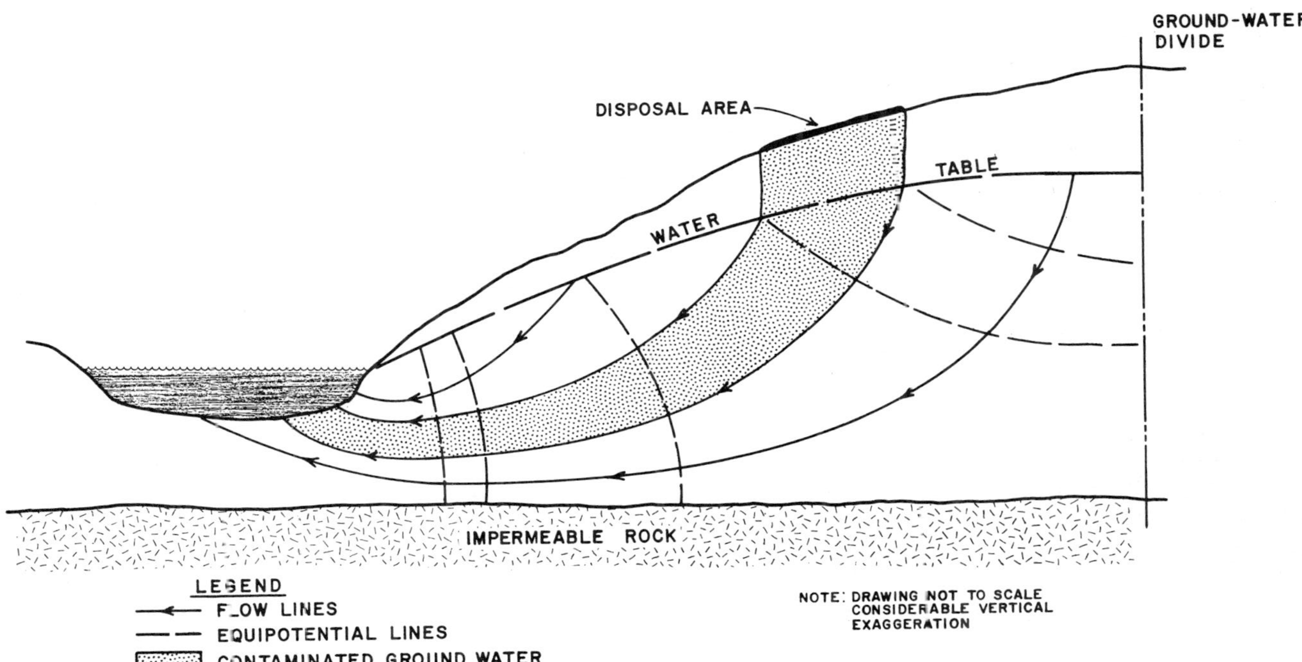

Figure 43. Flow in a water-table aquifer (humid region).

ies. The first, dispersivity, may be described as the inherent capability of the aquifer to cause dispersion. Dispersivity multiplied by ground-water flow velocity gives the dispersion coefficient, which is the dynamic equivalent under actual aquifer conditions. Both are given for longitudinal (in the direction of ground-water flow) and transverse directions.

The rate of ground-water movement within an aquifer is obviously of great importance. It is governed by the hydraulic gradient and aquifer permeability, the latter of which varies far more widely than any other physical property encountered in contamination studies. The U. S. Geological Survey 4) has determined permeabilities for a gravel through which, under a gradient of 10 ft/mi (2 m/km), water would move at the rate of 60 ft/day (18 m/day), and for a clay through which, under the same gradient, the rate of movement would be one ft (0.3 m) in about 30,000 years. Flow rates in most aquifers, however, range from a few feet per day to a few feet per year.

Theoretical solutions are available for the expression of dispersion phenomena. In digital models, these are usually combined with terms for molecular diffusion and adsorption isotherms. Unfortunately, these solutions are either restricted to relatively uncomplicated systems quite unlike those encountered in actual aquifers, or require the input of years of accumulated data to develop the values of otherwise undeterminable parameters. Mechanical dispersion, which is usually predominant in determining the shape of the plume of contamination, is so profoundly affected by heterogeneity that any attempt at detailed prediction is futile. Skibitzke [5] comments that "....the nature of the heterogeneous region can hardly be described through reference to the individual geometric discontinuities. Such a description would require an endless compendium of individual descriptions, a device so obviously impractical that it renders the region not amenable to description by measurement of any of the characteristics visible or accessible from the surface of the region."

One of the most informative studies on the spread of ground-water contamination, and the modeling thereof, is that carried out at the Idaho National Engineering Laboratory (INEL) and reported by Robertson and Barraclough [6], with additional background material in a report by Robertson, Schoen, and Barraclough. [7] Their findings show the state of the art of digital modeling for such purposes, and demonstrate clearly both the powers and the limitations of the method. The following discussion is directed to these ends, and technical details are limited to those necessary for a proper understanding.

The INEL site is on the Snake River Plain in southeast Idaho, overlying an aquifer consisting of thin basaltic flows and interbedded sediments, with a water table about 450 ft (137 m) below land surface. Industrial and low-level radioactive wastes have been discharged to the aquifer through seepage ponds since 1952, and since 1964 cooling tower blowdown has been injected directly into the aquifer through an injection well. The U. S. Geological Survey has monitored the facilities since their inception, and has analyzed the fate of the wastes, using data from about 40 observation wells. The complexity of the subsurface regime, however, is such that no explanation could be given for past behavior, and no predictions could be made about the future. To resolve these questions a digital model, simulating the aquifer, was developed. The modeling included a hydrology phase to solve the equation for ground-water flow, and a solute-transport phase to solve the equation for solute movement, both of which were verified on the basis of historical behavior. The verification procedure is used to adjust the values of various parameters, and Robertson and Barraclough note that the most speculative of these are the dispersivities and distribution coefficients, remarking that there is no effective and practical way of measuring coefficients in the field because of the large-scale aquifer inhomogeneities, and that it is therefore invalid to extend ordinary laboratory measurements to field conditions.

Simulations were made for chloride, a conservative ion; tritium, which is subject to radioactive decay; and strontium-90, which is strongly adsorbed. It was concluded that the model is a valid tool for estimating waste distribution in the aquifer. Even so, the authors warn that this is highly dependent upon future hydrologic conditions, which can only be assumed.

Note that this model (which still provides only a fair to good approximation) required the input of 20 years of data from about 40 observation wells. It would _not_ have been possible to predict the shape and extent of the plumes _a priori_ by means of this or any other model.

The transverse dispersivity value (450 ft or 137 m) required to give the best fit of the theoretical plume to the observed plume is much larger than had been expected from either classic theory or laboratory models. The actual chloride plume, after 16 years, extended about 5 mi (8 km) downgradient and had a maximum width of almost 6 mi (10 km). In contrast, Pinder [8] found a transverse dispersivity value of only 14 ft (4.3 m) in a case of chromium contamination in a glacial aquifer on Long Island. The shapes of the two types of plumes are shown in Figure 44. In this particular case, the shape of the plume of contamination could have been predicted with moderate accuracy from the time that contamination commenced, since the aquifer is fairly homogeneous in two dimensions. Drawing a three degree cone, as suggested by Danel (quoted by Todd [9]), along the flow lines, using the mound formed under the disposal ponds as an apex, gives nearly as good a fit as does the digital model. This approach does not, of course, involve the element of time. For practical purposes, however, it could be applied to similar aquifers to provide a general idea of what the area of contamination would be, but this by no means would eliminate the need for monitoring and periodic analysis of collected data.

Radioactive Decay [10]

Radioactive isotopes may be defined as forms of atoms that are characterized by spontaneous disintegration, with the release of energy. Some occur in nature (e.g., the isotopes of uranium), while hundreds more have been produced artificially. At least one radioactive isotope is known for every element. All of the radioactive and stable isotopes of an element are indistinguishable by chemical means, since they have the same atomic number. The differences are in the mass of the atomic nucleus, and the isotopes are identified by this mass number, as carbon-12 and carbon-14.

Radioactive contaminants of concern to ground-water systems can include waste materials produced from a variety of commercial and governmental activities. Both naturally occurring and so-called artificial or man-made radionuclides are included. By-products and wastes from uranium mining and milling activities contain uranium decay products, for example, which can enter ground-water systems. Ground-water contamination has occurred in conjunction with storage and disposal of nuclear fuel cycle wastes, including high-level liquid wastes leaking from steel tanks into the ground. The foremost example of this occurred at Hanford, Washington. Radioactive contaminants lose their radioactivity at a fixed and unalterable rate that is characteristic of the isotopes involved. This decay rate is expressed in terms of half-life, which is the time lapse required for the loss (per unit mass) of half the radioactivity. Half-lives range from fractions of a second to millions of years; but those of the isotopes of principal concern in ground-water contamination are mostly in the range of tens to thousands of years.

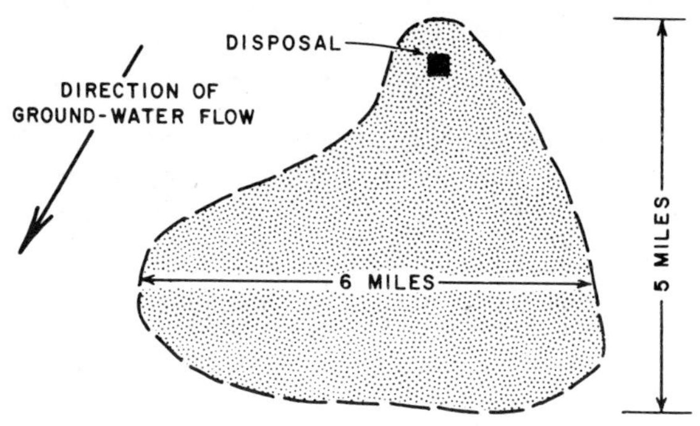

a) CHLORIDE PLUME, INEL, IDAHO
Transverse dispersivity: 450 feet
Time: 16 years

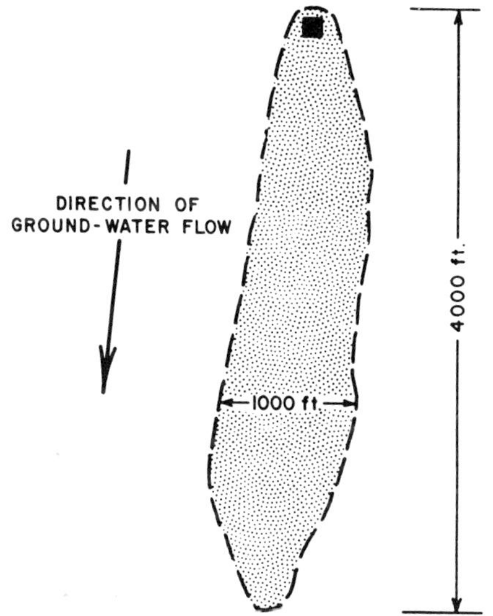

b) CHROMIUM PLUME, LONG ISLAND
Transverse dispersivity: 14 feet
Time: 13 years

Figure 44. Effect of differences in transverse dispersivity on shapes of contamination plumes.

Strontium-90, for example, has a half-life of about 28 years.

Many radioisotopes are members of radioactive decay chains or series wherein the daughters produced by decay are themselves radioactive. One example is the decay of strontium-90 to form yttrium-90, with a half-life of about 62 hours, which in turn decays to the stable zirconium-90. Thus, at any moment, all three isotopes will be present in any media containing strontium-90. Similarly, uranium-238 passes through 14 states of decay before arriving at lead-206, the stable end product.

In considering the rate of movement of radwaste materials into and through ground-water systems, the effects of radioactive decay, dispersion, and adsorption must be considered together. Within the ground-water system, the other mechanisms may be more effective than decay in reduction of radioactive contamination. For example, field data from the Idaho National Engineering Laboratory show a very small plume of strontium-90 as compared with tritium, from radioactive wastes which had entered the ground from various disposal operations. Because strontium-90 has a half-life over twice as long as tritium (28 years versus 12 years), one might expect the strontium to have migrated further than the tritium. The reason for the discrepancy is that strontium-90 is strongly adsorbed in the subsurface while tritium is not adsorbed at all.

Adverse water quality impacts from radionuclides are dependent upon numerous factors, chief of which are concentration, half-life, toxicity, hydrogeologic conditions, and biologic receptors (plants, animals, man). Attenuation in the environment also is dependent upon these factors, which must be mutually considered in evaluating the hazard of a given situation involving radioactive contaminants in ground water.

DISTRIBUTION OF CONTAMINATION UNDERGROUND

Specific statements cannot be made about the distances that contamination will travel because of the wide variability of aquifer conditions and types of contaminants. Also, each constituent from a source of contamination may follow a different attenuation rate, and the distance to which contamination is present will vary with each quality component. Yet certain generalizations which are widely applicable can be stated. For fine-grained alluvial aquifers, contaminants such as bacteria, viruses, organic materials, pesticides, and most radioactive materials, are usually removed by adsorption within distances of less than 328 ft (100 m). But most common ions in solution move unimpeded through these aquifers, subject only to the slow processes of attenuation.

A hypothetical example of a waste-disposal site is shown in Figure 45. Here ground water flows toward a river. Zones A, B, C, D, and E represent essentially stable limits for different contaminants resulting from the steady release of liquid wastes of unchanging composition. Contaminants form a plume of contaminated water extending downgradient from the contamination source until they attenuate to acceptable quality levels.

The shape and size of a plume depend upon the local geology, the ground-water flow, the type and concentration of contaminants, the continuity of waste disposal, and any modifications of the ground-water system by man, such as well pumping.[1] Where ground water is moving relatively rapidly, a plume from a point source will tend to be long and thin; but where the flow rate is low, the contaminant will tend to spread more laterally to form a somewhat wider plume. Irregular plumes can be created by local influences such as pumping wells and variations in permeability.

Plumes ordinarily tend to become stable in areas where there is a constant input of waste into the ground. This occurs for one of two reasons: (a) the tendency for enlargement as contaminants continue to be added at a point source is counterbalanced by the combined attenuation mechanisms, or (b) the contaminant reaches a location of ground-water discharge, such as a stream, and emerges from the underground. When a waste is first released into ground water, the plume expands until a quasi-equilibrium stage is reached. If sorption is important, a steady inflow of contamination will cause a slow expansion of the plume as the earth materials within it reach a sorption capability limit.

An approximately stable plume will expand or contract generally in response to changes in the rate of waste discharge. Figure 46 shows changes in plumes that can be anticipated from variations in waste inputs.

An important aspect of ground-water contamination is the fact that it may persist underground for years, decades, or even centuries. This is in marked contrast to surface-water pollution. The average residence time of ground water is on the order of 200 years; consequently, a contaminant which is not readily decayed or sorbed underground can remain as a degrading influence on the resource for indefinite periods. But the comparable residence time for water in a stream or river is on the order of 10 days; thus, contamination can be rapidly eliminated. Controlling ground-water contamination,

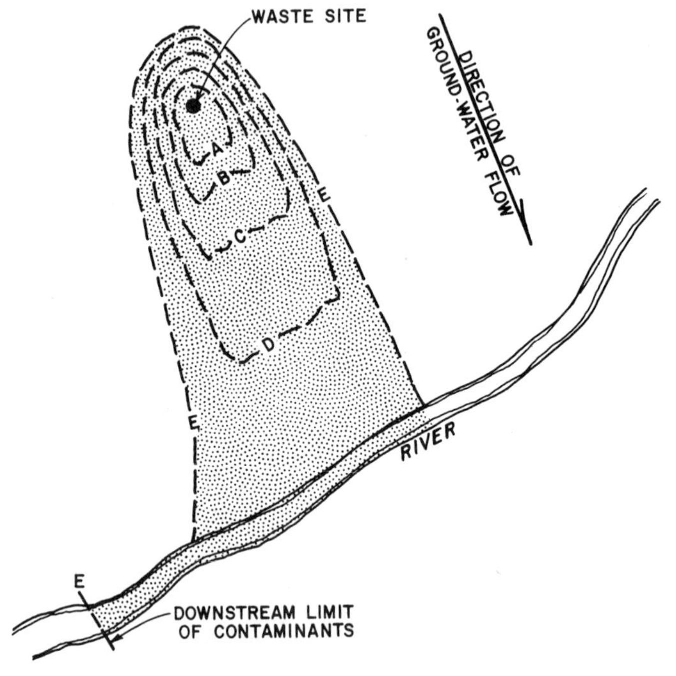

Figure 45. Plan view of a water-table aquifer showing the hypothetical areal extent to which specific contaminants of mixed wastes at a disposal site disperse and move. 1)

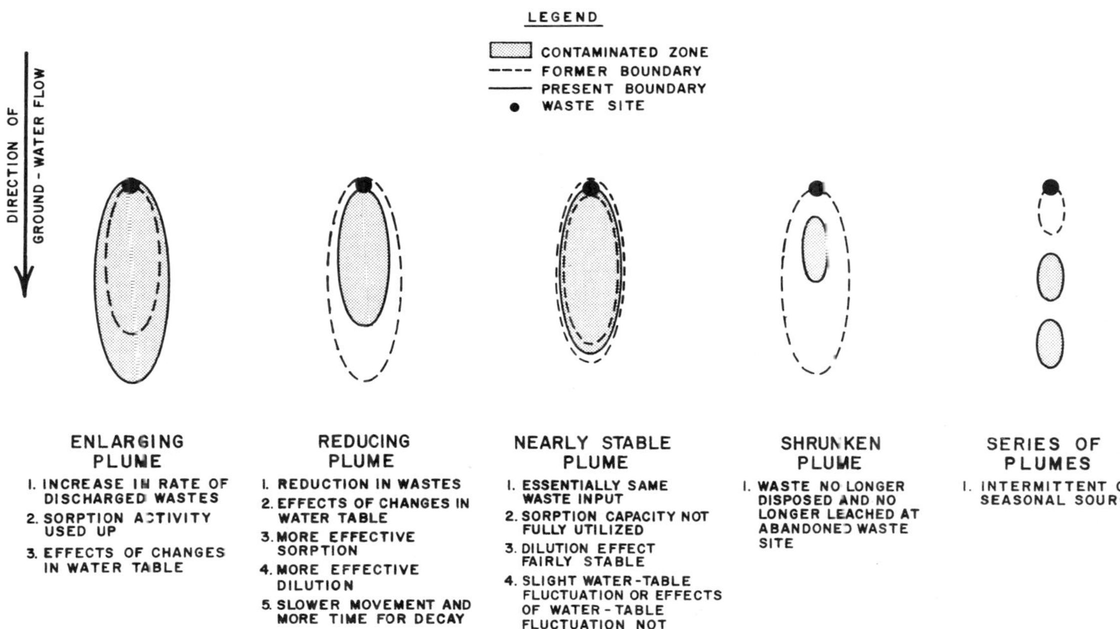

Figure 46. Changes in plumes and factors causing the changes. after 1)

therefore, is usually much more difficult than controlling surface-water contamination. Underground contamination control is best achieved by regulating the source of contamination, and secondarily by physically entrapping and, when feasible, removing the contaminated water from the underground.

REFERENCES CITED

1. LeGrand, H. E. 1965. Patterns of contaminated zones of water in the ground. Water Resources Research, Vol. 1. pp. 83-95.

2. Deutsch, M. 1963. Groundwater contamination and legal controls in Michigan. U. S. Geological Survey Water-Supply Paper 1691. 79 pp.

3. Jenne, E. A. 1968. Controls on Mn, Fe, Co, Ni, Cu, and Zn concentrations in soils and water: The significant role of hydrous Mn and Fe oxides. Pages 337-387 in Robert F. Gould, ed. Trace inorganics in water. Advances in chemistry series No. 73. American Chemical Society, Washington, D. C.

4. Wenzel, L. K. 1942. Methods for determining permeability of water-bearing materials. U. S. Geological Survey Water-Supply Paper 887.

5. Skibitzke, H. E. 1964. Extending Darcy's concept of ground-water motion. U. S. Geological Survey Professional Paper 411-F.

6. Robertson, J. B., and J. T. Barraclough. 1973. Radioactive and chemical-waste transport at National Reactor Testing Station, Idaho; 20-year case history and digital model. Underground Waste Management and Artificial Recharge. American Association of Petroleum Geologists, U. S. Geological Survey, and International Association of Hydrological Sciences. 1:291-322.

7. Robertson, J. B., Robert Schoen, and J. T. Barraclough. 1973. The influence of liquid waste disposal on the geochemistry of water at the National Reactor Testing Station, Idaho. 1952-1970. U. S. Geological Survey Open-file Report.

8. Pinder, George F. 1973. A Galerkin-finite element simulation of groundwater contamination on Long Island, New York. Water Resources Research. 9(6). 1657-1669.

9. Todd, D. K. 1959. Ground Water Hydrology. John Wiley & Sons, New York. 336 pp.

10. Glasstone, Samuel. 1958. Sourcebook on Atomic Energy. 2nd Ed. D. Van Nostrand, Princeton.

SECTION VI

INDUSTRIAL WASTE-WATER IMPOUNDMENTS

SUMMARY

Industrial waste-water impoundments are a serious source of ground-water contamination because of their large number and their potential for leaking hazardous substances which are relatively mobile in the ground-water environment. In some heavily industrialized sections of the nation, regional problems of ground-water contamination have developed where the areal extent and the toxic nature of the contaminants have ruled out the use of ground water from shallow aquifers. Contaminated ground water originating from impoundments at isolated industrial establishments can be even more important because of the potential for migrating to local water-supply wells with no warning.

Either by design, or by accident or failure, surface impoundments of industrial effluent can cause ground-water contamination because of leakage of waste waters into shallow aquifers. Potential contaminants cover the full range of inorganic chemicals and organic chemicals normally contained in industrial waste waters. Those documented as having degraded ground-water quality include phenols, acids, heavy metals, and cyanide.

United States' industries treat about 5,000 billion gal./yr (18 billion cu m/yr) of waste water before discharging it to the environment. Of this volume, about 1,700 billion gal. (6.4 billion cu m) are pumped to oxidation ponds or lagoons for treatment or as a step in the treatment process. Unknown quantities of industrial wastes are also stored or treated in other types of impoundments, such as basins and pits. Based on standard leakage coefficients and volumes of waste waters discharged, it is estimated that more than 100 billion gal./yr (380 million cu m/yr) of industrial effluents enter the ground-water system. This source of contamination is one of the most frequently reported, in spite of the almost complete lack of ground-water monitoring.

One option to correct leaking impoundments is the use of an impermeable barrier or liner. A second is to replace waste-water treatment operations now performed in ponds and lagoons with such alternatives as clarifiers, filtration or centrifugation equipment, and digestion (anaerobic, aerobic).

Impoundments of industrial wastes are normally not subject

to any special regulations unless it is shown that they may affect surface- or ground-water quality. In order to overcome this burden of proof, a few states have developed specific regulations covering such aspects as design of the facility to guard against or minimize leakage, reporting types and volumes of effluent, and installation of monitoring wells.

DESCRIPTION OF THE PRACTICE

In the literature, industrial waste-water impoundments are referred to as "lagoons," "basins," "pits," and "ponds." All these terms are used interchangeably. There is no typical design for an industrial impoundment. It may be a natural or man made depression, lined or unlined, and from a few feet in diameter to hundreds of acres in size. Oxidation lagoons are used for secondary treatment of waste water and are necessarily shallow. Pits are distinguished by a small ratio of surface area to depth. Any one of these can serve as a holding pond, and all are basically surface impoundments. Some lagoons, basins, and pits are intended to discharge liquid to the soil system, while others are designed to be leakproof. The former are unlined structures sited on good infiltrative surfaces; the latter are lined with clay, concrete, asphalt, metal, or plastic sheeting.

Industrial impoundments used for treatment are often unlined, although leakage is not a desirable feature. The nonferrous smelting and refinery industry, for example, utilizes predominantly unlined settling pits and basins for waste and scrubber waters, with some permanent sludge disposal in lagoons. These various types of surface impoundments range up to 40 acres (16 ha) in size. Either by design, or by accident or failure, surface impoundment of industrial effluent can cause serious ground-water contamination because of leakage of hazardous waste waters into shallow aquifers.

In industrial pollution control, ponds are most often used in some form of effluent treatment, frequently biochemical stabilization. Stabilization ponds can be aerobic, anaerobic, or both, and are used primarily to reduce biochemical oxygen demand levels. Aerating devices are used to improve oxygen transfer in many aerobic ponds.

Impoundments can be used for solids separation and are usually called settling ponds in these instances. Dewatering can be accomplished by a combination of evaporation and seepage. Impoundments thus utilized are referred to as evaporation pits, even though seepage of waste water to the ground-water system may be the principal mechanism of disposal. Other forms of treatment are occasionally accomplished in a lagoon, such as reducing the temperature of power plant cooling waters, ammonia reduction, pH neutralization, chemical coagulation and precipitation, and other forms of chemical treatment.

If the discharge rate is small enough, lagoons can be used for permanent disposal. With such a system, when one lagoon is full, wastes are simply diverted to other lagoons which are filled in turn. Such lagoons can be dredged to remove sediment buildup and then reused. Impoundments can also be used for temporary storage of wastes or other materials prior to treatment or use, in which case they are referred to as holding ponds.

A large variety of potentially hazardous wastes are deposited in industrial surface impoundments. Most of the substances contained in these wastes are complex, and many of the constituents that could find their way into ground waters are not normally included in routine analysis of water supplies. General guidelines regarding siting or designing of surface impoundments from the standpoint of protecting ground water from contamination by such wastes have not been enforced until recently. Consequently, industrial wastewater impoundments are frequently constructed to meet criteria such as convenience and lowest possible cost rather than to protect ground-water quality. Most impoundments operate on the principle that some leakage will occur. An evaporation pit may operate successfully in a humid region only if enough leakage takes place through the bottom and sides of the pond to create storage space for continued waste discharges.

Regional and local conditions which influence the contamination potential of a surface impoundment include soil permeability, height of the water table, rainfall, and evaporation. Also to be considered are the types of potentially hazardous materials contained in the wastes. For example, industrial wastes can contain toxic chemicals, such as heavy metals and synthetic organic compounds. If these chemicals enter ground-water supplies, serious health hazards may be created.

Industrial impoundments, therefore, require proper construction to provide protection against both surface-water and ground-water contamination. Artificial liners can be used or the structures can be located on naturally impervious soil. Except for dry climates, as found in parts of the western United States, lined impoundments without some form of discharge may occasionally overflow from rainfall accumulation or flood inundation of the area.

This section is directed toward ground-water contamination as a result of waste-water discharge to lagoons, ponds, pits, and basins. However, it should be noted that industrial waste-water impoundments along with landfills and dumps receive large volumes of sludge and other residuals in addition to liquid effluent. Sludge and other residuals disposed of in surface impoundments can be significant sources of ground-water contamination. The characteristics of sludge and their potential impact on ground-water quality are described in the sections on land spreading of sludges.

CHARACTERISTICS OF CONTAMINANTS

Four major industrial groups generated about 91 percent of the total volume of waste water put into ponds and lagoons in 1968: paper and allied products, 29 percent; petroleum and coal products, 22 percent; primary metals, 22 percent; and chemicals and allied products, 18 percent. [1)] While the number of individual chemical constituents contained in the waste waters of these major industrial categories is very large, general constituent groups that are useful in indicating contamination potential can be identified. Table 7 was modified from the EPA list of waste-water parameters having significant pollution potential by deleting those elements which do not represent a significant threat to ground-water quality. [2)] However, it must be kept in mind that no such table can be complete.

It is important to note that waste-water chemical characteristics are only one factor controlling the severity of ground-water contamination from impoundments. Volume of leakage is a second important consideration and depends upon the ability of the impoundment to seal itself, soil permeability, or the effectiveness of artificial sealing if the impoundment is lined. Geologic and hydrologic conditions along with the characteristics of the soil determine the degree of attenuation of contaminants in the waste water as it moves from the lagoon into and through the ground-water system. Natural water quality, ground-water use, and pumping patterns represent other key considerations.

EXTENT OF THE PROBLEM

Table 8 shows the volume of waste water discharged by the major industrial groups during the years 1959, 1964, 1968, and 1973. [1,3,4)] While there was an upward trend from 1959 to 1968, the total volume discharged in 1973 is nearly identical with the 1968 total. There were significant changes in volume discharged by several of the smaller industrial groups. However, the four key groups discussed in some de-

Table 7. INDUSTRIAL WASTE-WATER PARAMETERS HAVING OR INDICATING SIGNIFICANT GROUND-WATER CONTAMINATION POTENTIAL.[2)]

PAPER AND ALLIED PRODUCTS

Pulp and Paper Industry

COD	Phenols	Nutrients (nitrogen and phosphorus)
TOC	Sulfite	Total Dissolved Solids
pH	Color	
Ammonia	Heavy metals	

PETROLEUM AND COAL PRODUCTS

Petroleum Refining Industry

Ammonia	Chloride	Nitrogen
Chromium	Color	Odor
COD	Copper	Total Phosphorus
pH	Cyanide	Sulfate
Phenols	Iron	TOC
Sulfide	Lead	Turbidity
Total Dissolved Solids	Mercaptans	Zinc

PRIMARY METALS

Steel Industries

pH	Cyanide	Tin
Chloride	Phenols	Chromium
Sulfate	Iron	Zinc
Ammonia		

CHEMICALS AND ALLIED PRODUCTS

Organic Chemicals Industry

COD	TOC	Phenols
pH	Total Phosphorus	Cyanide
Total Dissolved Solids	Heavy metals	Total Nitrogen

Table 7 (Continued). INDUSTRIAL WASTE-WATER PARAMETERS HAVING OR INDICATING SIGNIFICANT GROUND-WATER CONTAMINATION OR POTENTIAL.[2)]

CHEMICALS AND ALLIED PRODUCTS (Continued)

Inorganic Chemicals, Alkalies and Chlorine Industry

Acidity/Alkalinity	Chlorinated Benzenoids and Polynuclear Aromatics	Chromium
Total Dissolved Solids	Phenols	Lead
Chloride	Fluoride	Titanium
Sulfate	Total Phosphorus	Iron
COD	Cyanide	Aluminum
TOC	Mercury	Boron
		Arsenic

Plastic Materials and Synthetics Industry

COD	Phosphorus	Ammonia
pH	Nitrate	Cyanide
Phenols	Organic Nitrogen	Zinc
Total Dissolved Solids	Chlorinated Benzenoids and Polynuclear Aromatics	Mercaptans
Sulfate		

Nitrogen Fertilizer Industry

Ammonia	Sulfate	COD
Chloride	Organic Nitrogen Compounds	Iron, Total
Chromium	Zinc	pH
Total Dissolved Solids	Calcium	Phosphate
Nitrate		Sodium

Phosphate Fertilizer Industry

Calcium	Acidity	Mercury
Dissolved Solids	Aluminum	Nitrogen
Fluoride	Arsenic	Sulfate
pH	Iron	Uranium
Phosphorus		

Table 8. WASTE-WATER DISCHARGE FOR ALL MAJOR UNITED STATES INDUSTRIAL GROUPS, 1959, 1964, 1968, and 1973. (Billions of gallons) 1,3,4)

Industrial Groups	1973	Total Waste Water Discharged 1968	1964	1959
Primary metals	4,756	4,696	4,093	3,142
Chemicals and allied products	3,911	4,175	2,866	2,470
Paper and allied products	2,300	2,078	1,888	1,774
Petroleum and coal products	1,158	1,217	1,130	1,118
Food and kindred products	744	752	545	470
Transportation equipment	227	293	206	204
Stone, glass and clay products	191	218	121	158
Machinery, except electrical	165	180	119	136
Textile mill products	160	136	86	79
Rubber and plastic products	142	128	120	89
Lumber and wood products	122	92	92	100
Fabricated metal products	100	65	35	28
Electrical equipment and supplies	97	118	71	77
Instruments and related products	34	36	23	19
Leather and leather products	7	14	9	7
Furniture and fixtures	6	3	2	2
Tobacco	4	4	2	2
Miscellaneous manufacturing industries	11	13	8	16
Total	14,135	14,218	11,416	9,891

tail within this section did not show substantial changes between 1968 and 1973. Because of this, and because the 1968 data provide more detail in certain categories, the 1968 data are primarily used in the following discussions.

Manufacturing industries used over 15,000 billion gal. (57 billion cu m) of water in 1968. This figure excludes hydroelectric power generation and those facilities that used less than 20 million gal./yr (76,000 cu m/yr). More than 14,000 billion gal. (54 billion cu m) were discharged after use. [1] Much of the discharge was returned to a stream or other surface-water body, but a substantial portion was sufficiently contaminated during use to require treatment prior to discharge. In 1968, United States industries treated 4,524 billion gal. (17 billion cu m) of waste water before discharging it to the environment. Of this volume, 1,668 billion gal. (6 billion cu m) was pumped to an oxidation pond or lagoon for treatment or as a step in the treatment process. [1] Table 9 lists the major industrial groups along with the volumes of water each used, discharged, treated, and lagooned in 1968. Figure 47 shows the total volumes of industrial waste water discharged to lagoons and ponds by region throughout the United States. The data for Figure 47 are given in Table 10. For historical comparison, Table 11 indicates the total industrial waste-water discharge for each state in 1959, 1964, 1968, and 1973.

Owing to the lack of data on either the individual or the large-scale impact of industrial waste impoundments on ground-water quality, the present discussion must be limited to the potential impact, based on what is known about leakage, chemical characteristics of industrial waste water, and concentrations of impoundments throughout the nation.

In addition to oxidation ponds and lagoons, primary sedimentation basins are used by industry to remove suspended solids from waste water prior to discharge or as a preliminary treatment step. However, since primary sedimentation basin area is estimated to be less than one percent of oxidation pond and lagoon area, the impact of basins is not included in the following summaries.

Paper and Allied Products Industries

Included within this category are pulp mills, paper mills, paperboard mills, converted paper products companies, paperboard container industries, and building paper and board mills. Altogether, 619 individual establishments representing this industry are considered. [1] Although the volume of water used by these plants was considerably less than that

Table 9 • VOLUMES OF WATER USE AND DISPOSAL FOR ALL MAJOR UNITED STATES INDUSTRIAL GROUPS. 1)
(Billions of gallons in 1968)

Industrial group	Intake	Percent of total	Discharge	Percent of total	Treated discharge	Percent of total	Discharge to pond or lagoon	Percent of total
Paper and allied products	2,252	15	2,078	15	915	20	484	29
Petroleum and coal products	1,435	9	1,217	9	917	20	364	22
Primary metals	5,005	32	4,696	33	1,430	32	362	22
Chemicals and allied products	4,476	29	4,175	29	674	15	304	18
Food and kindred products	811	5	752	5	184	4	72	4
Textile mill products	154	1	136	1	53	1	30	2
Machinery, except electrical	189	1	180	1	24	1	14	1
Stone, clay, and glass products	251	2	218	2	36	1	11	1
Electrical equipment and supplies	127	1	118	1	27	1	8	< 1
Lumber and wood products	118	1	92	1	18	0	7	< 1
Transportation equipment	313	2	293	2	22	0	5	< 1
Leather and leather products	16	0	14	0	9	0	2	< 1
Fabricated metal products	68	0	65	0	9	0	2	< 1
Rubber and plastic products	135	1	128	1	7	0	1	< 1
Instruments and related products	38	0	36	0	12	0	1	< 1
Furniture and fixtures	4	0	3	0	< 1	0	0	< 1
Tobacco	6	0	4	0	< 1	0	NA	-
Miscellaneous manufacturing industries	14	0	13	0	< 1	0	< 1	< 1
Total:	15,412	-	14,216	-	4,524	-	1,668	-

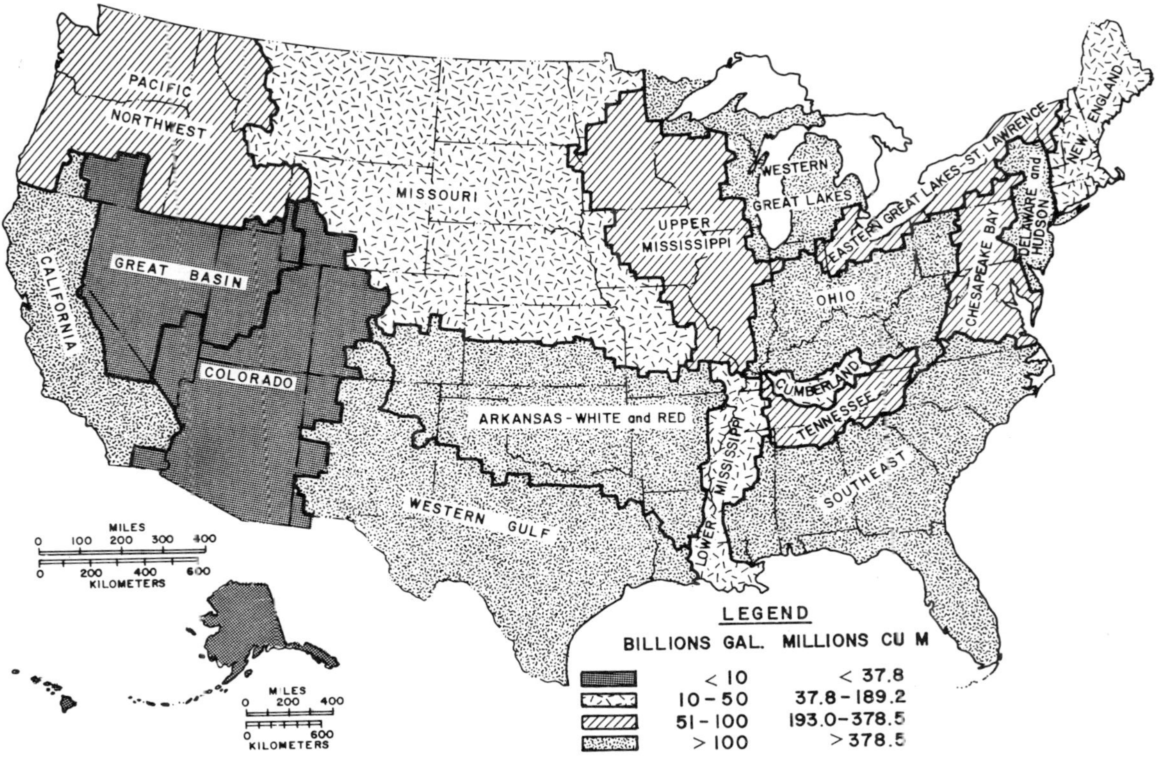

Figure 47. Total industrial waste water treated in ponds and lagoons, 1968.

Table 10. INDUSTRIAL LAGOON DISPOSAL BY REGION. [1]
(In billions of gallons for 1968)

Region	Paper	Petroleum	Metals	Chemicals	Total [a]
New England	3.5	0	0	0	11.7
Delaware and Hudson	8.4	92.7	66.9	66.3	239.2
Chesapeake Bay	21.8	NA	29.3	12.2	69.1
Eastern Great Lakes-St. Lawrence	4.6	0	44.9	16.2	78.9
Ohio River	4.3	7.9	51.8	27.9	104.1
Tennessee	0	None	6.8	58.0	78.8
Southeast	264.4	0	3.0	26.1	324.3
Western Great Lakes	25.0	16.1	115.6	2.6	163.1
Upper Mississippi	1.1	9.1	28.8	17.1	77.1
Lower Mississippi	0	0	NA	2.5	23.8
Missouri	0	14.8	0	1.7	25.7
Arkansas, White and Red	76.4	10.4	0	15.1	103.1
Western Gulf	23.8	119.2	0	21.5	166.4
Colorado Basin	NA	NA	0.6	NA	2.2
Great Basin	None	0	NA	NA	8.9
California	0	73.0	0	14.5	104.7
Pacific Northwest	33.3	0	0	11.9	81.4
Cumberland	NA	None	NA	NA	11.2
Alaska	NA	NA	None	None	0
Hawaii	None	NA	None	None	0

a): Total is for all industries, including the four listed
NA: Information was withheld to protect individual industries
None: No industries of this category in this area

Table 11. INDUSTRIAL WASTE-WATER DISCHARGE - 1959, 1964, 1968, AND 1973.[1] (Billions of gallons per year)

State	1973 Total water discharged	1968 Total water discharged	1964 Total water discharged	1959 Total water discharged
Texas	1,554	1,654	1,455	1,159
Pennsylvania	1,377	1,470	1,475	1,324
Louisiana	1,299	999	843	692
Indiana	1,178	1,072	830	629
Ohio	964	1,128	1,115	979
Michigan	803	738	739	780
West Virginia	511	610	690	540
Illinois	493	652	591	550
New York	491	519	569	587
Tennessee	406	445	287	228
California	387	307	313	277
New Jersey	384	391	395	415
Alabama	370	360	242	208
Maryland	341	404	401	266
Washington	328	333	341	261
Connecticut	297	187	118	128
Virginia	283	362	275	261
Florida	265	264	230	222
Wisconsin	244	272	236	192
Georgia	243	220	213	168
Delaware	182	179	164	155
North Carolina	175	141	146	93
Maine	167	153	163	135
Oregon	150	138	151	134
Massachusetts	148	141	144	169
South Carolina	134	133	101	84
Iowa	125	107	103	89
Mississippi	115	83	65	59
Kentucky	108	122	117	99
Minnesota	83	110	87	92

Table 11(continued). INDUSTRIAL WASTE-WATER DISCHARGE - 1959, 1964, 1968, AND 1973.[1] (Billions of gallons per year)

State	1973 Total water discharged	1968 Total water discharged	1964 Total water discharged	1959 Total water discharged
Missouri	81	88	82	66
Arkansas	73	50	42	33
Hawaii	50	93	102	-
Colorado	43	51	54	40
Montana	39	36	26	37
Idaho	38	53	47	43
Alaska	31	32	34	-
Kansas	30	36	24	22
Utah	27	21	27	22
Nebraska	24	21	24	22
New Hampshire	23	55	35	41
Oklahoma	18	9	10	9
Rhode Island	13	19	16	17
Arizona	10	10	10	5
Vermont	6	6	7	8
Nevada	5	10	4	2
Wyoming	4	3	7	4
New Mexico	2	1	1	1
South Dakota	1	2	5	-
North Dakota	1	1	1	-
Total:	14,124	14,255	13,157	11,736

- Not available

used by the metals and chemical industries, the volume of waste water discharged to ponds and lagoons was higher for the paper industry than any other. In 1968, the paper industry discharged 484 billion gal. (1.8 billion cu m) of waste water to ponds and lagoons, or 29 percent of the total of all impounded industrial waste water in the United States. This represents an increase of over five times the volume for 1954 and the quantity is expected to nearly quadruple by 1983 (Table 12). Figure 48 shows the distribution of impounded waste water from the paper industry in 1968 in various regions across the United States, based on data given in Table 10.

A number of investigators have calculated water seepage from waste-water lagoons, and an average value from these studies is approximately 30 in./yr (76 cm/yr). This value may vary by as much as a factor of 10, depending on local soil conditions, the ability of the surface impoundment to seal itself, and the amount of leakage that can take place before the surface impoundment seals itself. However, for a general estimate, based on 30 in./yr (76 cm/yr), leakage is approximately 6 percent of the total volume of waste water entering ponds and lagoons.

Thus, from Table 10 and Figure 48, the southeast region has by far the greatest impounded waste-water volume with 55 percent of the total, and leakage is estimated at 16 billion gal./yr (61 million cu m/yr). Other regions with significant percentages are Arkansas with 16 percent, Pacific Northwest with 7 percent, and Western Great Lakes, Western Gulf and Chesapeake Bay with about 5 percent each.

Petroleum and Coal Products

This industrial group includes several manufacturing processes; however, the use of lagoons and ponds to treat waste water is virtually confined to the petroleum refining industry. In 1968, 25 plants reporting water consumption over 20 million gal. (76,000 cu m) discharged a total of 1,217 billion gal. (4.6 billion cu m) of waste water. Of this total volume, petroleum refineries ponded or lagooned 363.8 billion gal. (1.4 billion cu m). The regions where impounding of waste water from this industry is most prevalent are shown on Figure 49, and the reported volumes by region are given in Table 10. Three regions dominate the United States in volume of refining waste water impounded: Western Gulf, Delaware and Hudson, and California account for 78 percent of the nation's total. Estimated leakage of petroleum refining lagoons and ponds for 1968 are: Western Gulf region, 7.2 billion gal. (27 million cu m); Delaware and Hudson re-

Table 12. INDUSTRIAL WASTE WATER TREATMENT IN PONDS AND LAGOONS OVER THE PERIOD 1954 TO 1968. [5)]
(In billions of gallons)

	1954	1959	1964	1968
Paper and Allied Products	86	191	311	484
Petroleum and Coal Products	90	168	277	342
Primary Metals	46	96	221	362
Chemical and Allied Products	-	-	227	304

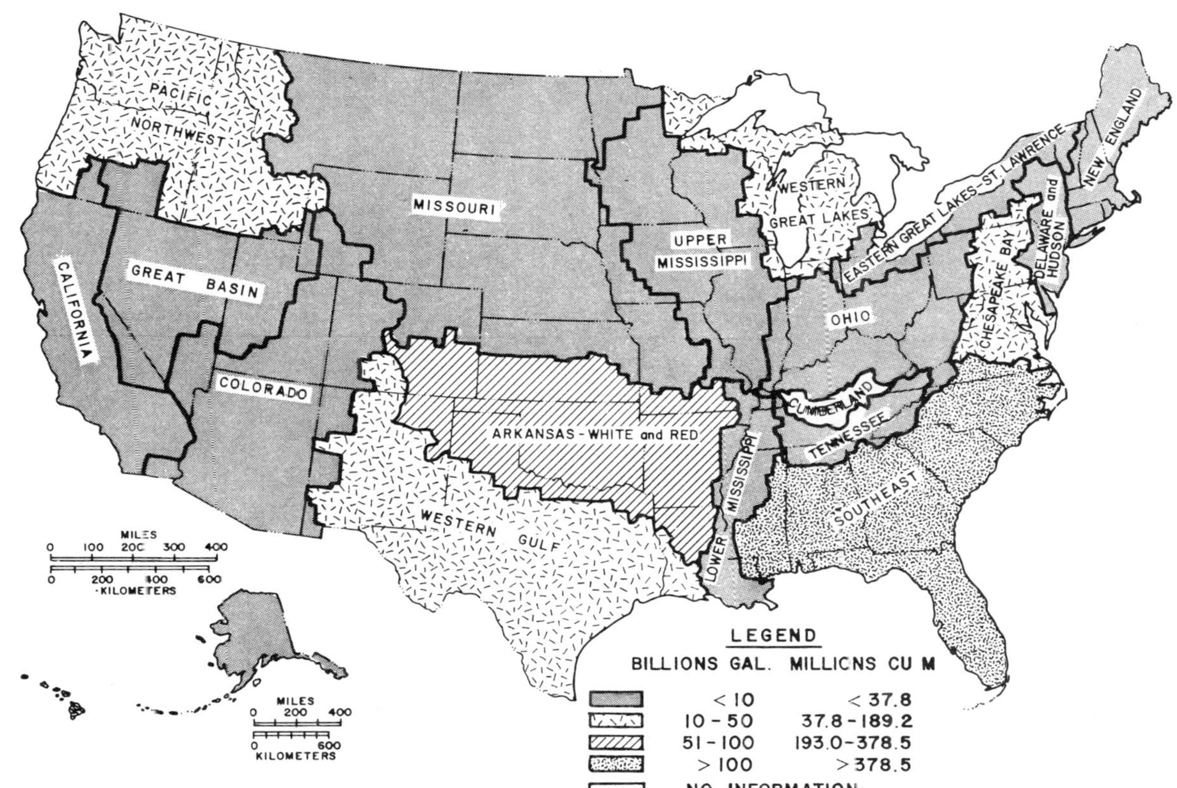

Figure 48. Paper and allied products industries - volume of industrial waste water treated in ponds and lagoons, 1968.

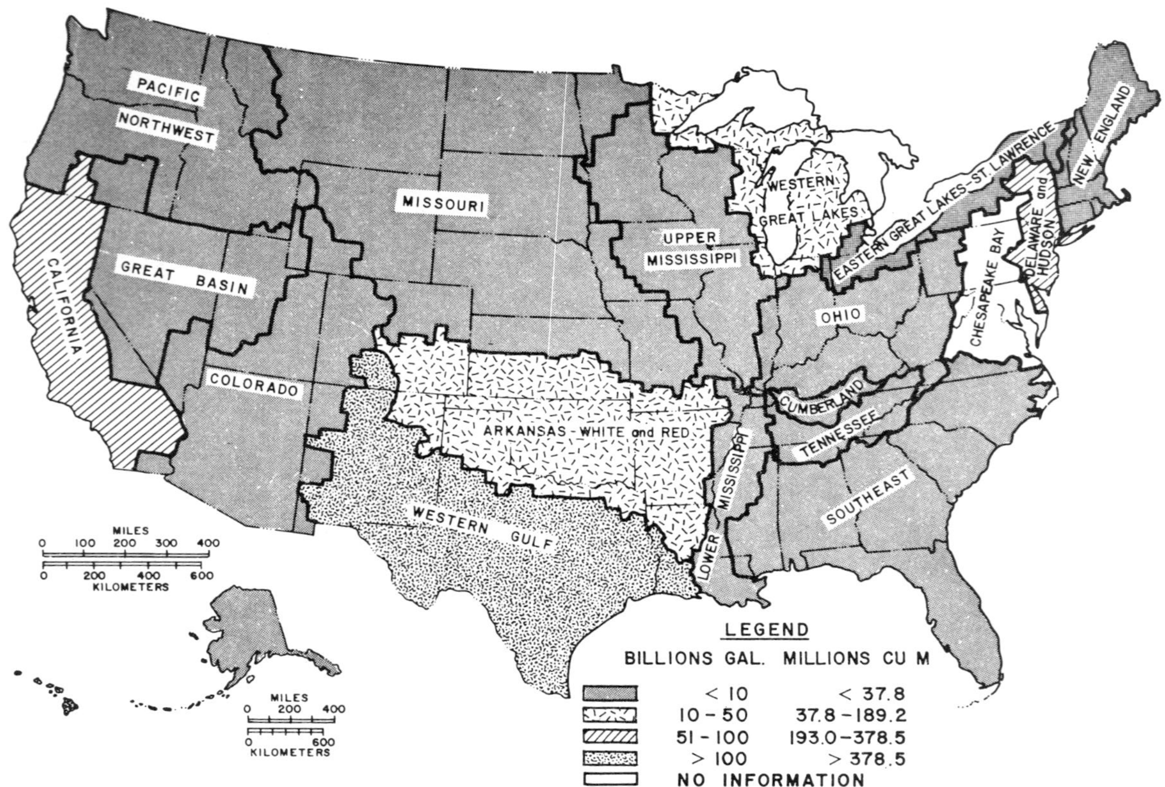

Figure 49. Petroleum and coal products industries - volume of industrial waste water treated in ponds and lagoons, 1968.

gion, 5.5 billion gal. (21 million cu m); and California region, 4.3 billion gal. (16 million cu m). The greatest concentration of pond and lagoon area per square mile would be in the Delaware and Hudson region.

Primary Metals

Of the primary metals industries, blast furnaces and basic steel production generate about 90 percent of the total volume of waste water discharged to ponds and lagoons. Most of the remaining 10 percent is produced by the electrometallurgical products and primary nonferrous metals industries.

As shown in Table 9, the primary metals industries use, discharge, and treat more water than any other industrial group. The volume of this waste water treated in ponds and lagoons in 1968, however, is almost exactly the same as the volume impounded by the petroleum industries and somewhat less than the paper industries.

The volume of waste water discharged to ponds or lagoons in 1968 was reported to be 362 billion gal. (1,375 million cu m). The regional distribution of this volume is shown in Figure 50. The Western Great Lakes region has the greatest volume with 116 billion gal. (440 million cu m) but the Delaware and Hudson region has about the same volume per unit area. Other regions with substantial percentages are the Eastern Great Lakes - St. Lawrence, Ohio River, Chesapeake Bay, and Upper Mississippi.

Estimation of leakage is as follows: Western Great Lakes region, 6.9 billion gal. (26 million cu m); Delaware and Hudson region, 5.6 billion gal. (21 million cu m); Ohio River region, 3.1 billion gal. (12 million cu m); Eastern Great Lakes - St. Lawrence region, 2.7 billion gal. (10 million cu m); Chesapeake Bay region, 1.8 billion gal. (7 million cu m); Upper Mississippi region, 1.7 billion gal. (6 million cu m). Total leakage for the industry in 1968 was about 22 billion gal. (84 million cu m). As shown in Table 12, the projected volume of waste water from the primary metals industry pumped into ponds and lagoons will be 1,029 billion gal. (3,910 million cu m) by 1983.[5] Regional volume distributions are expected to remain roughly proportional to what they were in the 1968 inventory.

Chemical and Allied Products

The chemical industry is the second largest user of water in the nation, but ranks fourth in volume of waste water treated in lagoons and ponds. However, the overall nature of

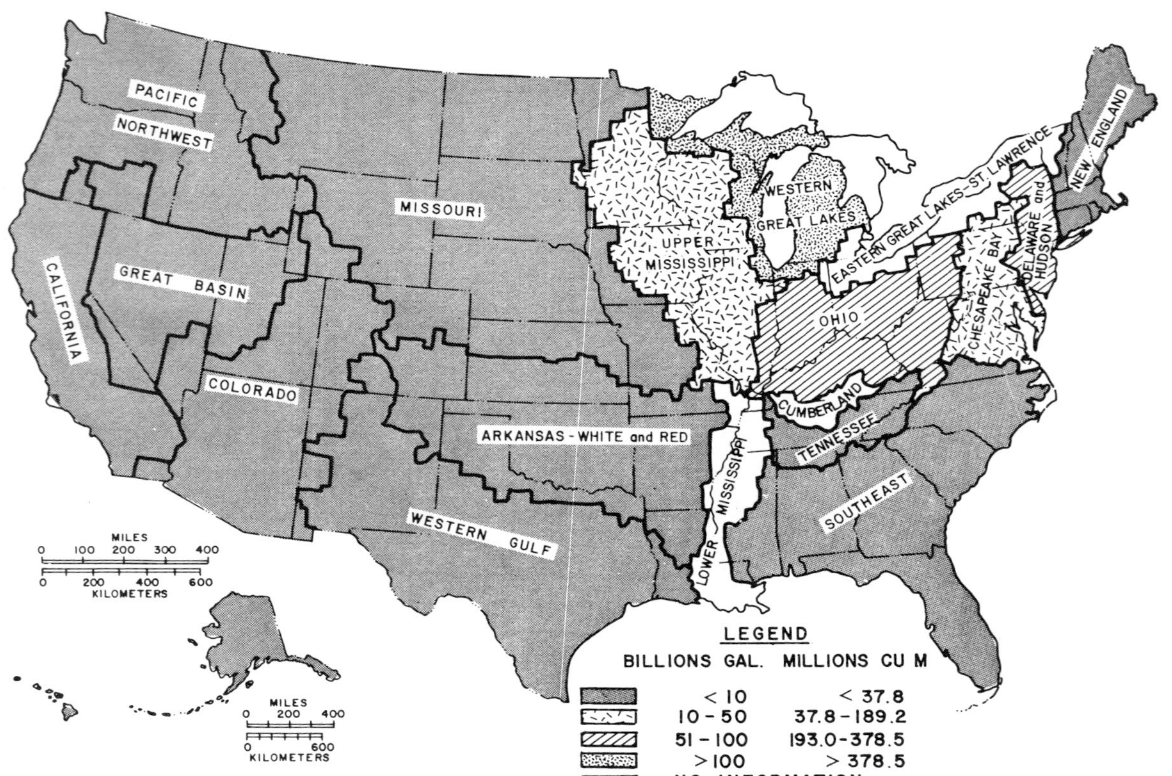

Figure 50. Primary metals industries - volumes of industrial waste water treated

this water is potentially more hazardous than the other industrial categories. The industrial chemicals industries generate 66 percent of the total chemical industries impounded waste water. Other substantial fractions are contributed by the following: plastic materials and synthetics, 20 percent; agricultural fertilizers, 6 percent; and explosives, 5 percent. Drugs, wood chemicals, adhesives and others contribute the remaining 3 percent.

The total volume of chemical industry waste water in treatment lagoons and ponds in 1968 was 304 billion gal. (1,155 million cu m). Figure 51 shows the regional distribution of this volume. The Delaware and Hudson, and Tennessee regions together contributed 41 percent of the total. Most of the remaining volume was fairly evenly divided among the nine regions indicated on Figure 51 as having 10 to 50 billion gal. (38 to 190 million cu m) each (see Table 10).

Estimated leakage from chemical industry impoundments in 1968 is 18 billion gal. (68 million cu m). The two relatively small regions of Delaware and Hudson, and Tennessee, with about 4 billion gal. (15 million cu m) each represent the areas of greatest concentration.

Based on the methods described above, the total estimated leakage from industrial lagoons in the United States in 1968 was 100 billion gal. (380 million cu m). If this volume were concentrated into one place, it would occupy an impoundment 10 ft (3 m) deep, one mi (1.6 km) wide, and 50 mi (80 km) long. If this quantity of liquid were placed in the ground, taking into account soil porosity, it might occupy as much as 5 times this volume. As ground water does not discharge quickly, accumulations of many years' input are likely to be found at any particular location.

Although only the major water-using industries have been discussed above, consideration must be given to the total number of manufacturing establishments that discharge waste water and probably use lagoons, basins, pits, and ponds. About 10,000 plants each used 100,000 gpd (380 cu m/day) or more of water in 1968 [1] and, assuming an average of two surface impoundments per plant, a minimum of 20,000 lagoons and ponds can be estimated. Taking into account that there are altogether about 250,000 water-using establishments in the United States, then a reasonable estimate for total surface impoundments of all types for the nation would be about 50,000. It is interesting to note that in an aerial survey of several industrialized portions of the state, personnel of the Pennsylvania Department of Environmental Resources counted 1,500 surface impoundments which contained indus-

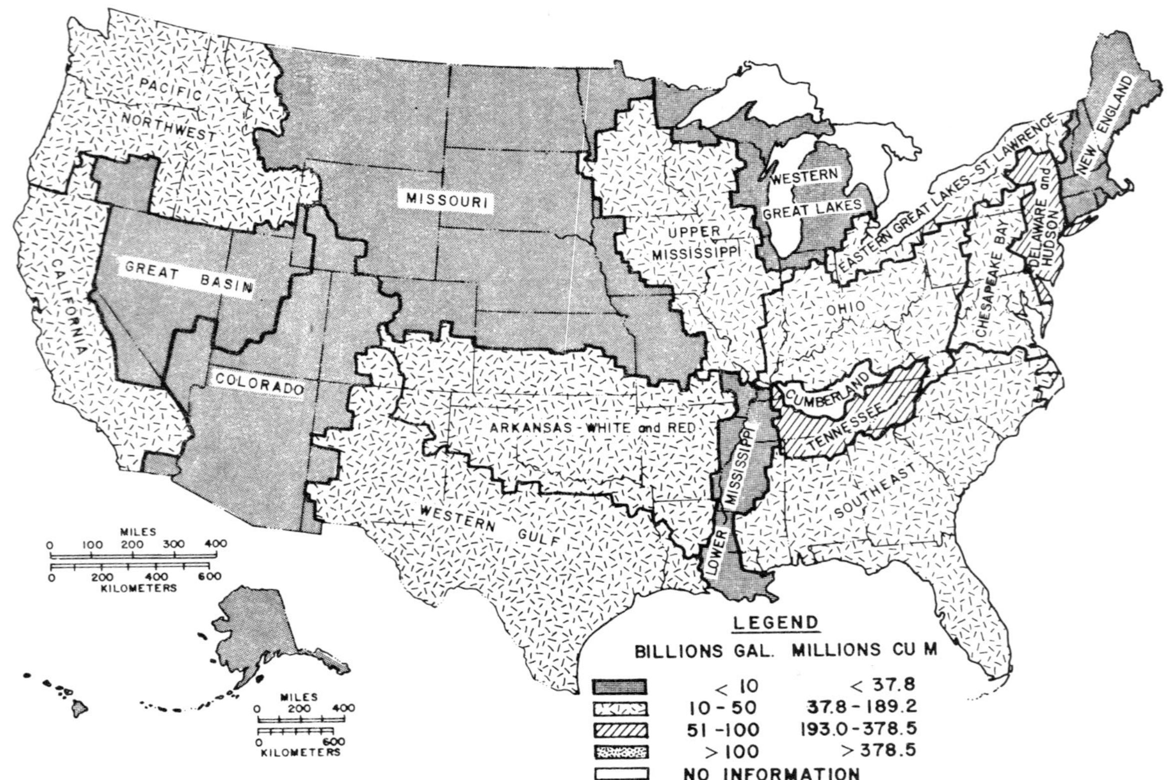

Figure 51. Chemical and allied products industries - volume of industrial waste water treated in ponds and lagoons, 1968.

trial effluent.[6]

Case Histories

The results of the inventory of ground-water contamination problems involving surface impoundments, carried out as part of an 11-state investigation in the northeastern part of the country, emphasize the variety of the contaminants and the diversity of the origins of waste water that can be encountered.[7] Table 13 is based on 57 cases of contamination taken from the files of public agencies and private organizations. Each involves a separate location where leakage of contaminants from some form of surface impoundment has entered the ground-water reservoir. In most cases, water-supply wells have been affected, and this is the only reason that the specific incident has been reported or investigated. In a few, simply observing operation of the impoundment has led officials of an environmental or health agency to investigate whether ground-water contamination has taken place. In others, the owner has noted the loss of a highly toxic substance to the ground and has brought this to the attention of authorities. In only a few cases had monitoring wells been installed specifically to detect degradation of ground-water quality. The types of surface impoundments represented in the 57 cases vary considerably, but lagoons and basins are listed most frequently.

One case in the northeast involved a 3-ft (1-m) wide, 48-ft (14.6-m) long, and 10-ft (3-m) deep concrete canal, used for storage of radioactive material at a private laboratory. An estimated 20,000 gal. (75.7 cu m) of slightly radioactive water leaked into a thin soil layer overlying shale and sandstone. The leak was reported to state authorities by the company, and to date, six monitoring wells have been installed in and around the facility to determine where the contaminant has traveled. Cobalt-60 activity has been picked up in some observation wells, and the investigation is continuing. Meanwhile, use of the canal has been curtailed.[7]

In Maryland, discharge of phenolic waste water to several clay-lined lagoons had been going on for 10 years before it was discovered that the lagoons were leaking. Contaminated ground water had migrated downslope to a fresh-water pond and a small stream. Geophysical surveys and monitoring wells installed under the direction of the state's Water Resources Administration revealed that an extensive zone of ground-water contamination exists in the water-table aquifer. Phenolic concentrations are 14.4 ppm. Discharge of this contaminated ground water has adversely affected the entire stream, from the industrial plant site to a marshy area two

Table 13. ORIGINS AND CONTAMINANTS IN 57 CASES OF GROUND-WATER CONTAMINATION IN THE NORTHEAST CAUSED BY LEAKAGE OF WASTE WATER FROM SURFACE IMPOUNDMENTS. 7)

Type of industry or activity	Number of cases	Principal contaminant(s) reported
Chemical	13	Ammonia Barium Chloride Chromium Iron Manganese Mercury Organic chemicals Phenols Solvents Sulfate Zinc
Metal processing and plating	9	Cadmium Chromium Copper Fluoride Nitrate Phenols
Electronics	4	Aluminum Chloride Fluoride Iron Solvent
Laboratories (manufacturing and processing)	4	Arsenic Phenols Radioactive materials Sulfate
Paper	3	Sulfate
Plastics	3	Ammonia Detergent Fluoride

Table 13. (continued). ORIGINS AND CONTAMINANTS IN 57 CASES OF GROUND-WATER CONTAMINATION IN THE NORTHEAST CAUSED BY LEAKAGE OF WASTE WATER FROM SURFACE IMPOUNDMENTS. 7)

Type of industry or activity	Number of cases	Principal contaminant(s) reported
Sewage treatment	3	Detergents Nitrate
Aircraft manufacturing	2	Chromium Sulfate
Food processing	2	Chloride Nitrate
Mining sand and gravel	2	Chloride
Oil well drilling	2	Chloride Oil
Oil refining	2	Oil
Battery and cable	1	Acid Lead
Electrical utility	1	Iron Manganese
Highway construction	1	Turbidity
Mineral processing	1	Lithium
Paint	1	Chromium
Recycling	1	Copper
Steel	1	Acid Ammonia
Textiles	1	Chloride

miles away. Because of the slow rate of movement of the contaminated ground-water body, it has been estimated that a century or more will be required before the stream can fully recover, even though the leaky lagoons are presently being removed. [8] It is interesting to note that in a recent survey by the State of Maryland, it was found that 75 percent of the liquid waste generated by industry is disposed of to the ground in "lagoons" and "pits" on site. [8]

In another case in the northeast, an abandoned sand and gravel pit was used by a paint manufacturer to contain liquid and sludge wastes removed from a stream during a clean-up operation. Monitoring wells installed later on the edges of the pit and driven to a depth of 15 ft (4.6 m) produced water with a chromium (hexavalent) concentration of as much as 7.2 ppm. [9]

Many of the contaminants reported in Table 13 are related to hazardous wastes, as indicated by the large number of heavy metals listed. The concentrations of these toxic substances can be very high at sites where the untreated industrial effluent is leaking from a surface impoundment and reaching the saturated zone almost unchanged in chemical composition. Concentrations of some of the heavy metals in water from a lagoon containing untreated industrial sludges and liquid wastes were: copper 5,250 ppm; chromium (trivalent) 1,380 ppm; and lithium 280 ppm. The site was investigated by a public agency after a stream near an abandoned plant property showed indications of contamination. The source of contamination in the stream was traced to the lagoon, which was leaking the waste effluent to the ground-water system. The contaminated ground water, in turn, was discharging into the stream. The problem is presently in litigation. [10]

The most grossly contaminated ground water encountered in the northeast investigation contained 10,000 ppm arsenic. [11] Liquids and sludges containing arsenate compounds had been deposited by a chemical company in unlined surface impoundments for many years, and the plume of contaminated ground water had reached a stream adjacent to the plant site, where arsenic concentrations as high as 40 ppm were observed. The lagoons were abandoned after the problem was recognized and the wastes were stored in plastic-lined drums. An attempt was made to pump out and treat the contaminated ground water. After 2.5 years of controlled pumping and monitoring, concentrations of arsenic in both ground water and surface water have been greatly reduced, but the condition is still dangerous.

To provide some insight into typical ground-water contamination cases related to surface impoundments, Table 14 has been prepared based on three detailed studies in the northeast region. The first is a well documented case of the dispersal of plating wastes in ground water in southeastern Nassau County, Long Island, New York. 12) The other two are based on investigations carried out in southern New Jersey and in central Connecticut. 13,14)

All three situations are related to industrial waste water having leaked out of surface impoundments. In two of the three cases, the plume of contaminated ground water had moved beyond the property limits of the industrial site before the problem became known and was defined. No monitoring had been carried out on the industrial plant property. The contaminated ground water migrated slowly toward an area of discharge. In two cases, major discharge is to streams draining the affected water-table aquifer. In the third, the pattern of ground-water movement was controlled by pumping from a series of water-supply wells, which were abandoned after contamination was discovered. Of significance is the small size of the impoundments as compared to the areal extent of the plumes of contaminated ground water. In two of the instances, the lagoon and basin areas represented 0.25 percent and 1.25 percent of the areal extent of the contaminated ground-water body. In only one case has removal or containment of the hazardous wastes been attempted.

An evaluation of ground-water pollution problems in the northwestern states has revealed similar instances in that region. 15) In Colorado, disposal of liquid chemical waste into unlined holding ponds at the Rocky Mountain Arsenal near Denver caused contamination of shallow ground water in a 12-sq mi (30-sq km) area of the South Platte River valley. 16,17,18) The problem was discovered through damage to crops that were irrigated with water from shallow wells. The contaminated water moved northwest in the normal direction of ground-water flow toward the South Platte River (Figure 52). Contaminants known to be present in the shallow aquifer included chloride, fluoride, arsenic, chlorate, the herbicide 2-4D, and the pesticides aldrin and dieldrin. 19) A total of 119 observation wells was installed and a systematic study of water quality was undertaken to map the extent of contamination by measuring chloride concentrations in shallow wells. These concentrations reached a maximum of 4,600 ppm in several areas. The approximate rate of ground-water movement was 13 ft (4 m) per day or about 4,800 ft (1,500 m) per year. Damage claims totaling $74,000 were paid by the government to five farmers that had suffered crop damage. 18)

Table 14. THREE CASE HISTORIES OF GROUND-WATER CONTAMINATION FROM LEAKAGE OUT OF SURFACE IMPOUNDMENTS. [7)]

Description of Surface Impoundment	Two disposal basins, 65x65x15 feet and one disposal basin, 130x54x15 feet	One storage lagoon approximately 50x50x10 feet	Series of lagoons and basins covering an area of about 15 acres and average about six feet in depth
Type of Waste	Aircraft manufacturing	Metal plating	Chemical
Principal Pollutant(s) Observed and Maximum Concentrations Reported in affected wells (mg/l)			
Chromium (Hexavalent)	40	2.3	-
Cadmium	10	-	-
Cyanide	-	0.4	-
Zinc	-	1.4	50
COD	-	-	5,000
Copper	-	-	135
Chromium (Total)	-	-	150
Nickel	-	-	10
Dimensions of Plume of Contaminated Ground Water			
Maximum Length (feet)	4,300	1,000	2,200
Maximum Width (feet)	1,000	200	1,200
Maximum Depth (feet below the water table)	70	60	30
Estimated Maximum Volume of Contaminated Ground Water in millions of gallons and year	200 (1962)	50 (1969)	20 (1972)
Year Reported	1949	1969	1971
Remedial Action(s) and Status of Problem	Periodic research and monitoring; affected wells abandoned; some treatment and reduction of waste effluent; concentrations of chromium and **cadmium** have declined but problem still present in 1974	Lagoon and affected wells abandoned; no further action; problem still present in 1974	Lagoons and basins sealed with cement and/or plastic liners; continuing program of monitoring; system of pumping wells installed to contain pollutants in area of plant site and in shallow aquifer zones; problem still present in 1974

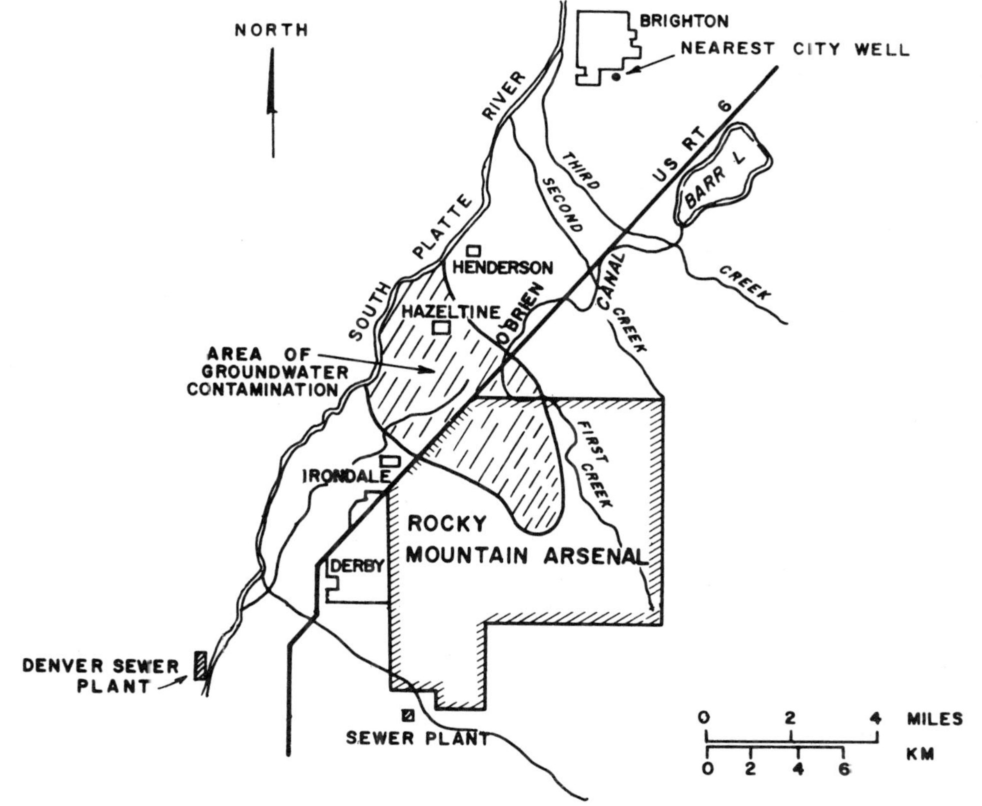

Figure 52. Area of ground-water contamination at Rocky Mountain Arsenal, Denver, Colorado. 17)

TECHNOLOGICAL CONSIDERATIONS

The primary contamination potential from impoundments is degradation of ground or surface waters from seepage of liquids. One countermeasure is to prevent seepage by installing an impermeable barrier. Another approach is to choose an alternative treatment method which can perform the function of the impoundment to be replaced, i.e., additional treatment, storage, or disposal. In the following paragraphs, these options and alternatives are briefly explored.

A wide range of materials are useful as barrier membranes for impounding liquids and sludges. Many are being used in the lining of ponds, reservoirs, lagoons and canals for reducing or eliminating the seepage of liquids into ground water. Today an increasing number of industries are installing synthetic liner materials, especially hypalon and polyvinyl chloride, to meet environmental quality standards. The number and location of lined lagoons and ponds is not known. Liners are not a guarantee against eventual leakage of contaminants into the ground-water system. They can fail mechanically or can be physically altered by the contained wastes.

Lagoons used primarily for storage can be replaced by leakproof facilities, such as above-ground tanks or concrete basins. The major criteria for storage tank selection involve quantity of the waste and the expected length of storage, and the physical and chemical properties of the waste. For example, the waste containing volatile contaminants should be stored in properly vented closed tanks, which could be either vertical or horizontal. In the cases in which volatility or odors pose no problem, wastes can be stored in open facilities.

A waste which is not corrosive can be stored in a concrete or steel tank; storage of wastes which are corrosive would require tanks made of other materials. Reinforced wall design is required for concrete basins, and the concrete must be water-proofed with a suitable paint or plastic coating. Short-term or temporary storage basins would have less stringent construction criteria than long-term or permanent storage.

More effective and environmentally-sound techniques are available to replace waste-water treatment operations now performed in ponds and lagoons or to reduce the volume of waste water now discharged to impoundments. Solids separation can be more effectively performed either in clarifiers, or by filtration or centrifugation. Another example is biological stabilization through use of activated sludge or trickling filtration rather than lagoons. Digestion (anaerobic, aerobic) can be used as an alternative treatment for sludges or wastes with high organic content. Chemical treatment is occasionally carried out in lagoons. The same reactions can be carried out in other facilities less prone to causing contamination.

An alternative to on-site treatment in general would be connection to a municipal treatment plant, assuming that it had the capacity and capability of treating the particular waste. In such a situation, a surcharge may be imposed by the municipality, the amount depending upon the volume and composition of the waste.

Sealing of lagoons, basins, pits, and ponds can have a major economic impact on industries. Lining can amount to several hundred thousand dollars in capital expenses at a particular industrial site. Also to be considered are the added costs involved in treating the additional quantity of waste effluent that previously would have been allowed to leak into the ground-water reservoir.

INSTITUTIONAL ARRANGEMENTS

In most states, impoundments of industrial waste are not subject to any special regulations but are considered simply as potential sources of contamination. If it can be shown that they affect or may affect surface- or ground-water quality, they may be maintained only under a permit. Application of these laws to industrial impoundments may simply be inferable from the general provisions of the statute, or specific, as in the Montana statute which specifies that the provisions of the water-pollution control law, including permit requirements, apply to:

> "...drainage or seepage from all sources including that from artificial, privately owned ponds or lagoons if such drainage or seepage may reach other state waters in a condition which may pollute the other state waters." 20)

A state with only general statutory provisions may find itself in the position of having to prove that contamination of ground water is occurring as a result of a storage or disposal activity, before it can prohibit or otherwise control the operation. It is to overcome this difficult burden that states have expanded upon their control authority in various ways.

A Pennsylvania regulation, for example, requires that impoundments for storage of industrial or other wastes be structurally sound, impermeable, protected from unauthorized acts of third parties, and that they maintain a 2-ft freeboard. If a person or municipality is operating or intends to construct an impoundment to contain more than 250,000 gal. (946 cu m) of waste, or where total capacity of several impoundments on any tract of land exceeds 500,000 gal. (1,893 cu m), or wherever the Department of Environmental Resources determines that a permit is necessary, a permit must be obtained. The Department then approves the location, construction, use, operation, and maintenance of the impoundment based upon a plan that the applicant must submit. [21]

The Pennsylvania statute allows regulation by permit of impounding, handling, storage, transportation, processing or disposal activities that create a danger of water contamination, or where regulation of the activity is necessary to avoid such contamination. [22]

Michigan, without a specific regulation for lagoons, controls industrial waste collections through its water pollution control law and regulations. A person who wants to dispose of wastes on the ground must file a "new use statement" (Statement of New or Increased Use of Water of the State for Waste Disposal Purposes), drill three initial observation wells, then file an application for a ground-water discharge permit, which is reviewed by the Water Resources Commission. The permit allows disposal of specified wastes under a specified monitoring program. It may require treatment of wastes. The permittee must sample and report each month, and the agency also checks monthly. The Commission does not follow general guidelines, but established requirements industry by industry as necessary to preserve U. S. Public Health Service standards. [23]

The Michigan statute requires that every industrial or commercial entity whether underground or on the ground, which discharges liquid wastes into any surface or ground waters, must have waste treatment or control facilities under the specific supervision and control of persons who have been certified by the Water Resources Commission as properly qualified to operate the facilities. This person must file monthly reports to the Commission showing the effectiveness of the treatment or control operation and the quantity of liquid wastes discharged, subject to revocation of his certificate if he makes a false statement. [24] In addition, the statute requires every person doing business within the state to annually report discharge of waste water (other than sanitary sewage) indicating quantities of "critical materials" used in its manufacturing processes. The Commission maintains a "Critical Materials Register" of organic and inorganic materials for this purpose.

The Michigan statute also requires a person engaged in removing liquid industrial wastes from the premises of others to be licensed and bonded, requires licensing and marking of his vehicles, and requires the licensee to keep records of materials transported and places of disposal. This law prohibits the licensee from disposing of wastes onto or into the ground except as approved by the Commission. 25) A few other states have similar laws. New York's law in addition applies to septic-tank cleaning firms.

An approach used by some states is to require a permit for construction of waste-creating facilities. A rule of the Florida Department of Pollution Control requires:

> "Any stationary installation which will reasonably be expected to be a source of pollution shall not be operated, maintained, constructed, expanded, or modified without an appropriate and currently valid permit issued by the Department, unless the source is exempted by Department rule. The Department may issue such permit only after it is assured that the installation will not cause pollution in violation of any of the provisions of Chapter 403, FS, or the rules and regulations promulgated thereunder." 26)

Applicable regulations are detailed, including standards for issuance of denial (the applicant must affirmatively provide the Department with reasonable assurance based on plans, test results, and other information that the activity will not cause pollution); revocation (including revocation for refusal to allow inspection); and detailed requirements for obtaining a permit, which includes:

> "An engineering report covering plant description and operations, types and quantities of all waste material generated whether liquid, gaseous or solid, and proposed waste control facilities, the treatment objectives and the design criteria on which the control facilities are based, and other information deemed relevant. Design criteria shall be based on the results of laboratory and pilot-plant scale studies whenever such studies are warranted. The design efficiencies of the proposed waste treatment facilities and the quantities and types of pollutants in the treated effluents or emissions shall be indicated...."

"Owners written guarantee to meet the design criteria as accepted by the Department and to abide by Chapter 403, FS, and the rules and regulations of the Department as to the quantities and types of materials to be discharged from the plant. The owner may be required to post an appropriate bond to guarantee compliance with such conditions in instances where the owner's financial resources are inadequate or proposed control facilities are experimental in nature."

Delaware's regulation requiring a permit prior to construction, instead of applying only to installations which may be a source of pollution, includes "any structure or facility the occupancy or use of which will generate liquid waste." It specifies four types of permits: (1) septic tanks, (2) liquid waste treatment systems, (3) bulk storage, transfer, and pipelines, and (4) sewers or pipelines carrying liquid waste. 27)

In a few states, landfill laws apply to liquid industrial wastes. Generally, state sanitary landfill laws specifically prohibit disposal of liquids in the landfill. California's is an exception. The State Water Resources Control Board has established site classifications, with restrictions that must be observed for disposal of certain types of waste.

California's system is unique in that requirements for each waste disposal site, whether solid or liquid, are established by Regional Water Quality Control Boards which issue "discharge orders" which must be consistent with a water management plan adopted for the region. These requirements vary from one region to another. Orders are "tailored" to a particular site, and are adopted after a public hearing held by the board. 28)

REFERENCES CITED

1. Bureau of the Census. 1971. 1967 census of manufactures, water use in manufacturing. U. S. Department of Commerce, Washington, D. C. 361 pp.

2. U. S. Environmental Protection Agency. 1973. Handbook for monitoring industrial waste water. U. S. Environment Protection Agency, Washington, D. C.

3. Bureau of the Census. 1975. 1972 census of manufactures, water use in manufacturing. U. S. Department of Commerce, Washington, D. C. 194 pp.

4. Bureau of the Census. 1966. 1963 census of manufactures, water use in manufacturing. U. S. Department of Commerce, Washington, D. C. 174 pp.

5. Karubian, J. F. 1974. Polluted ground water: estimating the effects of man's activities. Office of Research and Development, U. S. Environmental Protection Agency, Washington, D. C. EPA 600/4-74-002. 99 pp.

6. Westlund, C. W. 1973. Personal communication. Pennsylvania Department of Environmental Resources, Harrisburg, Pennsylvania.

7. Miller, D. W., F. A. DeLuca, and T. L. Tessier. 1974. Ground-water contamination in the northeast states. U. S. Environmental Protection Agency, Washington, D. C. EPA 660/2-74-056.

8. Schiffman, Arnold. 1975. Disposal of hazardous and industrial wastes in Maryland. Report to Legislative Council, Maryland General Assembly. Maryland Department of Natural Resources, Annapolis, Maryland.

9. Schiffman, Arnold. 1973. Personal communication. Ground Water Technical Services, Maryland Department of Natural Resources, Annapolis, Maryland.

10. Confidential communication. 1973.

11. Wright, J. F. 1973. Administrative and legal considerations: an interstate viewpoint. University of California Water Resources Engineering Education Series.

12. Perlmutter, N. M., and Maxim Lieber. 1970. Dispersal of plating wastes and sewage contaminants in ground water and surface water, South Farmingdale-Massapequa area, Nassau County, New York. U. S. Geological Survey Water-Supply Paper 1879-G.

13. Geraghty & Miller, Inc. 1972. Consultant's report. Port Washington, New York.

14. Geraghty & Miller, Inc. 1969. Consultant's report. Port Washington, New York.

15. van der Leeden, Frits, L. A. Cerrillo, and D. W. Miller. 1975. Ground-water pollution problems in the northwestern United States. U. S. Environmental Protection Agency, Washington, D. C. EPA 660/3-75-018.

16. Petri, L. R. 1961. The movement of saline ground water in the vicinity of Derby, Colorado, in Robert A. Taft Sanitary Engineering Center. Proceedings of 1961 symposium on ground water contamination. Technical Report W 61-5.

17. Walton, Graham. 1961. Public health aspects of the contamination of ground water in the vicinity of Derby, Colorado, in Robert A. Taft Sanitary Engineering Center. Proceedings of 1961 symposium on ground water contamination. Technical Report W 61-5.

18. Gahr, W. N. 1961. Contamination of ground water - vicinity of Denver. Paper presented at 128th meeting of the American Association for the Advancement of Science, Denver, Colorado.

19. Division of Water Supply and Pollution Control. 1965. Ground water pollution in the South Platte River valley between Denver and Brighton, Colorado. U. S. Department of Health, Education and Welfare. Published Report 4.

20. Montana Revised Code, Section 69-4804.

21. Pennsylvania Department of Environmental Resources. Rules and Regulations, Chapter 101.

22. Pennsylvania Clean Streams Law, Section 402.

23. Michigan Department of Natural Resources, Water Resources Commission. General Rules, Part 21.

24. Michigan Water Resources Commission Act. Public Acts of 1929.

25. Michigan Public Acts of 1969. Act 136.

26. Florida Department of Pollution Control. Rules, Chapter 17-4.

27. Delaware Water Pollution Control Regulations, Section 4.

28. California State Water Resources Control Board. Waste discharge requirements for waste disposal to land.

SECTION VII

LAND DISPOSAL OF SOLID WASTES

SUMMARY

Solid waste land disposal sites can be sources of ground-water contamination because of the generation of leachate caused by water percolating through the bodies of refuse and waste materials. Precipitation falling on a site either becomes runoff, returns to the atmosphere via evaporation and transpiration, or infiltrates the landfill. Contamination problems are more likely to occur in humid areas, where the moisture available exceeds the ability of the waste pile to absorb water.

Leachate is a highly mineralized fluid containing such constituents as chloride, iron, lead, copper, sodium, nitrate, and a variety of organic chemicals. Where manufacturing wastes are included, hazardous constituents are often present in the leachate (e.g., cyanide, cadmium, chromium, chlorinated hydrocarbons, and PCB). The particular makeup of the leachate is dependent upon the industry using the landfill or dump. Another problem is the disposal of low-level radioactive wastes.

There are about 18,500 land disposal sites which accept municipal wastes, of which only about 20 percent are "authorized." Most are open dumps, or poorly sited and operated landfills, and most receive some industrial wastes. There is no national inventory available on privately owned industrial land disposal sites. However, it is estimated that 90 percent of industrial wastes that are considered hazardous are landfilled, mainly because it is the cheapest waste-management option.

Problems presently associated with existing or abandoned dumps and landfills should not be considered in the same category as potential problems at new, properly designed sanitary landfills because there are methods available for minimizing environmental effects and managing leachate production. Proper siting in locations where potential contamination of ground water is limited is one method. Reduction of leachate formation by use of selected cover materials and surface grading of the refuse pile is another. Also promising but costly are such techniques as pre-treatment capable of reducing the volume or solubility of the waste, detoxification of hazardous wastes prior to disposal, and collection of the leachate by means of impermeable barriers or liners followed by treatment.

There is no effective Federal regulatory control of land disposal of solid waste except as it may enter navigable waters. Forty-four states have statutes which prohibit the disposal of solid waste without a permit. The range of requirements for state permit systems extends from simple notification that a facility exists to detailed site descriptions including the results of soil borings and sampling of baseline ground-water quality. About 15 states have regulations limiting land disposal of hazardous wastes.

DESCRIPTION OF THE PRACTICE

A wide variety of wastes from industries, residences, and municipalities is disposed of on the land. This practice ranges from simple dumping of refuse on a readily available piece of property to controlled disposal of processed wastes on sites which are designed to minimize the potential for contamination of local water resources.

According to EPA, the term "solid waste land disposal site" refers to the following types of operation which may accept garbage, refuse, municipal and industrial sludges and liquid waste; discarded solid materials resulting from agricultural, industrial, and commercial operations and from community activities; and hazardous wastes:

Dump: An uncovered land disposal site where solid and/or liquid wastes are deposited with little or no regard for pollution control or aesthetics. Dumps are susceptible to open burning and are exposed to the elements, vectors, and scavengers.

Landfill: A land disposal site located without regard to possible effects on water resources, but which employs intermittent or daily cover to minimize scavenger, aesthetic, vector, and air pollution problems.

Sanitary Landfill: A land disposal site employing an engineered method of disposing of solid wastes on land in a manner that minimizes environmental hazards by spreading the solid wastes in thin layers, compacting the solid wastes to the smallest practical volume and applying and compacting cover material at the end of each operating day. 1)

Secured Landfill: A land disposal site that allows no hydraulic connection with natural waters, segregates the waste, has restricted access, and is continually monitored. Landfills and dumps are very common. True sanitary landfills are rare, and, to date, secured landfills are experimental only.

Landfills or dumps invariably are located on land that is considered to have little or no value for other uses, for example: marshlands, abandoned sand and gravel pits, old strip mines, or limestone sinkholes, all of which are susceptible to ground-water contamination problems. In one eastern state, 85 percent of the existing landfills were originally designed as "reclamation" projects to fill marshlands and abandoned sand and gravel pits. Control procedures for minimizing contamination of ground water were not instituted.

Solid waste land disposal sites can be sources of ground-water contamination because of the generation of leachate caused by water percolating through the refuse. Precipitation falling on a site either becomes runoff, returns to the atmosphere via evaporation and transpiration (water use by plants), or infiltrates the refuse. This infiltrating water ultimately will form leachate (water that has percolated through the wastes and picked up soluble and suspended contaminants).

The process of leachate formation and subsequent ground-water contamination is dependent upon the amount of water which passes through the refuse. Water which infiltrates the surface of the cover will first be subject to evaporation and plant transpiration. Any water in excess of field capacity will percolate into the layers of solid waste. Additional surface runoff from the surrounding land, moisture contained in the solid or liquid waste placed in the fill, moisture from solid-waste decomposition, and water entering through the bottom or sides of the site also contribute to the generation of leachate.

Problems associated with existing or abandoned dumps and landfills are not in the same category as potential problems at new, properly designed sanitary or secured landfills where modern methods are employed for minimizing environmental effects and managing leachate production. Proper design and operation of a new site and the use of liners and diversion trenches can essentially eliminate input water from surrounding surface runoff and native ground water. Input water from precipitation can be minimized by controlling such surface conditions as (1) the steepness of slope, (2) the permeability of the material used for cover, and (3) the type and amount of vegetation.

CHARACTERISTICS OF CONTAMINANTS

The largest component of municipal waste is paper, but substantial food wastes, yard wastes, glass, metals, plastics, rubber, and liquid wastes are also included. Many municipal sites also receive industrial process residues and pollution control system sludges in addition to septic tank pumpings, sewage sludge, bulky wastes, street sweepings, and construction/demolition wastes.

The constituents in leachate from a municipal site result from simple solution of compounds in the wastes and from the decomposition of the refuse contained in the land disposal site. The latter is a biological, chemical, and physical process and is affected by the degree of microbial activity that is proceeding within the fill. Microbial activity is generally dependent on refuse composition, temperature, moisture content, and the availability of free oxygen.

The types and concentrations of contaminants in leachate are of great importance in determining potential effects on the quality of ground and surface water. Table 15 shows ranges in concentration of chemical constituents and physical parameters of typical leachate generated from municipal solid wastes.

In addition to refuse generated by residences and commercial establishments, a wide variety of industrial wastes are disposed of on the land. Examples of some of those identified by EPA studies, in approximate order of volume, are as follows:

Inorganic Chemicals - process sludges, cell tear-down rubble, waste-water treatment sludges, dry residues and dusts. 3)

Organic Chemicals - liquid tars, still bottoms, filter residue sludges, residual pitch solids, filter cakes, spent catalysts, pesticides. 4)

Petroleum Refining - tank bottoms, cooling tower sludges, air flotation float, slop oil, biological sludge, storm silt, spent lime, filter clays, API separator bottoms, Fluidized Catalytic Cracker (FCC) fines, coke and coke fines, hydrofluoric acid, alkylation sludges, cleaning sludges. 5)

Primary Metals Smelting and Refining - slags, dusts, pollution-control sludges. 6)

Electroplating - waste-water treatment sludges, air pollution control sludges, organic solvents, concentrated plating

Table 15. SUMMARY OF LEACHATE CHARACTERISTICS BASED ON 20 SAMPLES FROM MUNICIPAL SOLID WASTES. [2)]

Components	Median value (ppm) a)	Ranges of all values (ppm) a)
Alkalinity ($CaCO_3$)	3,050	0 - 20,850
Biochemical Oxygen Demand (5 days)	5,700	81 - 33,360
Calcium (Ca)	438	60 - 7,200
Chemical Oxygen Demand (COD)	8,100	40 - 89,520
Copper (Cu)	0.5	0 - 9.9
Chloride (Cl)	700	4.7 - 2,500
Hardness ($CaCO_3$)	2,750	0 - 22,800
Iron, Total (Fe)	94	0 - 2,820
Lead (Pb)	0.75	< 0.1 - 2.0
Magnesium (Mg)	230	17 - 15,600
Manganese (Mn)	0.22	0.06 - 125
Nitrogen (NH_4)	218	0 - 1,106
Potassium (K)	371	28 - 3,770
Sodium (Na)	767	0 - 7,700
Sulfate (SO_4)	47	1 - 1,558
Total Dissolved Solids (TDS)	8,955	584 - 44,900
Total Suspended Solids (TSS)	220	10 - 26,500
Total Phosphate (PO_4)	10.1	0 - 130
Zinc (Zn)	3.5	0 - 370
pH	5.8	3.7 - 8.5

a) Where applicable

baths, process wastes. 7)

Paints - raw materials packaging, spills and spoiled batches, waste-water treatment sludges, air pollution control dusts, organic cleaning solvents. 8)

Battery - process scrap, reject and scrap batteries, battery processing slags. 9)

Pharmaceuticals - waste solvents, still bottoms, miscellaneous organics, heavy metals, filter cakes, chemical muds, returned goods, reject materials. 10)

Low-Level Radioactive Waste - contaminated paper and plastics (70 percent by volume), dead laboratory animals, broken equipment and glassware, protective clothing, evaporator residues, ion exchange resins, organic cleaning solvents, pesticides, and chemical processing residues.

A partial listing of the potentially hazardous constituents of these industrial waste categories is given in Table 16. Some of the contaminants detected in ground water pose an acute threat to public health because of their toxicities, (e.g., cyanide, arsenic, phenols, etc.). Others are hazardous because of chronic effects, such as carcinogens or teratogens (e.g., some chlorinated hydrocarbons, vinyl chloride, chromium, lead). Many contaminants have both characteristics.

Public health and environmental effects are normally correlated with the concentration of and duration of exposure to the specific contaminant. This has been better documented for acute effects resulting from high concentrations over a short period of time than for chronic effects resulting from exposure to low concentrations for a long period. 11)

EXTENT OF THE PROBLEM

According to the latest available estimates, 135 million tons (122 million tonnes) of residential and post-consumer commercial wastes are disposed of in the United States annually. 12) This amounts to 3.5 lb (1.6 kg) per person/day. Not included in this figure are sewage sludge, industrial processing wastes, air pollution control wastes, demolition and construction residue, street sweepings, discarded automobiles and automotive parts, and bulky tree and landscaping wastes, all of which are found in varying quantities throughout the nation at municipal land-disposal sites.

It is estimated that approximately 260 million tons (236 mil-

Table 16. COMPONENTS OF INDUSTRIAL WASTE.

	Metals mining	Primary metals	Pharmaceuticals	Batteries	Inorganic chemicals	Organic chemicals	Pesticides	Explosives	Paints	Petroleum refining	Electroplating
Ammonium salts		X								X	
Antimony	X				X				X		
Arsenic	X	X	X		X					X	
Asbestos					X				X		
Barium									X		
Beryllium	X									X	
Biological waste			X								
Cadmium	X	X		X	X				X	X	X
Chlor. hydrocarbons					X	X			X		X
Chromium		X	X	X	X				X	X	X
Cobalt									X	X	
Copper	X	X	X	X					X	X	X
Cyanide		X			X					X	X
Ethanol waste, aqueous			X								
Explosives (TNT)								X			
Flammable solvents						X			X		
Fluoride		X			X						
Halogenated solvents			X								
Lead solvents	X	X		X	X				X	X	X
Magnesium	X										
Manganese		X									
Mercury		X	X	X	X				X	X	
Molybdenum										X	
Nickel		X		X	X					X	
Oil		X								X	X
Organics, misc.						X					
Pesticides (organo-phosphates)							X				
Phenol		X								X	X
Phosphorus					X						X
Radium	X										
Selenium	X	X	X							X	
Silver				X						X	X
Vanadium										X	
Zinc	X	X	X	X	X				X	X	X

lion tonnes) of industrial wastes of all types are generated and disposed of annually (on a dry basis), which is almost twice as much each year as is produced by residential and commercial sources. Many of these wastes are potentially hazardous. Preliminary data from industrial waste surveys conducted by EPA indicate that over 90 percent of all hazardous industrial wastes are disposed of on the land, mainly because it is the cheapest waste management option. A similar percentage of the non-hazardous portion of industrial wastes is most likely disposed of on land, as opposed to undergoing treatment (chemical, thermal, etc.) and/or recovery. Thus, about 240 million tons (218 million tonnes) of industrial wastes end up in land disposal sites each year.

There are currently about 18,500 land disposal sites which accept municipal wastes in the United States. 13) Only about 20 percent of these are "authorized." About 20 sites are lined, and about 60 sites have leachate treatment facilities. Most of the 18,500 sites are open dumps, or poorly located and operated landfills; very few are truly "sanitary landfills." Most receive some industrial wastes.

There is no national inventory of industrial land disposal sites, which are generally located on private property. The locations, or even the existence of these exclusively industrial dumps and landfills are rarely recorded with public agencies, and the few that are, are generally not inspected on a routine basis. Therefore, ground-water contamination problems resulting from such sites are not normally discovered until water from nearby supply wells has been noticeably affected.

Some indication of the concentrations of industrial wastes disposed of in different regions of the United States can be developed from surveys conducted by EPA over the past few years. One conclusion of these surveys is that industrial land disposal contamination problems are widespread. Preliminary data indicate that the highest percentage of industrial hazardous waste is generated in EPA Regions V (Midwest) and VI (Gulf Coast), followed by Regions II and III (Mid-Atlantic) and IV (Southeast). Each of these regions contains heavy concentrations of industrial production. In the study, the total industry wastes (and hazardous portion) are presented by total amount generated per industry by state.*

* The data developed during these industrial surveys are not provided by industrial waste disposal processes but rather by industrial category. Thus, the critical areas can only be illustrated by industry category or waste type, and not by treatment or disposal method.

The number of industrial establishments in a state generally provides an indication of the relative magnitude of concern for a particular waste category and disposal technique combination. For instance, the inorganic chemicals industry landfills approximately 55 percent of its concentrated, potentially hazardous wastes. 3) States with 50 or more inorganic chemicals manufacturing establishments are Texas, California, New Jersey, Ohio, Louisiana, and Illinois. Thus, land disposal of industrial wastes containing inorganic chemicals should be of greater concern in these states than in any of the others.

Figure 53 indicates the geographical distribution of industrial establishments on the basis of value of production (value added). Although each specific industrial category has its own geographical distribution pattern, Figure 53 is indicative of the generally high industrial concentrations along the northeast seaboard, in the Great Lakes region, and in California and Texas.

As noted above, the amount of infiltration from precipitation that falls on a disposal site is the major factor affecting the quantity of leachate that can be generated. Therefore, the extent of the potential problem of groundwater contamination resulting from leachate entering aquifers is greatest in areas where average annual precipitation exceeds the potential water losses by evaporation and transpiration. As shown on Figure 54, such areas are found east of the Mississippi River and in the coastal region of the Pacific Northwest. About 71 percent of the municipal refuse disposal sites found in the United States are located in these water surplus areas.

It should be emphasized, however, that no matter where they are located geographically, disposal sites which are placed in wetlands, in flood plains, or in other areas where the water table is close to land surface, can produce leachate and can be the cause of ground-water contamination. In some places, such as low lying coastal areas, the water table is so high that all disposal sites constructed without sufficient natural or artificial barriers will contaminate ground water.

While the most common economic damage resulting from leachate is the contamination of domestic, industrial, and public supply wells, there are numerous cases where leachate has directly contaminated surface waters. In confined, slow moving, or relatively low-volume surface waters, leachate has killed vegetation and fish, wiped out spawning areas, and ruled out the use of existing and planned recreational

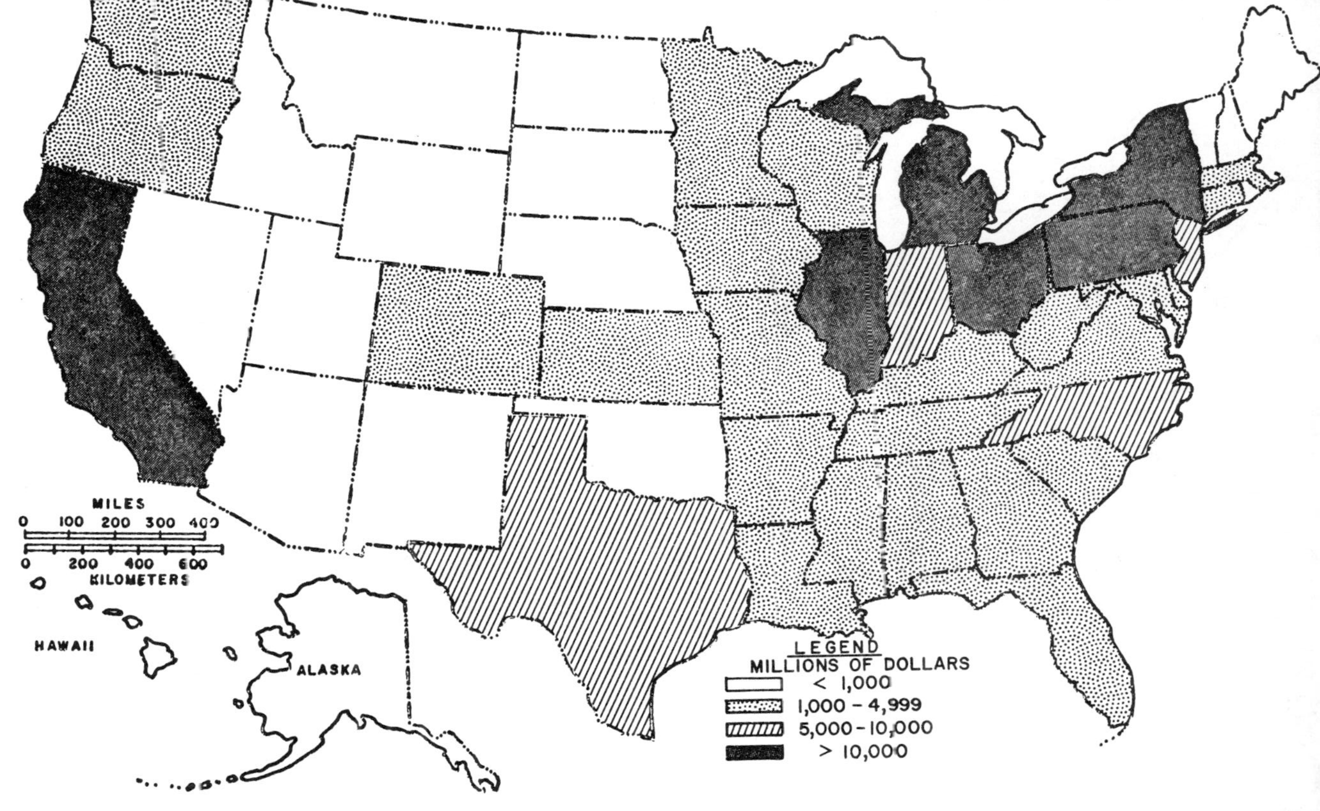

Figure 53. Value added by manufacture (in millions of dollars) 1972. 14)

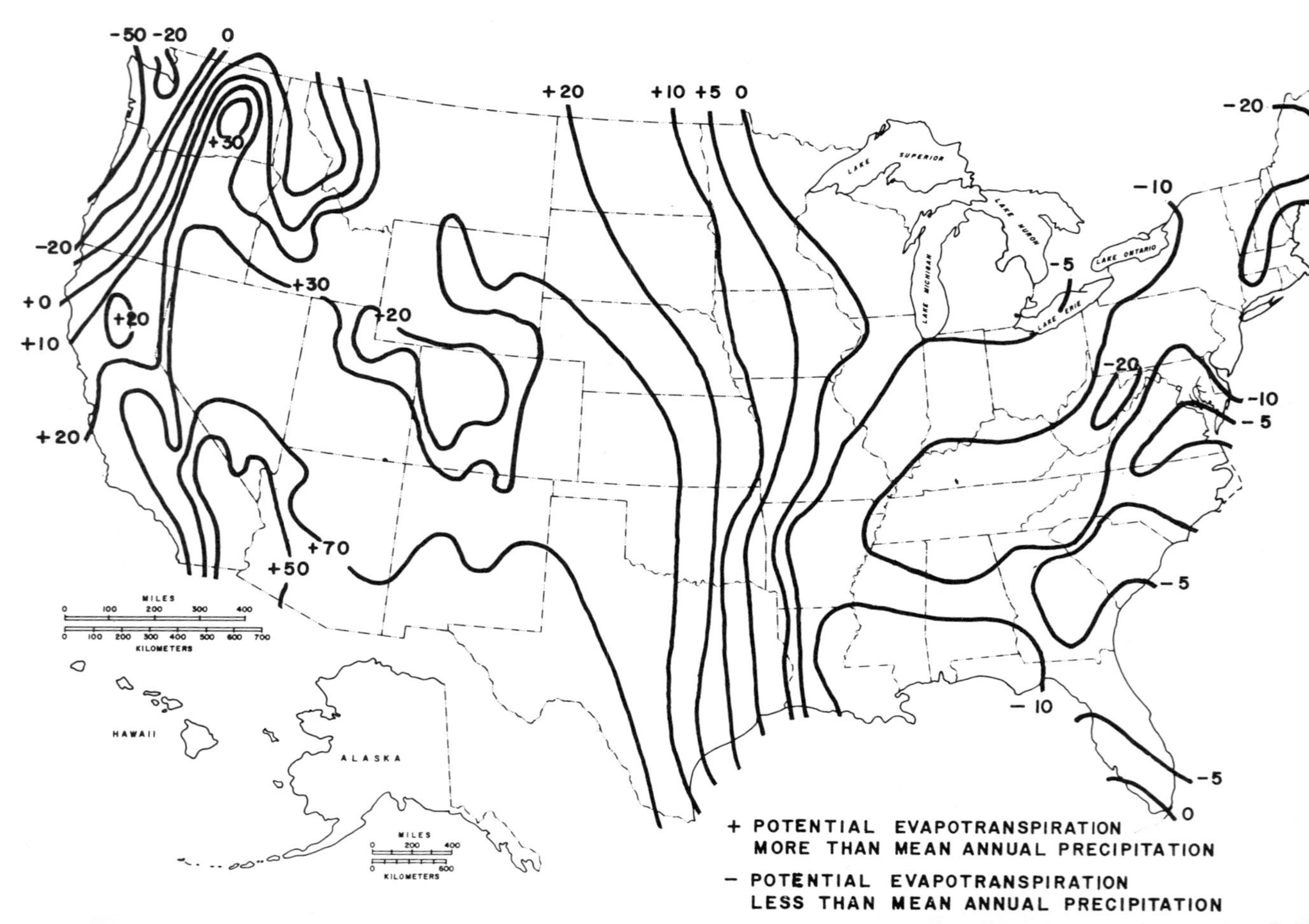

-50 -20 0
+30
-20
+0
+20
+10
+20
+30
+20
+70
+50
+20
+10 +5 0
-20
-10
-5
-20
-10
-5
-5
-10
-5
0
LAKE SUPERIOR
LAKE MICHIGAN
LAKE HURON
LAKE ERIE
LAKE ONTARIO
MILES
0 100 200 300 400
0 100 200 300 400 500 600 700
KILOMETERS
HAWAII
ALASKA
MILES
0 200 400
0 600
KILOMETERS
+ POTENTIAL EVAPOTRANSPIRATION MORE THAN MEAN ANNUAL PRECIPITATION
– POTENTIAL EVAPOTRANSPIRATION LESS THAN MEAN ANNUAL PRECIPITATION

areas.

Table 17 lists possible types of damages that can result from the introduction of leachate to surface- and ground-water resources. The damages that have actually been reported are underlined.

On a local level, in addition to the possibility of grave health effects as a result of chronic exposure, leachate contamination frequently causes severe economic hardships, distresses, inconveniences, and inequities to damaged land owners. In no case recorded to date have the victims been fully compensated for their losses. Due to the number of years (50 to 100 is possible) that land disposal sites may produce leachate, it is difficult to assess the impact of possible damages to future generations.

Case Histories

EPA is currently compiling a national inventory of damage incidents that have been caused by improper land disposal of municipal and industrial wastes. Table 18, based on 391 industrial damage incidents documented as of February 1976, indicates that ground-water contamination is the most common type of damage reported, followed by surface-water contamination. Table 18 also shows that landfills, dumps, and "other land disposal" (i.e., promiscuous dumping), account for the large majority of reported damage incidents. It should be noted that twice as many damage incidents have been associated with "other land disposal" than with true landfills and dumps. The data summarized in the table are not nationally representative since 63 out of the 391 cases studied were obtained from an incomplete survey of one state that already has a permit system for landfills and surface impoundments. The most flagrant environmental offenses generally occur in those states that do not have regulatory programs for industrial waste disposal.

In the 1974 EPA study of ground-water contamination in the northeast states, 60 case histories were tabulated, including 18 pertaining to industrial landfills.[17)] Examination of Table 19, which has been reproduced from that study, will convey some of the dimensions of the problem, at least for the Northeast. The study uncovered about 100 cases in which landfills were pinpointed as the contamination source. The 60 cases on Table 19 were selected from the 100 on the basis of adequacy and reliability of information.

Although Table 19 indicates that municipal sites account for more cases of ground-water contamination than do industrial,

Table 17. TYPES OF POSSIBLE LEACHATE DAMAGE.[16)]

DAMAGE TO LIFE [a)]

I. Humans

acute and chronic health effects (e.g. illness, skin damage, partial paralysis, brain damage, death)

II. Domestic Animals

III. Wild Animals

terrestrial (e.g. mammals, birds)
aquatic (e.g. fishkills, spawning areas, shellfish [b)], crabs)

IV. Farm Crops

V. Other Vegetation

grasses, shrubs, trees

PHYSICAL DAMAGE [a)]

I. Water Resources Contaminated

springs, lakes, streams, rivers, ground water, marshlands

II. Drinking Water Supplies

surface water (reservoirs, springs, lakes, rivers)
ground water (domestic, industrial and public supply wells)

III. Land/Water Use

sport/recreation: fishing, shellfish, swimming, parks
decreased property value or facility utilization
loss of future use of water resource (surface and ground water)

IV. Material Destruction

damage to laundry, pipes, sinks, water heater, food, etc.
damage to industrial equipment or products

a) Underlined portions are leachate damages identified to date.
b) Shellfish concentrate accumulations of toxic metallic ions.

Table 18. PRELIMINARY ESTIMATE OF THE RELATIONSHIP BETWEEN DISPOSAL METHOD AND DAMAGE MECHANISM, EXPRESSED AS PERCENT OF CASES STUDIED. [a]

Disposal Method	Surface Impoundments	Landfills, dumps	Other land disposal [b]	Storage of wastes	Smeltings, slag, mine tailings	Unknown
Percentage of cases studied	21	23	44	3	4	5
Damage mechanism (percentage of cases studied)						
Ground water (61)	13	15	25	2	3	4
Surface water (39)	9	12	15	-	2	<1
Air (3)	-	1	2	-	-	<1
Fires, explosions (3)	-	2	<1	-	-	-
Direct contact poisoning (13)	<1	1	10	1	-	-
Unknown (2)	<1	<1	<1	<1	<1	<1
Wells affected [c] (34)	7	6	16	1	<1	3

a) All numbers refer to percent of 391 cases studied thus far. The total percentages in the matrix add up to more than 100, because several damage incidents involved more than one damage mechanism. All percentages have been rounded to the nearest integer.

b) Haphazard disposal on vacant properties, on farmland, spray irrigation, etc.

c) Not included as a damage mechanism.

Note: The data presented in this table have been derived solely from case studies associated with land disposal of industrial wastes.

Table 19. SUMMARY OF DATA ON 42 MUNICIPAL AND 18 INDUSTRIAL LANDFILL CONTAMINATION CASES.[17)]

Findings	Type of Landfill: Municipal	Industrial
Assessment of principal damage		
Contamination of aquifer only	9	8
Water supply well(s) affected	16	9
Contamination of surface water	17	1
Principal aquifer affected		
Unconsolidated deposits	33	11
Sedimentary rocks	7	3
Crystalline rocks	2	4
Type of pollutant observed		
General contamination	37	4
Toxic substances	5	14
Observed distance traveled by pollutant		
Less than 100 feet	6	0
100 to 1,000 feet	8	4
More than 1,000 feet	11	2
Unknown or unreported	17	12
Maximum observed depth penetrated by pollutant		
Less than 30 feet	11	3
30 to 100 feet	11	3
More than 100 feet	5	2
Unknown or unreported	15	10
Action taken regarding source of contamination		
Landfill abandoned	5	6
Landfill removed	1	2
Containment or treatment of leachate	10	2
No known action	26	8
Action taken regarding ground-water resource		
Water supply well(s) abandoned	4	5
Ground-water monitoring program established	12	2
No known action	26	11
Litigation		
Litigation involved	8	5
No known action taken	34	13

this is somewhat deceptive. As previously noted, very little information is on file with regulatory agencies regarding the location and operation of industrial landfills. Furthermore, ground-water contamination problems have been studied to a greater extent at municipal than at industrial sites.

Because of the large volumes of waste material involved, removing the source of contamination when dealing with landfills obviously is almost impossible. Thus, most of the cases included in Table 19 are listed under the category "no action taken." In a few, involving small quantities of toxic wastes, the material causing the problem was excavated. In others, the site was closed, but this alternative is difficult to accomplish because an acceptable new location must be found and approved, or new facilities must be designed and constructed for handling the waste, such as recovery, treatment, or incineration. In addition, simple abandonment of the site does nothing to remedy the problem. In cases where well supplies have been affected, abandonment of the wells is sometimes a last resort because of the high costs involved in developing and piping a new source of water supply.

A few contamination cases are known to have resulted in litigation. This procedure normally takes the form of a local or regional regulatory agency using existing laws to force the contaminator to take action to modify the operation of or to abandon the site.

As discussed in other sections of this report, the major perils inherent in ground-water contamination are the elusive nature and the long duration of the problem. Almost all of the case studies reported to date were discovered after the damage to the ground water had already occurred. Also, the subsurface migration of contaminants is a very slow process, which means that most of the damages caused by the disposal of huge quantities of hazardous wastes during the past decades are still to be evidenced. This point is well illustrated by the following case study. 18)

In May 1972, a local building contractor occupied a new office and warehouse structure at the outskirts of Perham, a town of 1,900 residents in western Minnesota. At that time, a well was drilled to supply drinking water for about 13 people who worked on the premises.

Early in May, five employees became ill with gastrointestinal symptoms. Following this, and continuing throughout the next 10 weeks, other employees also became ill. Arsenic poisoning was determined to be the cause, which affected a total of 11 out of 13 persons exposed to the water. Two required hospitalization and treatment. One of the victims lost the use of his legs for about six months due to severe neuropathy. The medical aspects of this ground-water contamination incident have been well documented by Dr. E. J. Feinglass. [19)]

Chemical analysis of samples taken from the affected well established arsenic concentrations of up to 21,000 ppb. (The EPA drinking water standard for arsenic is 50 ppb.) As Dr. Feinglass pointed out in his article, the particularly serious consequences of chronic arsenic poisoning were probably avoided in this instance because of the extremely high concentration of arsenic in the drinking water. The acute course of the illness allowed early recognition of the problem.

The source of the well water contamination has been traced back to the mid-1930's, at which time grasshoppers had constituted a serious problem to farmers in the area. Some old-timers recalled that excess grasshopper bait had been buried at the former County Fairgrounds, in a corner which was used as the village dump in those days. That area is now directly adjacent to the new facilities of the building contractor whose well became contaminated.

The exact area of disposal was located approximately 20 ft (6 m) from the well. The well is 31 ft (9 m) deep and the arsenic trioxide was buried at a depth of about 7 ft (2 m). Analysis of soil samples established a maximum arsenic concentration of 40 percent at the spot where a white crystalline material was found. The Minnesota Department of Agriculture has estimated that less than 50 lb (23 kg) of grasshopper bait was disposed of in the trench about 40 years ago.

The association of public health effects with the contamination of a drinking water supply through hazardous waste disposal, as illustrated by the previous case study, is extremely difficult. Nevertheless, EPA is aware of numerous incidents where waste disposal practices have impaired the use of domestic and public drinking water supplies. The following case study illustrates the wide-scale impairment of domestic water supplies and the associated economic damages. [20)]

In March 1971, a major chemical company hired an independent waste hauler to remove drums containing chemical wastes from one of its plants in New Jersey. The wastes consisted of organic wash solvents and still bottoms and residues from the manufacturing of a variety of organic chemicals and plastics.

The waste drums were to be taken to a landfill for disposal, and some of the drums which were removed initially from the plant probably reached their intended destination. However, in December 1971, about 4,000 of the drums were discovered on a former chicken farm in a sparsely populated suburban area of New Jersey. About 10 percent of the drums were empty. There were trenches on the property into which the contents of some drums had been emptied. Chemical wastes and some of the full drums were also buried in other sections of the property. On the grounds that the storage of drums containing flammable and explosive chemicals presented fire and explosion hazards in the area, and that the storage of chemical waste by the lessee did not have the approval of the property owner, a court order was obtained to have the chemical company remove all drums and chemical wastes from the property and dispose of them in an approved manner.

Early in 1974, about two years after the discovery of chemical waste storage/disposal on the leased land, some of the residents in the area discovered unusual taste and odor in their well water. Subsequent chemical analysis of water samples from these and other wells in the area indicated the presence of petrochemical contaminants. On the basis of these analytical results, the very strong and persistent taste and odor problem associated with the water from some of the wells, and the documented case of earlier waste chemical storage and burial on the nearby land, the New Jersey Department of Environmental Protection concluded that the ground water in at least the immediate vicinity of the disposal site was contaminated with hazardous organic chemicals.

To protect the health of the area's residents, the local Board of Health passed an ordinance forbidding the use of well water from 148 wells for any purpose. For a period of about 6 months while steps were being taken to extend the services of a public water supply company to the area on a permanent basis, emergency water supply was provided to the residents using water tanks which were stationed at strategic locations in the area. Some residents and public facilities used bottled water for drinking and cooking purposes. In some sections of the area where construction of new wells was still allowed, the wells had to be drilled deeper than was formerly necessary in order to obtain uncontaminated water.

Preliminary estimates indicate the immediate monetary damage resulting from the incident was in excess of $400,000. The major items of cost include: the extension of the public water supply to the area ($249,000), 20 new wells drilled to a deeper aquifer ($46,000), interim emergency water for area residents ($5,000, minimum), and sampling and analysis of water ($35,500). Costs associated with litigation, removal of wastes from the property, and salaries for many of the professionals in the state and local agencies investigating the incident are not included in the above estimates.

The stockpiling of hazardous waste materials without adequate precautions can seriously impair ground-water quality. This is illustrated by the following incident which affected a public water supply system.[20] In 1967, an industrial operation recovering metals from waste materials moved near the well field of a New Jersey municipality. The company stockpiled materials containing zinc, lead, and cadmium in the open, and the metals leached into adjacent surface and ground waters. Subsequently, some public water supply wells had to be closed in 1971 and 1972 due to high concentrations of zinc. Other wells in the same field are in jeopardy. In addition, a surface stream flows into a pond near the well field; zinc and lead concentrations in that pond have been analyzed at 12,250 ppm and 600 ppm, respectively. (The U. S. Public Health Service recommended drinking water limit for zinc is 5.0 ppm; the EPA mandatory drinking water limit for lead is 0.05 ppm.)

Prolonged disposal of hazardous wastes at landfills and dumps can result in the accumulation of extremely large amounts of material, the effects of which may not be observed for many years. From 1953 to 1973, a laboratory company in Iowa utilized a dump site for solid-waste disposal. Over 250,000 cu ft (7,000 cu m) of arsenic-bearing wastes had been deposited there. Monitoring wells around the dump have established over 175 ppm arsenic in the ground water. (The U. S. Public Health Service drinking water limit for arsenic is 0.05 ppm.) The dump site is located above a limestone bedrock aquifer, from which 70 percent of the nearby city's residents obtain their drinking and crop irrigation water. Although there is no evidence that the drinking water is being affected, the potential for future contamination is significant.[20]

In 1975, EPA conducted an assessment of leachate damages at five municipal disposal sites located in the northeast and midwest (Table 20). All five sites contaminated the ground water and polluted residential, industrial, or public supply wells. Up to 5,000 ft (1,500 m) and 0.42 sq mi (1.1 sq km) of ground water were contaminated.

In all but one case, all the wells were abandoned and public water piped in. In this case the residential wells were replaced by piped water, but the public supply and industrial

	Site 1	Site 2	Site 3	Site 4	Site 5
Type of operation	open dump, landfill	open dump, landfill incinerator residue	landfill	open dump, landfill	open dump, landfill incinerator residual
Location	swampy area, stream	sand pit	gravel pit	on bedrock	sand pit
Years of operation	1945 - present	1933 - present	1960 - 1968	1961 - 1972	1947 - 1972
Size of operation					
Acres	8	17	56	22	40
Peak annual tonnage	2,500	?	200,000	68,000	94,000
Depth (ft.)	25	50	40	55	55
Annual precipitation (inches/yr)	42	42	42	37	37
Damages	25 residential wells	3 residential wells home fixtures	33 residential wells 3 industry wells 8 public supply wells	7 residential wells home fixtures	4 residential wells 4 industry wells 1 public supply well
Remedial actions	wells abandoned public water supplied	filters on wells wells abandoned public water supplied fixtures replaced	resid. wells abandoned public water supplied public wells cut back counterpumping	wells abandoned public water supplied	all wells abandoned public water supplied new public supply cover soil on landfill
Litigation	No	No	No	Yes	No
Costs to date					
Value of damaged resources	31,200	3,600	39,600	7,000	94,800
Damage costs (clothes, fixtures)	0	852	?	?	0
Corrective costs [a]	0	0	1,371,762	0	12,000
Avoidance costs [b]	500,000	6,032	628,238	88,000	115,000
Total cost	531,200	10,484	2,039,600	95,000	221,800
Status	Costs continuing	Concluded	Costs continuing	Concluded	Concluded

[a] Costs necessary to clean up or control contamination, so that a ground-water or surface-water scurce may be used again.

[b] Costs necessary to avoid use of a contaminated aquifer and to develop an alternative water supply

wells have been forced to cut back production. Counterpumping wells are being used to retard the leachate migration. Since counterpumping is so expensive, serious consideration is being given to the removal of the landfill. This would allow the aquifer to eventually cleanse itself and once again permit full use of the public supply and industrial wells.

In none of the five cases were the damaged well owners fully compensated for their losses, including their inconvenience, lost time, and water consumption charges. The domestic well owners in one case assumed all the cost themselves. They used temporary water for four years before donating their private road to the city and connecting to a public supply. In another case, the domestic well owners sued the landfill operator. In winning the suit they recovered only the legal fees, public supply costs, and some house fixture replacements. Unrecovered were such costs as temporary water, inconvenience, lost time, road repair, and water consumption charges. They also were required to become annexed to the town which supplied the water.

Litigations against contaminating land disposal sites take two forms: (1) action by the state health or environmental authority to force corrective actions to prevent continued contamination, and (2) damage suits by the impacted well or land owners. Both types of litigation have severe constraints. A major problem with both types is the expense and effort of proving the contamination was indeed caused by the disposal site and by no other source. Such proof typically requires drilling of monitoring wells and analyses of water samples to determine the direction of ground-water flow and the constituents and extent of contamination. This means a major expense before litigation can start. At one site, such an examination was estimated by the U. S. Geological Survey to cost $130,000.

A second and greater constraint on state litigation is that, in the absence of a disposal alternative, the state cannot realistically shut down a disposal site regardless of the contamination it causes. Coercion must be used with or without litigation. Further, any effective corrective action is generally prohibitively costly (a private owner could declare bankruptcy first). The damaged party usually cannot afford the professional expense to conclusively prove the source of contamination. In addition, the owner of the contaminated well usually has no alternative water-supply source other than bottled water, so time is of the essence in obtaining relief.

Radioactive Wastes

Radioactive wastes are disposed of by shallow-land burial at a number of facilities across the United States. 21) Six of these sites are commercial, and are required to be situated on Federally- or state-owned land, even though they are managed by private industry. The commercial sites are located in Washington, Nevada, Illinois, Kentucky, New York, and South Carolina, and in general, are regulated by the state's Environmental Protection Agency.

Energy Research and Development Administration (ERDA) has five principal burial sites at their facilities in New Mexico, Idaho, Tennessee, Washington, and South Carolina. All are located on Federal land, and they are operated by ERDA contractors.

The sources of waste, which are in a solid form, are derived from the nuclear power industry (fuel fabrication, reactor operation, and fuel reprocessing); the university and industrial research centers, medical diagnostic and treatment centers; military laboratories and service facilities, including the shipyard servicing of nuclear-propulsion naval vessels; and certain ERDA operations which do not have on-site waste burial capability. The low-level radioactive wastes which are disposed of by shallow-land burial can be defined as those which neither originate from the first extraction process of the nuclear fuel reprocessing operation nor generate sufficient heat or radiation so as to require special cooling or shielding.

Inventory data accumulated by EPA during the past few years reveal that the volume and quantity of low-level waste is growing rapidly. As of 1973, some 9.2 million cu ft (260,000 cu m) of low-level radioactive wastes have been buried, and it is estimated that by 1990 this volume will have accumulated to some 175 million cu ft (5 million cu m). It is estimated that the ERDA facilities will have accumulated about 40 million cu ft (1.1 million cu m) by 1990. Based on EPA data, estimates, and projections concerning the existing commercial burial sites, if all sites are used to capacity and no changes to the present practices or trends occur, all six commercial sites could be filled by 1992.

Examples of problems that can occur if such wastes are not handled properly are illustrated by the following case histories. Between 1963 and 1974, approximately 3.7 million cu ft (100,000 cu m) of solid "low-level" radioactive waste were buried at the commercial radioactive waste disposal facility at the Maxey Flats, Kentucky site. 22) These wastes contained more than 175 lb (80 kg) of plutonium-239, a large undetermined quantity of other plutonium isotopes, and 1,600,000 Ci of by-product material. The burial media at Maxey Flats is an unsaturated jointed shale with relatively low permeability which is underlain by a series of jointed shales, siltstones, and sandstones. The main water table is approximately 150 ft (45 m) below land surface. In 1972, it was noted that some trenches had filled or were filling with leachate and radioactivity was detected in monitoring wells and in the environment around the burial facility.

Geotechnical, operational, and regulatory factors or problems observed at Maxey Flats which appear to have a direct bearing on the land burial of hazardous wastes follow:

- The intent of disposing of plutonium and other radionuclides at Maxey Flats was absolute containment but contamination has migrated hundreds of metres from the site in less than 10 years. While EPA scientists are confident that, at the present time, this movement of plutonium and other radioactive materials does not present a public health hazard, the potential long-range impact of these contaminants is not known.

- The primary source of the leachate is precipitation infiltrating through the trench caps. If the caps had been impermeable (rather than of "very low permeability"), there may not have been a problem.

- The permeability of the burial media was sufficiently low to cause the trenches to fill like bathtubs with leachate and overflow; yet, the joints in the burial media allowed subsurface migration of the leachates.

- There is no information on the physical and chemical character of the wastes.

- There is insufficient information on the hydrogeology of the site and the wastes buried there to predict where or how the radioactivity will migrate or what the impact on man or the environment will be. An estimated $1,300,000 will be required to evaluate the problem.

- Plutonium is thought to be insoluble in water and unable to move more than a few centimetres after coming in contact with the ground. It is suspected that the reaction of plutonium in the leachate-saturated environment of the burial trenches may have mobilized it so that it has been able to migrate hundreds of metres through the subsurface. If true, this could have serious implications on the land disposal of other heavy metals.

- The site is located over a series of jointed and fractured rock and the burial media itself is jointed. This jointing and fracturing: (1) prevents prediction of where contaminants will move; (2) allows channeling and unexpectedly rapid movement of contaminants through the subsurface; and (3) allows the ion exchange mechanism, a major secondary safety factor in land disposal, to be bypassed.

- Mitigating actions such as dewatering the trenches, evaporating the leachates, and improving the caps have been taken. They do not, however, ensure that the spread of contamination has been stopped. They are on-going, interim actions which may well be required for hundreds of years. Once dewatering stops, leachates probably will again fill the trenches and migration of contaminants will continue. No solution to the problem has yet been developed; however, final corrective action may well cost several millions of dollars.

- The burial facility is regulated by the state. The regulations are, in general, comparable to regulations for other state- and Federally-regulated radioactive waste disposal sites.

A similar situation has developed at a second commercial radioactive waste burial facility in West Valley, New York.[23)] Although the hydrogeology is different, precipitation has infiltrated the trench caps, causing the formation of radioactive leachates. These leachates have filled the trenches and overflowed, allowing the release of radioactivity. Again insufficient information was available on the geology, hydrology, depth to water, and physical and chemical character of the wastes.

At Oak Ridge National Laboratory, buried wastes from two of four burial grounds annually contribute 1.8 to 2.8 curies of strontium-90 to adjacent ground water and ultimately to the Clinch River. From 1951 to 1973, approximately 35 million gal. (130,000 cu m) of waste containing over one million curies of mixed fission products were placed into four pits and seven trenches excavated into the Conasaug Shale. About half of the activity consisted of strontium-90 and cesium-137 with most of the balance being ruthenium-106. Breakthrough and transport (in ground water) was commonplace and, in part, led to subsequent disposal efforts involving hydrofracturing of the shale and injection of slurries.

Future Trends

Current or probable trends that will tend to reduce the problems of ground-water contamination from land disposal sites are:

1. Many open dumps and other contaminating disposal sites will be closed, and larger regional facilities, properly located and designed to minimize contamination, will be developed.

2. More sites will require permits from regulatory agencies and, thus, will be constructed in compliance with state controls and regulations.

3. More wastes will be processed (incinerated or shredded) and/or recovered (materials and energy), resulting in less waste volume.

It should be emphasized that the foregoing favorable trends relate only to municipal and not industrial wastes.

Current or probable trends that will tend to increase the problem of ground-water contamination from land disposal sites are:

1. The volume of waste is continuing to increase, both absolutely and on a per capita basis.

2. An increasing amount of industrial wastes will require land disposal as a result of the implementation of the Federal Water Pollution Control Act Amendments of 1972, the Marine Protection Research and Sanctuaries Act of 1972, and the Clean Air Act of 1970. This is illustrated in Figure 55, which shows that residues of pollution control efforts account for the greatest increase in wastes destined for land disposal in the case of select industries that were examined by EPA.

3. Increasing numbers of municipal sites will refuse to accept industrial wastes, resulting in more on-site disposal facilities, which are difficult to regulate, inventory, and monitor.

4. Disposal sites now operating or currently closed may continue to produce leachate and cause severe damages for the next 50 to 100 years.

5. The demand for ground and surface waters will continue to increase, forcing utilization of water resources that

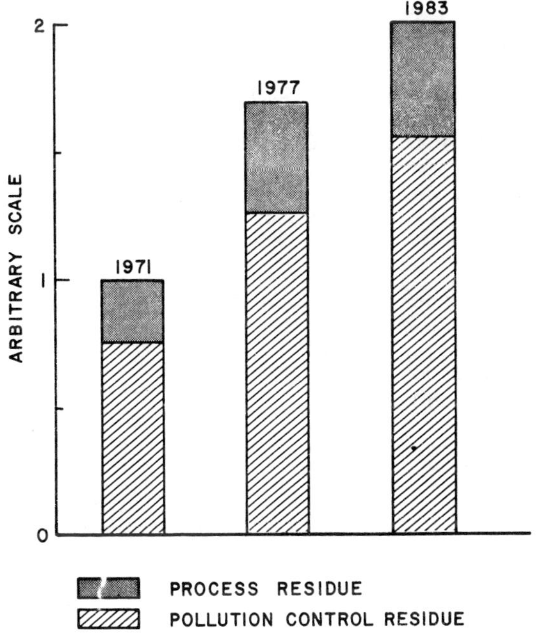

re 55. Projected growth of combined waste quantities for four representative industries (inorganic chemicals, paper, steel, and non-ferrous smelting).

may have been affected by leachate.

The net effect of these favorable and unfavorable trends can be summarized as follows:

1. The increase in the recovery of material and energy will not offset the increases in waste generation (Table 21).

2. The greatest problems will be controlling the disposal of hazardous industrial wastes, and developing effective permitting and monitoring programs for all land disposal sites.

3. In populous areas, it may become necessary to recycle water, or treat brackish or highly mineralized water to replace water which has been contaminated.

4. Extensive, sophisticated land and ground-water use planning and surveillance programs will have to be developed to avoid placing wells in contaminated aquifers, or locating disposal sites in areas where contamination could occur.

TECHNOLOGICAL CONSIDERATIONS

Methods for preventing, reducing, or managing leachate are (1) natural attenuation, (2) prevention of leachate formation, (3) collection and treatment, (4) pretreatment capable of reducing the volume or solubility of the waste, and (5) detoxification of hazardous wastes prior to disposal.

Descriptions of these processes are given below, including the effectiveness of protecting ground-water resources. The current status of the various leachate control methods is given in Table 22.

Natural Attenuation

As leachate migrates through soil, it undergoes natural attenuation by various chemical, physical, and biological processes. The ability of a proposed sanitary landfill site to attenuate the leachate generated should be estimated on a site-by-site basis using available technology. If natural attenuation appears inadequate, it may be desirable to line the site and collect and treat the leachate.

Prevention

The second control method involves preventing leachate generation. If water is restricted from entering the site,

Table 21. U.S. BASELINE POST-CONSUMER SOLID WASTE GENERATION PROJECTIONS.[a]

	Estimated			Projected	
	1971	1973	1980	1985	1990
Total Gross Discards					
Million tons per year[b]	133	144	175	201	225
Pounds per person per day[c]	3.52	3.75	4.28	4.67	5.00
Less: **Resource Recovery**					
Million tons per year	8	9	19	35	58
Pounds per person per day	0.21	0.23	0.46	0.81	1.29
Equals: **Net Waste Disposal**					
Million tons per year	125	135	156	166	167
Pounds per person per day	3.31	3.52	3.81	3.86	3.71

a) Resource Recovery Division, Office of Solid Waste Management Programs, U.S. EPA, revised December 1974. Projections for 1980 to 1990 based in part on contract work by Midwest Research Institute.

b) Tons x 0.9078 = tonnes
c) Pounds x 0.454 = kilograms

Table 22. STATUS OF LEACHATE CONTROL METHODS.

Method	Effectiveness	Degree of use	Cost (examples)
NATURAL ATTENUATION			
Clay	promising research	unknown	natural
Silt	unknown	unknown	natural
Sand	unknown	unknown	natural
PREVENTING LEACHATE GENERATION	ranges from complete to partial control	limited	not available
COLLECTION AND TREATMENT			
Liners	promising research	limited	$1.50 to $4.00/sq yd
Biological treatment	promising research	very limited	not available
Physical-Chemical	promising research	very limited	not available
Recirculation	promising research	very limited	not available
Spray irrigation	promising research	very limited	not available
IMMOBILIZATION			
Chemical Stabilization	research progressing looks promising	limited but growing	$10 to $20/ton
Encapsulation	research progressing looks promising	very limited	$16/ton
Fixation and encapsulation	research progressing looks promising	not in use	$40/ton
VOLUME REDUCTION			
Dewatering	effective	widely practiced in water pollution	$5 to $20/ton
Incineration	effective for organics	moderate	$20 to $100/ton
DETOXIFICATION	varies widely by process and waste	limited to specific wastes	varies widely

then the amount of leachate generated will be greatly reduced. Water cannot be completely prevented from entering in some locations, but through proper design and operation, the quantity can be minimized.

Control measures, such as diversion of upland drainage; use of relatively impermeable soils for cover material; compacting, grading, and sloping of the daily and final cover to allow runoff; rapid attainment of final elevations; planting of high-transpiring vegetation; use of impermeable membranes overlying the final lift of solid waste; maintenance of final grades; and use of subsurface drains and ditches to control ground water are available to the design engineer and operator. Use of impermeable membranes and soil cover requires vents to control gases and drains to manage the intercepted leachate. However, there presently is a general lack of quantitative information on the use of these controls.

Collection and Treatment

The third control method is to collect and treat the leachate. An impermeable liner may be employed to prevent the movement of leachate into the ground. This is a relatively new technique, and because impermeable liners have not been used for long periods, their long-term durability has not been established.

An impermeable liner can be made from different types of materials, including: natural clay, soil additives, conventional paving asphalt, hot sprayed asphalt, polyethylene (PE), polyvinyl chloride (PVC), butyl rubber, hypalon, chlorinated polyethylene (CPE), and ethylene propylene rubber (EPDM).

EPA research currently underway provides for the evaluation of common liner materials for resistance to various types of municipal and industrial leachates. The evaluation will take two years, and results will become available beginning in May 1977. The costs of lining sanitary landfills will also be evaluated. [24,25]

Where sanitary landfills use collection for control of leachate, provisions must be made to treat it prior to discharge to the surrounding environment. Several researchers have studied the treatment of leachate, and promising results were obtained with a number of methods. It was found that biological treatment methods are effective when treating organics generated in a new site. [2] Physical-chemical treatment methods showed better results than biological methods when treating leachate from an old fill. Industrial leachate may require more specific treatment techniques, depending upon the wastes involved.

Three projects have been awarded by the EPA to demonstrate leachate treatment processes. The first is an activated sludge plant in Tulleytown, Pennsylvania. The treatment plant will use lime pretreatment to remove heavy metals prior to entering the activated sludge units. Because this is the first full-scale activated sludge plant to treat landfill leachate, accurate cost data are presently not available. This plant was designed to treat leachate from a 50-acre (20-ha) sanitary landfill in a moderately high-rainfall area. The reported capital cost for the treatment plant is $350,000.

The second demonstration project is an anaerobic filter at Enfield, Connecticut. At the present time, EPA knows of no cost data that are available on the capital and operating cost of a full-scale anaerobic filter used to treat landfill leachate; these will be provided by the Enfield project. The third project tests the use of soils and vegetation for reducing contaminants in leachate and measures the environmental and ecological impact of leachate applied by spray irrigation.

A laboratory study is currently underway at the University of Illinois to test the different alternatives and combinations of physical-chemical treatment methods to polish the effluent of the extended aeration, activated sludge, and anaerobic filter units used in treating leachate. Preliminary results indicate that a very good quality effluent can be obtained (COD of less than 5 ppm and a total dissolved solids content of less than 50 ppm). Similar treatment schemes can be used to treat well-stabilized leachate from an old landfill.

Leachate can be spray irrigated on certain land and is treated through reactions in the soil mantle. Research work already carried out by EPA has shown that spray irrigation can remove contaminants from leachate. The test results from the study showed COD removal rates of 85 to 99 percent; iron, 88 to 99 percent; and similar results for zinc. Lower removal rates were observed for potassium, calcium, magnesium, and sodium.

Managing Industrial Wastes

While much industrial waste has characteristics somewhat similar to municipal refuse, process wastes tend to be composed of higher concentrations of more homogeneous materials. Often these are liquids, slurries, or sludges, and typically they are toxic, flammable, or otherwise hazardous to health and the environment. Properties of such wastes vary widely, and the techniques used to manage them adequately vary accordingly. Many of the techniques discussed above are applicable to certain classes of hazardous wastes, or can be made suitable if modifications are made, or proper precautions are taken.

Although the preferable alternative for many waste disposal problems is the conventional sanitary landfill, certain wastes should probably never be land disposed because of the extreme hazards posed by the escape of even small quantities of a hazardous constituent. Due largely to the hazardous nature of some industrial wastes, the potential for leachate contamination of surface and ground waters is large even in a well-run sanitary landfill. Also, waste-materials handling procedures at a landfill may be less than adequate to ensure the landfill operators' safety.

EPA has awarded a five-year demonstration grant to the Minnesota Pollution Control Agency. The overall project goal is to conduct a complete demonstration of a chemical waste land disposal site, which examines the technological, economic, organizational, and social/institutional issues involved in establishing and managing an environmentally acceptable site designated for hazardous wastes.

Specific objectives include:

1. Demonstration of site selection methods.

2. Demonstration of appropriate site preparation techniques to prevent ground-water infiltration.

3. Demonstration of waste preparation techniques.

4. Demonstration of monitoring and surveillance techniques.

5. Evaluation of waste handling and operational procedures.

6. Determination of costs.

7. Evaluation of social and institutional issues.

The facility will become operational in late 1977 and continue to serve industrial waste generators in the Minneapolis-St. Paul area after the demonstration period ends in 1980.

The broad categories of hazardous waste handling are chemical stabilization and encapsulation. The purpose of these techniques is to render the waste less soluble or less available for transport to the water related environment. In the chemical fixation process, additives are mixed with waste sludges, and the resulting mixture is deposited in a land disposal site. Depending on the technique employed, some wastes, such as heavy metals, may be complexed to produce insoluble compounds. Chemical fixation costs are typically in the range of 5 to 10 cents/gal. (approximately $10 to $20/ton or $11 to $22/tonne), not including the cost of land at the final disposal site.

Although its application is somewhat limited, numerous incineration techniques are currently available which are appropriate to a variety of disposal situations. The primary advantages of proper incineration are listed below:

1. Incineration technology is relatively well developed. Facilities exist that are currently utilized in both municipal and industrial waste disposal.

2. Incineration can destroy or detoxify a very wide variety of hazardous organic materials. With the proper incinerator temperature, residence time, and effluent scrubbers, many dangerous compounds are oxidized to harmless combustion products. In this case, some residual materials will still remain for land disposal, but these will often be less toxic than the original material.

3. Incineration can greatly reduce the volume of waste. A concentrated ash is more suitable for materials recovery and requires less space for land disposal.

4. Incineration has the potential for energy recovery. With the increasing cost of fuel, the heat value of waste materials represents a potentially valuable energy source.

Incineration does have disadvantages. The equipment is relatively expensive and unless proper control devices are installed, air pollution problems can result.

Although many industrial wastes are currently being incinerated in commercial incinerators, the combustion conditions and the effluents are not always adequately controlled or monitored. Also, information which is gathered is generally kept in confidence. In order to provide the desired information base to the government and the public, a demonstration program has been initiated by EPA.

This program will demonstrate the environmental, operational, and economic feasibility of destructing industrial wastes via incineration. The work will be accomplished through a series of test burns conducted on various types of existing commercial-scale incinerators, using actual samples of industrial waste streams.

INSTITUTIONAL ARRANGEMENTS

Levels of Regulatory Control

There is no Federal regulatory control of land disposal of solid waste except as it may enter navigable waters. The use of the Refuse Act of 1899, which prohibits discharging solid waste to navigable waters, has been awkward and generally ineffective.

A recent district court ruling stated that the Federal Water Pollution Control Act (PL 92-500) specifically excluded ground water from Federal regulatory control (U.S. vs. GAF 2-5-75). The National Pollutant Discharge Elimination System (NPDES), established under this Act to control groundwater discharges, has yet to be successfully employed. Further obstacles to Federal control are inherent in the characteristics of "sanitary landfills." "Sanitary landfills" are not designed to discharge leachate. Therefore, the application of an NPDES-style permit is infrequent.

Traditionally, regulation and control of solid-waste disposal have been concerns of the states. While there are dramatic differences in perceived responsibilities and objectives among the 50 state regulatory agencies, there is agreement among state laws that solid-waste disposal shall not pose a threat to health. All 50 states have addressed the problem of solid-waste disposal and have passed legislation to control, to varying degrees, health and environmental impacts associated with solid-waste disposal.

Permits

Forty-four states have statutes which prohibit the disposal of solid waste without a permit. Typically, the permit is issued by a state agency, but in a number of states, permits are handled by local or regional agencies, with state oversight.

The range of requirements for state permit systems ext
from the minimal requirements of four states for noti
tion that a facility exists, to the requirements of
like Pennsylvania, that applicants engage engineers

pare thorough site descriptions, including soil borings, water analyses, and general aquifer descriptions, for submission to the State Geological Survey for review. Under a typical system, an applicant for a permit must supply detailed information concerning his proposed site and disposal plan. The agency determines whether the site is acceptable, and if so, what conditions would be placed in the permit. Special requirements must be met for hazardous wastes.

Many statutes also specifically address the prevention of water contamination, including ground water. In some states, these are strictly enforced, but regulation of the disposal of solid waste is, in practice, frequently shaped by a balancing of competing interests, of which ground water is the least apparent.

As conditions for issuance of a permit, regulations typically require:

1. Plans and specifications for the proposed site and facility; some states require that these be prepared by a registered professional engineer.

2. A map or aerial photograph of the area showing land use within the adjoining area. Locations of nearby water bodies may be required. Delaware requires wells within one mile of the site to be identified, [26] but the requirement in other regulations is usually less than this -- down to as little as 500 ft (150 m).

3. A report on geologic formations and soil conditions, including depth to ground water. The Wisconsin regulation specifies 3 borings for a site up to 5 acres (2 ha) in size, one boring for each additional 5 acres up to 50 acres (20 ha), and one boring for each additional 50 acres. [27] Many states, such as Florida, now require that hydrogeologic factors be considered prior to permitting new sites. A hydrogeologic survey of a site can show type and permeability of soil, height of water table, quality of ground water, and direction of ground-water movement.

   Illinois requires data describing soil classification, grain size distribution, permeability, compactability, and ion-exchange properties of the subsurface materials for those strata essential to design of the land disposal site; comprehensive analyses of water samples from on-site and nearby wells; a description of ground-water conditions including flow below and adjacent to the proposed site; and an appraisal of the effect on ground and surface waters.[28]

The importance given to hydrogeological analysis in the site review process varies widely among the states. Converted dumps and landfills, which make up the majority of sites, usually are not required to undergo such analyses.

4. A description of surface drainage patterns. The California regulation, for example, requires calculations for the flooding frequency of streams within or adjacent to the site.[29]

5. A report of (a) population and area to be served by the facility, (b) anticipated type, quantity, and source of wastes, (c) source and characteristics of cover materials, and (d) type and amount of equipment, and operating plans.

6. Information concerning measures proposed for prevention of water contamination and for control of drainage, leachate, and gases.

In addition, some regulations require a statement or plan as to ultimate use of the site after closing.

The statute or regulation may require that a representative of the regulatory agency inspect the site prior to issuance of a license. Most states either automatically require a hearing prior to issuance of a permit, or require one only if requested by a person who believes he may be adversely affected by the disposal operation.

Every state has implemented some type of control over its municipal sites, but control over industrial sites is much less common. Where the state solid-waste authority does require permits, there may be a permit to construct, a permit to operate, or some combination of the two.

Regulations Limiting Hazardous Waste Disposal at Landfill Sites

Alabama, California, Florida, Hawaii, Illinois, Indiana, Massachusetts, Mississippi, Montana, Nevada, New Mexico, Oklahoma, South Carolina, Tennessee, and Texas have regulations limiting hazardous waste disposal at solid-waste sites. However, defining which wastes are hazardous presents a difficult problem.

Some states, such as Massachusetts, have published detailed

regulations specifically addressing hazardous waste management. However, a number of state personnel have argued that one or two paragraphs contained in general solid-waste management (or other) regulations equally constitute "having hazardous waste regulations." Their belief is that the whole of their regulations, of which the hazardous waste section (toxic waste, etc.) may be only a small part, is adequate authority to accomplish the same goals as the more detailed regulations in other states.

Many states have published regulations for land disposal of hazardous wastes, even though only a few states have hazardous waste legislation. State solid-waste management or water contamination control laws sometimes give the state authority to control certain aspects of "hazardous," "toxic," "liquid," "industrial," or "special" waste management. Any of these terms can be construed in such a way as to give the state authority to regulate most of the wastes usually included in the term "hazardous." In nearly every case, the state has been given (or chosen to exercise) authority over hazardous wastes only at the disposal site. Many of these states choose to issue proscriptive regulations, such as, "special wastes may not be placed in landfills without prior approval."

As mentioned above, commercial burial facilities for radioactive wastes are managed by private industry, are located on public lands with limited access, and are generally controlled by the state in which they are situated. Exceptions are sites in Illinois, Nevada, and Washington, where NRC (National Research Council) retains direct regulation and/or licensing of the handling of special nuclear material.

The licenses for the burial of radioactive waste generally contain provisions for site maintenance, inventory control, health and safety, and environmental monitoring. The former AEC (Atomic Energy Commission) and agreement states issued 29 other licenses to companies for the purpose of collecting, packaging, storing, and transporting radioactive wastes from the users of radioactive materials to the commercial disposal facilities. These companies are not involved with the actual disposal of the waste, but act only as intermediaries.

Published Criteria for Designating Hazardous Wastes

Three states have published criteria for determining which wastes are "hazardous": California, Massachusetts, and South Carolina. The California criteria are considerably more detailed than those of the other two, and (also unlike the other two) the California criteria are embodied in a

"Hazardous Substances Act." Minnesota and Oregon are reportedly in the process of developing their criteria.

Publishing criteria has several advantages over publishing a list of hazardous wastes. Most importantly, it allows the state to describe what is being regulated without naming it. The state may unintentionally omit a flammable compound from its list of hazardous wastes, but if the state has published its criterion for measuring flammability, the omitted substance would be covered. Additionally, wastes can be an almost infinite variety of compounds and mixtures. For example, it would be nearly impossible for the state to anticipate every form and combination of waste which might be flammable. By publishing criteria, however, the generator would have a reliable method for determining whether or not he had a hazardous waste regardless of the composition.

<u>Program Limitations</u>

Evaluation of state solid-waste programs is very difficult without first hand experience of each particular state. Reliance on staff size, budget, legislation, or regulations as measures of effectiveness can be deceptive. The enthusiasm with which a particular state solid-waste agency fulfills its responsibilities may be contingent on administrative leadership, staff professionalism, the agency's perception of problem potential, or historical precedent. As previously mentioned, the regulation of disposal sites may be subject to the forces of competing interests. Additionally, a large percentage of each state's disposal sites was developed prior to the current regulations, and these old sites generally are not subject to rigorous examination by the state agency.

Current state solid-waste budgets range from $0.01 to $0.32/capita/year. One of the possible reasons for the wide range of these budgets is that those states which have delegated some of the regulatory functions to sub-state governments would not need as large a state budget or staff as those states that maintain a greater degree of centralized control. Also, Federal EPA grants have often been used to help supplement state solid-waste programs. The size of a solid-waste budget may reflect the efficiency of a program, but they have not been a reliable measure of its effectiveness.

The number of employees in a state solid-waste program ranges from one to 50. Many states are hampered by inadequate staff, resulting in fewer inspections and a backlog of permit requests. A number of state programs have sufficient staff to operate only under authority legislated in general environmental acts. For state solid-waste programs on the average, 27 percent of staff is in administration, 27 percent in enforcement, 26.5 percent in technical assistance, and 19.5 percent in planning. The high priority of technical assistance reflects the positive incentive approach many states are using to improve solid-waste facilities and regulatory programs.

Federal Guidelines

The Administrator of the U. S. Environmental Protection Agency has prepared Solid Waste Management Guidelines for the disposal of municipal waste under directive of the 1970 amendments to the Solid Waste Disposal Act of 1965 (PL 89-272). The guidelines represent the judgment of the EPA regarding what is acceptable design and operation of land-disposal facilities to insure protection of the environment. They are recommended for adoption by state and local governmental agencies, and are mandatory for Federal agencies and for solid-waste disposal on any Federal lands.

REFERENCES CITED

1. U. S. Environmental Protection Agency. 1974. Rules and regulations, title 40, chapter 1, part 241; guidelines for the land disposal of solid waste, section 241.101: definitions.

2. U. S. Environmental Protection Agency, Solid and Hazardous Waste Research Laboratory, National Environmental Research Center. 1974. Summary report: gas and leachate from land disposal of municipal solid waste. Cincinnati, Ohio. 62 pp.

3. Shaver, R. G. 1975. Assessment of industrial hazardous waste practices: inorganic chemicals industry. Final report. Versar, Inc. Contract No. 68-01-2246. Springfield, Virginia.

4. Gruber, G. I., and M. Ghasseni. 1975. Assessment of industrial hazardous waste practices: organic chemical, pesticides and explosives industries. Unpublished draft. TRW Systems and Energy. Contract No. 68-01-2919. Redondo Beach, California.

5. Weisberg, G. 1975. Assessment of industrial hazardous waste practices: petroleum refining. Unpublished draft. Jacobs Engineering Co. Contract No. 68-01-2288. Pasadena, California.

6. Leonard, R. P. 1975. Assessment of industrial hazardous waste practices in the metal smelting and refining industry. Unpublished draft. Calspan Corporation. Contract No. 68-01-2604. Buffalo, New York.

7. Hallowell, J. B. 1975. Assessment of industrial hazardous waste practices: electroplating and metal finishing industries. Unpublished draft. Battelle Columbus Laboratories. Contract No. 68-01-2664. Columbus, Ohio.

8. Levin, J. 1975. Assessment of industrial hazardous waste practices: paint and coatings manufacturer, solvent reclaiming, and factory-applied coatings operations. Unpublished draft. Wapora, Inc. Contract No. 68-01-2656. Washington, D. C.

9. Shaver, R. G. 1974. Assessment of industrial hazardous waste practices: storage and primary batteries industries. Versar, Inc. National Technical Information Service, Report PB 241-204. Springfield, Virginia.

10. McMahan, J. R. 1975. Hazardous waste generation, treatment and disposal in the pharmaceuticals industry. Unpublished draft. Arthur D. Little, Inc. Contract No. 68-01-2684. Boston, Massachusetts.

11. U. S. Environmental Protection Agency. 1974. Report to Congress: disposal of hazardous wastes. Environmental Protection Publication SW-115.

12. U. S. Environmental Protection Agency, Office of Solid Waste Management Programs, Resource Recovery Division. 1973. 1973 estimates. Report in preparation.

13. Anonymous. 1975. Exclusive Waste Age survey of the nation's disposal sites. Waste Age. January 1975.

14. U. S. Department of Commerce, Bureau of the Census. 1975. 1972 census of manufactures, water use in manufacturing. Washington, D. C. 194 pp.

15. Flach, K. W. 1973. Land resources. Pages 113-120 *in* University of Illinois. Proceedings of joint conference on recycling municipal sludges and effluents on land. Champaign, Illinois.

16. Shuster, K. A. 1975. Leachate damage assessment. Interim Report. Unpublished report. Office of Solid Waste Management Programs, U. S. Environmental Protection Agency.

17. Miller, D. W., F. A. DeLuca, and T. L. Tessier. 1974. Ground water contamination in the northeast states. U. S. Environmental Protection Agency, Report 660/2-74-056.

18. U. S. Environmental Protection Agency, Office of Solid Waste Management Programs. 1975. Hazardous waste disposal damage report. Environmental Protection Publication SW-151.

19. Feinglass, E. J. 1973. Arsenic intoxication from well water in the United States. New England Journal of Medicine 288(16):828-830.

20. U. S. Environmental Protection Agency, Office of Solid Waste Management Programs. Internal files.

21. O'Connell, M. F., and W. F. Holcomb. 1974. A summary of low-level radioactive wastes buried at commercial sites between 1962-1973, with projections to the year 2000. Radiation Data and Reports, Vol. 15, No. 12. Pages 759-767.

22. Kaufman, R. F., G. G. Eadie, and C. R. Russell. 1975. Summary of ground-water quality impacts of uranium mining and milling in the Grants Mineral Belt, New Mexico. Technical Note ORP/LV-75-4. U. S. Environmental Protection Agency, Office of Radiation Programs, Las Vegas, Nevada. 71 pp.

23. U. S. Environmental Protection Agency, Radiation Source Analysis Branch. 1975. Personal communication.

24. Haxo, H. E., Jr., and G. M. White. 1974. Evaluation of liner materials exposed to leachate: first interim report. Matrecon, Inc. Oakland, California. 58 pp.

25. Geswein, A. J. 1975. Liners for land disposal sites. Environmental Protection Publication (SW-137). U. S. Government Printing Office, Washington, D. C. 23 pp.

26. Delaware Water Pollution Control Regulations.

27. Wisconsin Department of Natural Resources. 1973. Regulations, Chapter NR 151: solid waste management.

28. Illinois Pollution Control Board. 1973. Rules and regulations, Chapter 7, rule 316: application.

29. California State Water Resources Control Board. Waste discharge requirements for waste disposal to land.

SECTION VIII

SEPTIC TANKS AND CESSPOOLS

SUMMARY

Septic tanks and cesspools rank highest in total volume of waste water discharged directly to ground water and are the most frequently reported sources of contamination. However, most problems are related to individual homesites or subdivisions where recycling of septic fluids through aquifers has affected private wells used for drinking water. Except in situations where such recycling is so quick that pathogenic organisms can survive, the overall health hazard from on-site domestic waste disposal is only moderate, with relatively high concentrations of nitrate representing the principal concern.

Twenty-nine percent of the population, representing about 19.5 million single housing units, dispose of their domestic waste through individual on-site disposal systems. Almost 17 million of these housing units use septic tanks or cesspools. Regional ground-water quality problems have been recognized only in those areas of the greatest density of such systems, primarily in the northeast and southern California. Across the United States, there are four counties (Nassau and Suffolk, New York; Dade, Florida; and Los Angeles, California) with more than 100,000 housing units served by septic tanks and cesspools, and there are 23 counties with more than 50,000. Data on discharge to industrial septic tanks are not available.

Where the density of on-site disposal systems has created problems, collection of domestic waste water by public sewers and treatment at a central facility is the most common alternative. Other alternatives, which are generally limited to special situations where natural conditions or restrictive codes rule out conventional septic tank systems, include aerobic treatment tanks, sand filters, flow reduction devices, evapotranspiration systems, and artificial soil (mounds) disposal systems.

Where sewer systems are not economically feasible, prevention of ground-water quality problems has normally been attempted by low density zoning at the local government level, although increased regulation of septic tank siting, construction and design is emerging at the state government level. More than half the states now participate in septic tank permitting or regulation of some type, and a large number are providing local agencies with data to aid in land-use planning as applied to septic tank density.

DESCRIPTION OF THE PRACTICE

There are three methods of on-site domestic waste disposal in wide use today. The most sophisticated and acceptable of these is the septic tank and its associated subsurface disposal system. Septic tanks are installed for most new housing when public sewer service is not available. The cesspool, common in many older installations with deep permeable soils, is less satisfactory than the septic system and no longer approved for new installations in many areas. And, finally, the pit privy is common in many rural areas, where pressurized water systems are not available.

Septic Tank System

The septic tank system is composed primarily of two components: the septic tank which traps the settleable solids, and floating grease and scum contained in the raw sewage, and the subsurface disposal system (trench bed, leach field, etc.) which receives the liquid effluent from the septic tank. When acceptable soil conditions exist, the two units will effectively treat the sewage generated by a household. Maintenance of the system requires periodic pumping and removal of solids collected in the tank. The material removed, termed septage, is transported to a central treatment facility, dumped, lagooned, spread on land, or deposited at a landfill. A typical septic system is shown in Figure 56. [1)]

Because the liquid effluent from a septic system has not been "purified" within the septic tank, the soil to which it is discharged is relied upon to perform this function. This is primarily achieved through aerobic decomposition at the soil interface (decomposition by bacteria in the presence of free oxygen) and by physical and chemical removal of suspended and dissolved solids (filtering and sorption). The ability of a soil to perform these latter functions is primarily related to the sizes of the individual soil grains; the larger the grains, the less filtering and sorption.

The Cesspool

Because the cesspool was once commonly used as an independent unit, performing the functions of both the septic tank and the subsurface disposal system, this method of sewage disposal is mentioned separately here. The cesspool is typically a 5- to 6-ft (1.5- to 2-m) diameter sump, buried several feet below ground surface. The facility receives raw

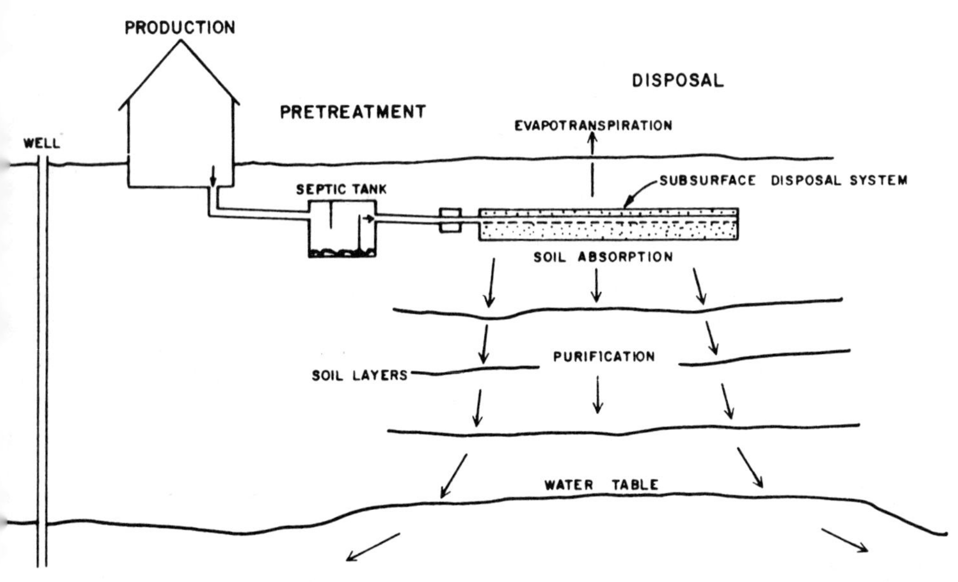

Figure 56. Diagram of a typical domestic septic tank system. 1)

sewage directly from the house drain. The larger solids settle to the bottom of the sump or are otherwise trapped inside while the liquid fraction seeps out through openings in the sides and bottom. Separation of solids from the liquid waste water is poor and cesspools will work only in very coarse or highly fissured materials. Thus, essentially raw sewage may move directly into ground water, and the contamination potential of these systems is considered very high. While new installations of cesspools are widely prohibited, their popularity in the past has left hundreds of thousands currently in operation.

The Privy (Outhouse)

This form of human waste disposal is common in rural areas where lack of indoor pressurized water systems precludes the use of other systems. The pit privy is typically a small, shallow pit or trench which normally receives only human waste and paper. Properly constructed pit privies allow effective decomposition and treatment of human wastes. The volume of water introduced to such facilities is relatively small; thus problems associated with odors and disease-carrying insects are more closely associated with pit privies than is the problem of ground-water contamination.

CHARACTERISTICS OF CONTAMINANTS

Sewage from individual homes consists of about 99.9 percent water (by weight), 0.02 to 0.03 percent suspended solids, and other soluble organic and inorganic substances.[2] Also present in domestic sewage are bacteria, viruses, and other microorganisms from the digestive tract, respiratory tract and skin. Domestic sewage composition is not uniform, but rather, it varies from day to day, even from hour to hour, and from house to house.

The volume of waste water directed to the septic system from a typical household ranges from 40 to 45 gpd/person (150 to 170 l/day/person).[3,4] The portions of this total from various sources within a particular house are related to such factors as the use of automatic washing machines and personal habits of the occupants. Ranges reported in the literature are as follows:

Source	Percentage of Total Domestic Waste Load
Toilet	22 - 45
Laundry	4 - 26
Bath	18 - 37
Kitchen	6 - 13
Other	0 - 14

A wide variety of household contaminants may enter the waste water flow from the major sources within a house. The organic chemical content of these wastes comes primarily from human wastes, soaps, detergents, and food wastes. 5)

Domestic sewage is a complex mixture, and some of its specific substances have yet to be fully identified. However, certain collective characterizations can be made as illustrated in Table 23. The constituents which present the greatest threat to ground-water quality and some of the problems which arise are as follows:

- Excessive concentrations of nitrate in drinking water produce a bitter taste and may cause physiological distress. Water from wells containing more than 45 ppm nitrate as NO_3 has been reported to cause methemoglobinemia in infants.

- Discharge of ground water with high phosphate concentrations to surface-water bodies can cause eutrophication.

- Lead, tin, iron, copper, zinc, and manganese (from household pipes and human waste) are toxic in excessive concentrations.

- Sodium, chloride, sulfate, potassium, calcium and magnesium, can create health hazards to some individuals, ranging from laxative effects to aggravated cardiovascular or renal disease, if concentrations exceed recommended limits.

- Aquifers being recharged by large volumes of septic tank effluent can contain water which exceeds the U. S. Public Health Service recommended limit of 0.5 ppm of MBAS (a nonbiodegradable detergent constituent and an indicator of contamination).

- Excessive BOD in septic fluid discharged to surface water from clogged drain fields can deplete dissolved oxygen supplies necessary to aquatic life.

Table 23. TYPICAL COMPOSITION OF DOMESTIC SEWAGE.[6]
(All values except settleable solids are expressed in ppm)

Constituent	Concentration Strong	Medium	Weak
Solids, total	1,200	700	350
Dissolved, total	850	500	250
Fixed	525	300	145
Volatile	325	200	105
Suspended, total	350	200	100
Fixed	75	50	30
Volatile	275	150	70
Settleable solids, (ml/l)	20	10	5
Biochemical Oxygen Demand, 5-day, 20°C (BOD5 20°)	300	200	100
Total Organic Carbon (TOC)	300	200	100
Chemical Oxygen Demand (COD)	1,000	500	250
Nitrogen, (total as N)	85	40	20
Organic	35	15	8
Free ammonia	50	25	12
Nitrite	0	0	0
Nitrate	0	0	0
Phosphorus (total as P)	20	10	6
Organic	5	3	2
Inorganic	15	7	4
Chloride [a]	100	50	30
Alkalinity (as $CaCO_3$) [a]	200	100	50
Grease	150	100	50

a) Values should be increased by amount in carriage water.

- Fecal coliform, a non-pathogenic species of bacteria, indicates the potential presence of pathogenic micro-organisms.

EXTENT OF THE PROBLEM

Twenty-nine percent of the United States population, or about 19,500,000 individual housing units, dispose of their domestic waste through individual on-site disposal units. 7) These are primarily (about 85 percent) septic tanks and cesspools, although in some southern states there are almost as many privies as other types of on-site units.

Ground-water contamination problems created by on-site domestic waste disposal systems can be classified as follows: individual, local, or regional. An individual problem is created when one disposal system on a particular piece of property contaminates one or more wells in the immediate vicinity. This type of problem can occur almost anywhere. A local problem exists when a high density of individual disposal systems in a definable housing development contaminates an aquifer which is used to supply water for that area. Such instances have been experienced throughout the country. A regional problem is created when many individual disposal units contaminate extensive aquifers which supply water over a broad area such as one or more counties. Only regional problems are considered in the following discussion.

The most important parameter influencing regional ground-water contamination from on-site domestic waste disposal systems is the density of these facilities in an area. While geology, depth to water and climate affect the nature and degree of the contamination problem, density is the principal factor. Regional problems are extremely difficult to correct because of the complexity and high cost of eliminating the source and the persistence of some contaminants in the ground-water system long after the septic tanks and cesspools are eliminated by replacement with community sewer systems.

Figure 57 shows three density ranges of housing units using on-site domestic waste disposal facilities: less than 10/sq mi (3.8/sq km), between 10 and 40/sq mi (15 sq km), and more than 40 sq mi. These ranges can be considered low, intermediate, and relatively high, respectively. Data for Figure 57 were obtained from the 1970 Census of Housing and mapped on a county by county basis. Adjoining counties falling into the same range form regions of varying ground-water contamination potential. A few large counties with numerous on-site disposal units, which may be concentrated in limited

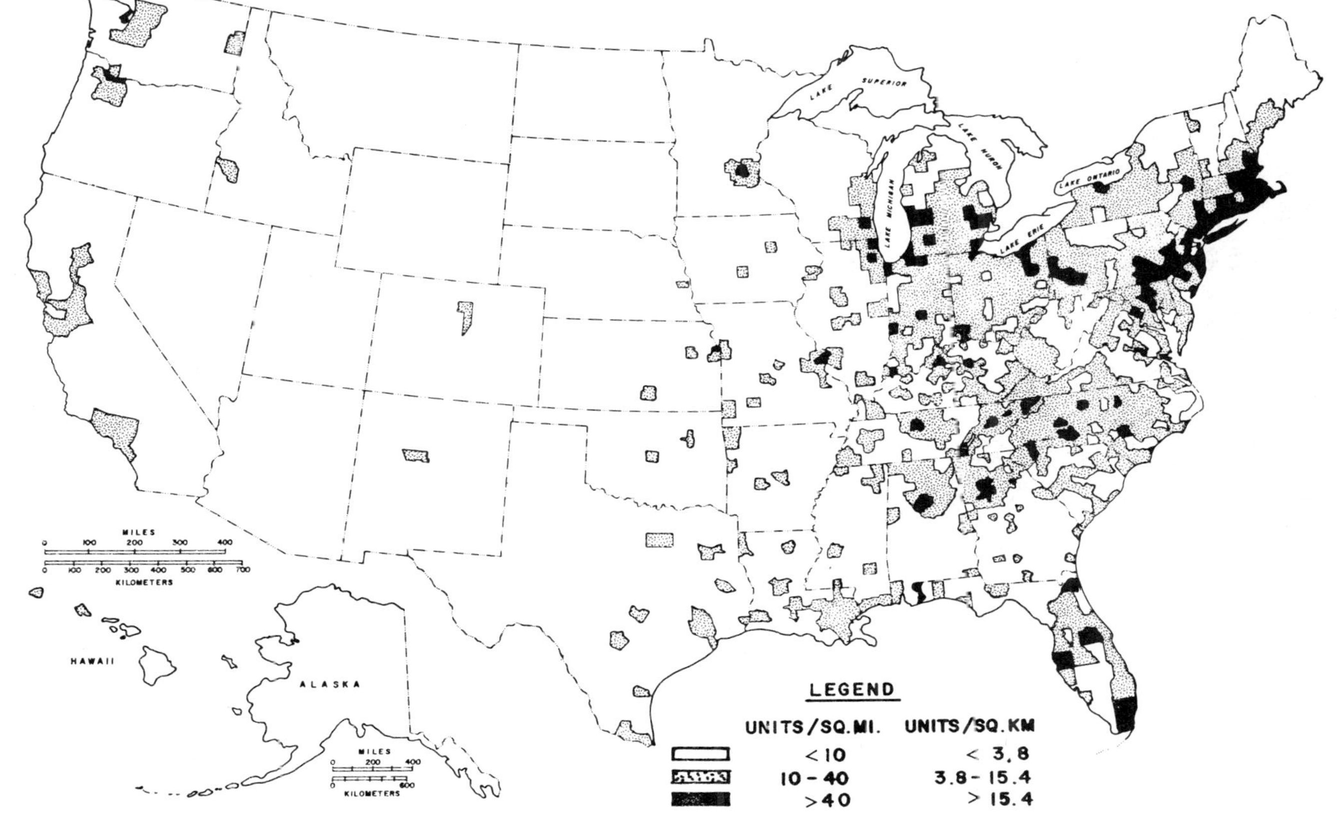

Figure 57. Density of housing units using on-site domestic waste disposal systems (by county).

areas of the county and thus create potential regional problems, do not appear in the relatively high range on Figure 57. To rectify this anomaly, Table 24 lists all counties with more than 50,000, and those with more than 100,000 housing units using on-site domestic waste disposal systems.

The potential for ground-water contamination in a region is suggested by the relative density of on-site domestic waste disposal units shown on Figure 57. A calculation of the volume of waste water discharged to the ground from these units in any particular location cannot be used to determine the existence or magnitude of a ground-water contamination problem without consideration of the other parameters previously mentioned (i.e., hydrology, geology, soils). However, the actual volume of domestic waste water discharged to the underground in high density areas can be very large and in some instances represents a significant form of recharge or replenishment to the local aquifers, even in humid areas. For example, the combined discharge from septic tanks and cesspools from the adjoining counties of Nassau and Suffolk, New York, is approximately 60 mgd, or about 50,000 gpd/sq mi (95 cu m/sq km).

The impact of this amount of waste discharge has become very obvious in Nassau County. Effluent from cesspools and septic tanks has been a major contributing factor, along with leachate from chemical fertilizers, to nitrate contamination of major aquifers in a 180-sq mi (466-sq km) area. 8) Nitrate enriched water has penetrated hundreds of feet into the principal artesian aquifer, and the water from 16 public supply wells serving thousands of residents has been degraded beyond current health standards.

Figure 57 indicates that there is one major region of high contamination potential along the northeast coast extending from about Washington, D. C. to north of Boston, Massachusetts. There are 40 other isolated regions, principally scattered over the eastern third of the country, in which high densities of septic tanks and cesspools also are present. The Los Angeles-San Bernardino-Riverside group of counties are of special note, even though they are not shown as high density areas on Figure 57, because of the great number of disposal units concentrated in urban areas of these very large counties.

Again, it should be noted that a septic tank density of greater than 40/sq mi designates a region of potential contamination problems. Actual densities in documented problem areas are considerably higher than 40/sq mi.

Table 24. COUNTIES WITH MORE THAN 50,000 AND COUNTIES WITH MORE THAN 100,000 HOUSING UNITS USING ON-SITE DOMESTIC WASTE DISPOSAL SYSTEMS.

More than 50,000

Jefferson, Alabama
Riverside, California
San Bernadino, California
Fairfield, Connecticut
Hartford, Connecticut
New Haven, Connecticut
Broward, Florida
Duval, Florida
Hillsborough, Florida
Jefferson, Kentucky
Bristol, Massachusetts
Middlesex, Massachusetts
Norfolk, Massachusetts
Plymouth, Massachusetts
Worcester, Massachusetts
Genesee, Michigan
Oakland, Michigan
Monmouth, New Jersey
Multnomah, Oregon
Westmoreland, Pennsylvania
Davidson, Tennessee
King, Washington
Pierce, Washington

More than 100,000

Los Angeles, California
Dade, Florida
Nassau, New York
Suffolk, New York

The second range of on-site disposal unit density (10 to 40 units per square mile) is again more or less confined to the eastern third of the country. In almost all of the western two-thirds of the country, plus extensive areas in northern New England and the rural portions of Georgia, Alabama and Mississippi, the density of septic tanks and cesspools is so low as to almost rule out the potential for significant regional contamination problems.

The uncontrolled disposal of septage represents a significant potential threat to ground-water quality. For example, continuing studies of this problem by EPA Region I indicate that about 400 million gal. (1.5 million cu m) of domestic sewage waste is pumped from septic tanks and cesspools in New England each year. Much of this volume is dumped in abandoned sand and gravel pits, along roadways and streams, or landfilled or lagooned at unapproved refuse disposal sites.

Septic tanks and seepage systems at industrial facilities operate on the same principal as those serving single and multiple dwellings. Of course, the volume of effluent can be greater, but what is more important is the possibility of wastes from the manufacturing process being incorporated with the domestic waste and ultimately migrating into an aquifer used for drinking water supplies. The hazard involved is magnified if the industrial wastes happen to be toxic.

Industrial septic tank discharge is an individual problem and not of a regional nature. Very little data exist on a national basis regarding the density of industrial septic tanks, and they probably number in the tens of thousands as compared to the millions of septic tanks and cesspools serving housing units across the country.

Case Histories

It has already been pointed out that the northeast represents the region of greatest density for on-site domestic waste disposal. A number of studies of ground-water contamination related to this type of waste disposal practice have been carried out. The results of two such investigations are described below to illustrate the typical nature of regional problems.

Boston Suburban Area, Massachusetts -

In a study carried out by the U. S. Geological Survey in the Ipswich and Shawsheen River basins of Massachusetts, north of Boston,[9] the investigators concluded that "development of housing beyond the reach of the municipal sewer systems of metropolitan areas has lowered the quality of the environment in many of the (housing) developments and has created health hazards in others." Chloride and specific conductance were used as tracers. Correcting for highway deicing salts, which are the only known contaminants other than septic tank discharge contributing significantly to this particular water-quality degradation situation, the investigators were able to develop a correlation between the relationship of housing density to residual conductance, and accretion of dissolved solids in the baseflow of streams.

Seventeen small drainage basins, all but one less than one square mile in area, were selected for study. All basins are served by public water supplies, but none has municipal sewer systems, and individual houses are served by on-site disposal systems. Housing density ranges from 0 to 900 units/sq mi (0 to 347/sq km). The concentration of chloride is about 50 ppm higher in the septic tank effluent than in the tap water entering the home. Septic tank flow per house is estimated to be 200 gpd (757 l/day). The results of the investigation indicated that the reduction of chemical contaminants during travel of the septic tank effluent through the soil and bedrock aquifer is slight.

<u>State of Delaware</u> -

In a 1972 report, John C. Miller of the Delaware Geological Survey states, "inspection of water analyses on file at the Delaware Geological Survey revealed that 25 percent of the shallow wells (less than 50-ft or 15-m deep) in the state yield water with nitrate (as NO_3) levels above 20 ppm."[10] Natural nitrate (NO_3) levels in ground water are less than 10 ppm. This indication of the potential for widespread ground-water contamination has led to an evaluation of some of the principal sources of nitrate enrichment of ground water in the state, including septic tank discharges.

Two suburban areas in the coastal plain were chosen for analysis of potential problems of ground-water quality degradation due to septic tanks.[11] The first area was selected on the basis that it is characterized by an extremely high water table and poorly drained soils. In addition, there had been numerous reports of overflowing septic-tank systems during rainy periods. For comparison purposes, the second area selected is underlain by deep, well-drained soils on uplands. In both areas, homes are situated on one-quarter to one-half acre lots, each of which has its own septic tank and shallow well-water system.

The results of the study showed that in the first area of poorly drained soils, nitrate (as NO_3) levels averaged 6.9 to 11 ppm during the period of sampling. A number of wells were contaminated by coliform bacteria. In the second area with well-drained soils, nitrate content ranged from 22 to 136 ppm, and concentrations in water from many wells were above the recommended EPA interim drinking water standards of 45 ppm nitrate (as NO_3). No wells were found to be contaminated with coliform bacteria.

The state investigators concluded that "the standard percolation test is not a suitable means for determination of the acceptability of a site for septic-tank effluent." Percolation tests in the first area were conducted during dry periods, and the favorable results led to installation of septic tanks. After installation, the systems overflowed during wet periods, and bacteriological contamination of domestic wells took place because of the introduction of sewage effluent from the land surface around well casings. On the other hand, the movement of the effluent through the fine soils has minimized the buildup of nitrate concentrations in the ground water. In the second area, the physical operation of the septic tanks has been successful because of the permeable soil sediments, which also apparently filtered out pathogenic organisms. However, nitrate contamination of ground water in that area is severe because of the favorable environment for oxidation of nitrogen compounds, and the rapid movement of septic-tank and tile-field effluent to the water table.

TECHNOLOGICAL CONSIDERATIONS

The manual of septic tank practice [12)] describes a soil as being suitable for the absorption of septic tank effluent if it has an acceptable percolation rate, without interference from ground water or impervious strata below the level of the absorption system. For a septic tank system to be approved by a local health agency, several criteria normally must be met: a specified percolation rate, as determined by a percolation test; and a minimum 4-ft (1.2-m) separation between the bottom of the seepage system and the maximum seasonal elevation of ground water. In addition, there must be a reasonable thickness, again normally 4 ft, of relatively permeable soil between the seepage system and the top of a clay layer or impervious rock formation. For what they are intended to insure, i.e., keeping the sewage below the ground, these criteria have been adequately successful.

Keeping the sewage below ground may have been sufficient under the low density, rural conditions for which the septic system was originally designed, but their widespread use after World War II in subdivisions with small lot sizes is largely responsible for the degradation of ground water that has occurred in many of these areas.

Appraisal of the potential contamination of ground water by septic tank systems in these high density areas requires an understanding of the ground-water system into which the effluent is discharged. First, ground-water recharge areas and flow patterns should be delineated. Second, the quantity of ground-water recharge must be estimated to establish the degree of natural dilution of the effluent. And third, the capability of the soil system to renovate the effluent should be known.[13)] While these concepts go beyond the widely established septic system siting criteria, their institution is essential if ground-water quality is to be protected in high density septic tank areas.

Collection of domestic waste water by public sewers and treatment at a central facility is the most common alternative to septic tank disposal systems. Where standard gravity sewers are not practical, an alternate method of domestic waste disposal is a septic tank-pressure sewer combination.

Aerobic treatment devices are an available alternative to septic (anaerobic) tanks. Aerobic treatment devices are generally scaled-down versions of activated sludge plants and most employ the extended aeration mode. Some investigators feel that aerobic tanks, under proper conditions of design, installation and operation, can achieve a significantly higher quality effluent than can septic tanks.

An approach for providing soil treatment where the native soils are not suitable for the disposal of effluents from septic tanks or other treatment devices -- where, for example, shallow soils are underlain by till, creviced or channeled rock, or there is a high ground water table -- is the use of artificially constructed above-ground mounds. Another alternative is the recirculating sand filter treatment system which consists of a septic tank, a recirculation tank and an open sand filter. This system has proven to be economical, and with disinfection, the effluent can meet surface-water discharge standards of regulatory agencies.

Also available are several flow reduction and completely self contained devices such as incinerating toilets, composting toilets, biological toilets, and vacuum system toilets. While these devices may hold promise for the future, their present use is generally restricted to special situations where normal on-site waste disposal methods are not feasible due to natural conditions or highly restrictive codes.

INSTITUTIONAL ARRANGEMENTS

Three essential phases of septic tank life are subject to regulation: installation, operation-maintenance, and failure detection and correction. Regulation of the installation, both design and siting, exists in all but a few states. Site inspection and issuance of a permit to make the installation is handled variously by state or local governments. Operation-maintenance is largely not regulated and left to the discretion of the homeowner. Failure detection and correction is difficult to regulate and is typically handled on an individual complaint basis or when a health hazard arises. Protection of ground-water quality is best accomplished by regulation of installation of the system and only this phase of the septic tank life is considered in this dicussion. Cesspools, where regulations exist, are generally not approved for new installations. Privies probably do not constitute a significant threat to ground water and their regulation is not considered here.

Regulation of septic tank installation is, in most cases, either by state, county, town, regional authority or a joint effort by two or more of these entities. For example, a state may regulate all septic tank installations; or it may regulate only installations serving something other than a single family residence; or it may regulate only installations in certain critical areas. The state may delegate regulation responsibilities to local governments or there may be no regulations at all. Where regulations exist, inspection may be comprehensive or spotty.

Some states are now restricting septic tank installation to non-subdivision situations (scattered lots). Mississippi, for example, will not approve individual residential sewage disposal systems of any type in new subdivisions, additions to existing subdivisions, or undeveloped portions of existing subdivisions unless the establishment of a community sewage system is economically unfeasible. [14)] Louisiana has resolved that every effort should be made to prevent the use of individual sewage disposal facilities in land development involving urban sized lots, unless it can be clearly demonstrated that the individual facilities are temporary and will be replaced with proper community facilities within a short period of time. [14)]

As states begin to discover ground-water contamination problems resulting from on-site domestic waste disposal systems, new regulations and controls are emerging. For example, the Oregon Department of Environmental Quality is in the process of updating its on-site disposal facility regulations based on engineering studies of problems in that state.[15] The new regulations will take into account, on a state-wide basis, such factors as regional soil conditions and climate.

The State of Maine included alternative systems such as composting, incinerating, chemical, recirculating and vacuum toilets and above-ground mounds in its regulations for use under certain problem conditions.[16]

Some states, such as Minnesota, while not actually regulating on-site domestic waste disposal systems, establish standards which are recommended to local authorities. Enforcement is then at the discretion of the local governments, some of which adopt the state regulations, while others either establish their own or have no regulation system.[17]

The pattern of regulation of installation of septic tanks and cesspools that has emerged in many states is for responsibility to be exercised at the local or regional level, principally through requirements of health departments. However, the laws of most states provide for enforcement intervention at the state level in cases where the activity is judged to be in conflict with existing state requirements. Controls are moving towards clearer definition of responsibility as states revise their statutes to provide for a greater degree of water protection, usually through amendments covering sewage disposal facilities. These mandate enforcement of procedures to protect potable water supplies. Public officials contacted during the course of this investigation were generally of the opinion that a more uniform enforcement policy is desirable and that present practice has evolved more by default than by design.

Figure 58 illustrates the present division of activity between state and local agencies in permitting and inspecting the installation of on-site domestic waste disposal systems across the country.[14,15,16,17,18,19] The map is generalized since it is not always clear from available descriptions, exactly which agency does the regulating or how thoroughly it is carried out. For example, some states regulate only specific items of location, design, etc., and in some instances the inspector responsible for enforcing the regulations may be employed by two cooperating governmental agencies. State regulation and inspection of septic tank installation is generally considered to be more effective than local regulation.

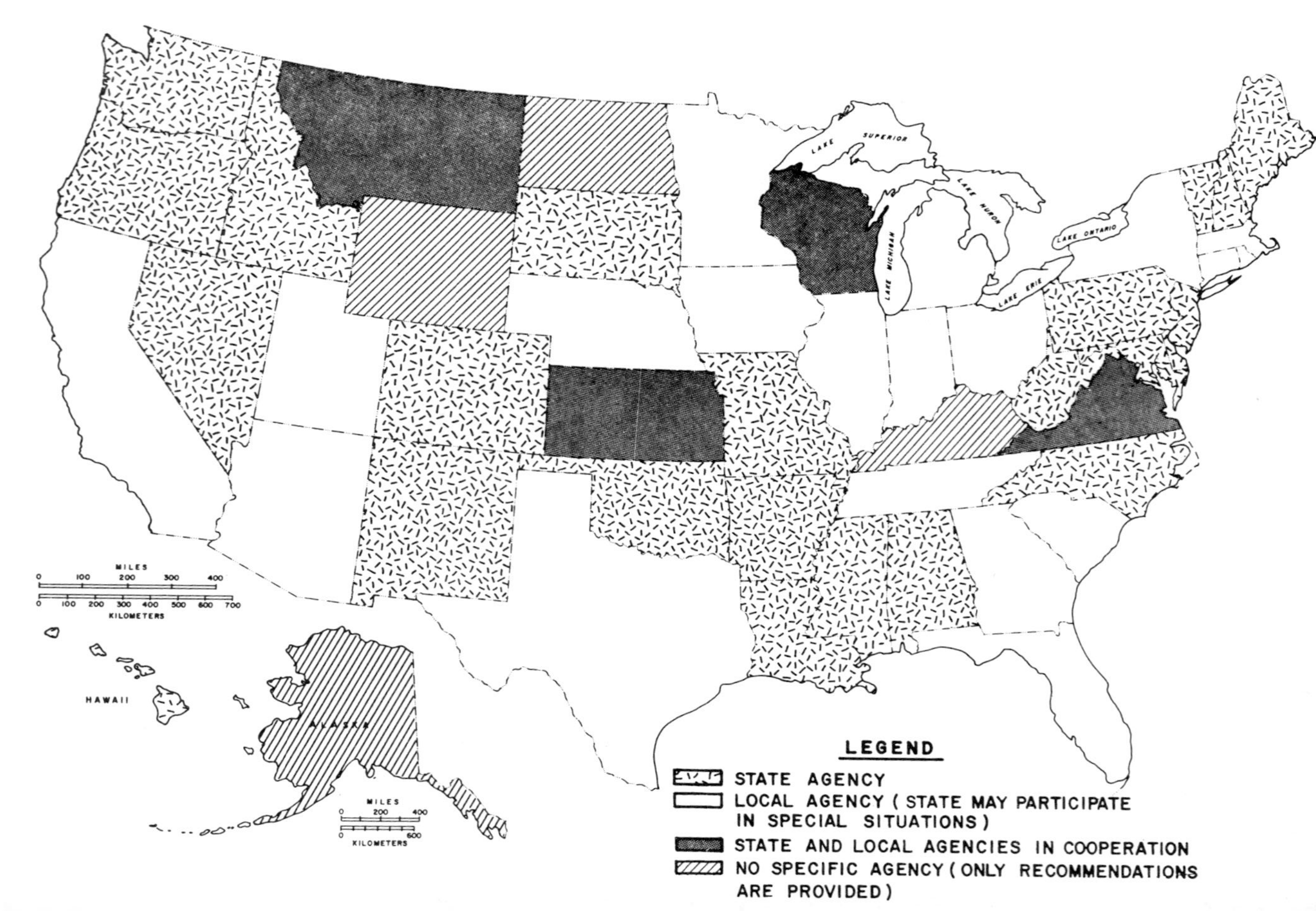
LAKE SUPERIOR
LAKE MICHIGAN
LAKE HURON
LAKE ERIE
LAKE ONTARIO
MILES
0 100 200 300 400
0 100 200 300 400 500 600 700
KILOMETERS
HAWAII
ALASKA
MILES
0 200 400
0 600
KILOMETERS
LEGEND
STATE AGENCY
LOCAL AGENCY (STATE MAY PARTICIPATE IN SPECIAL SITUATIONS)
STATE AND LOCAL AGENCIES IN COOPERATION
NO SPECIFIC AGENCY (ONLY RECOMMENDATIONS ARE PROVIDED)

An effective approach for alleviating many domestic waste related ground-water contamination problems by regulatory agencies is the establishment of a broad physiographic data base for use in a general land use planning effort. One specific aspect of this would be regulating septic tank density. Some state and local agencies are currently applying this approach. A data base can be developed to include such variables as soils, geology, physiography, hydrology, vegetation and climate. Analysis of these data for a location results in an overall physical capability rating for that location. These ratings can then be reduced to map form useful to planners and regulating agencies. For example, a residential capability map might be constructed that would indicate what specific site limitations exist for the construction of housing units. The state does not use the data to establish regulations on land use but rather makes the data available to local governments which perform this function.

By applying this type of wide range planning data to potential problem areas, maximum densities could be established for any regions throughout the country. Regulations could then be established to restrict the installation of new septic tank disposal systems to areas which have not yet reached critical densities. As areas reached the critical septic tank density, further residential development would be required to use alternate methods of sewage disposal. Areas which have already exceeded the established critical septic disposal system density could be individually evaluated to determine what corrective measures might be taken.

REFERENCES CITED

1. Bouma, J., et al. 1972. Soil absorption of septic tank effluent. University of Wisconsin, Soil Survey Division, Information Circular Number 20.

2. Pelczar, M. J., and R. D. Reid. 1965. Microbiology. 2nd ed. McGraw-Hill Book Company, New York. 512 pp.

3. Bennett, E. R., et al. 1974. Rural home waste water characteristics. Home Sewage Disposal. Proceedings of the national home sewage disposal symposium, Chicago, Illinois. American Society of Agricultural Engineers, St. Joseph, Michigan. pp. 74-78.

4. Witt, M., et al. 1974. Rural household waste water characteristics. Home Sewage Disposal. Proceedings of the national home sewage disposal symposium, Chicago, Illinois. American Society of Agricultural Engineers, St. Joseph, Michigan. pp. 79-88.

5. Goldstein, S. N., et al. 1972. A study of selected economic and environmental aspects of individual home waste water treatment systems. The Mitre Corporation. Washington, D. C. 90 pp.

6. Metcalf & Eddy, Inc. 1972. Waste water engineering: collection, treatment, disposal. Published by McGraw-Hill Book Company, New York. 782 pp.

7. Detailed housing characteristics. 1970 Census of Housing. Department of Commerce. Bureau of the Census. 53 Volumes.

8. Perlmutter, N. M., and Ellis Koch. 1972. Preliminary hydrogeologic appraisal of nitrate in ground water and streams, southern Nassau County, Long Island, New York. U. S. Geological Survey Professional Paper 800-B, Geological Survey Research.

9. Morrill, G. B., III, and L. G. Toler. 1973. Effect of septic-tank wastes on quality of water, Ipswich and Shawsheen River basins, Massachusetts. U. S. Geological Survey Journal of Research, Vol. 1, No. 1.

10. Miller, J. C. 1972. Nitrate contamination of the water-table aquifer in Delaware. Delaware Geological Survey Report of Investigation No. 20.

11. Miller, J. C. 1975. Nitrate contamination of the water-table aquifer by septic-tank systems in the coastal plain of Delaware. Pages 121-134 in W. J. Jewell and R. Swan, editors. Water pollution control in low density areas. University Press of New England, Hanover, New Hampshire.

12. U. S. Department of Health, Education and Welfare. Revised 1967. Manual of septic tank practice. Public Health Service Publication No. 526. 92 pp.

13. Holzer, T. L. 1975. Limits to growth and septic tanks. Pages 65-74 in W. J. Jewell and R. Swan, editors. Water pollution control in low density areas. University Press of New England, Hanover, New Hampshire.

14. Patterson, J. W., et al. 1971. Septic tanks and the environment. Illinois Institute for Environmental Quality, Chicago, Illinois.

15. Jackman, R. July 1975. Personal communication. Oregon State Department of Environmental Quality.

16. State of Maine. 1974. Plumbing code, Part II, Private sewerage disposal regulations. Department of Health and Welfare, Bureau of Health, Division of Health Engineering.

17. Minnesota State Department of Health. July 1975. Personal communication.

18. Commission on Rural Water. 1974. Guide to state and Federal policies and practices in rural water-sewer development. Information Clearinghouse, Commission on Rural Water, Chicago, Illinois. 223 pp.

19. State Board of Health, State of Washington. July 1975. Personal communication.

SECTION IX

COLLECTION, TREATMENT, AND DISPOSAL OF MUNICIPAL WASTE WATER

SUMMARY

Municipal waste water follows one of three direct routes to reach ground water: leakage from collecting sewers, leakage from the treatment plant during processing, and land disposal of the treatment-plant effluent. In addition, there are two indirect routes: effluent disposal to surface-water bodies which recharge aquifers, and land disposal of sludge, which is subject to leaching. Although the volume of waste water entering the ground-water system from these various sources may be substantial, there have been few documented cases of hazardous levels of constituents of sewage or storm water affecting well-water supplies. However, the impact on ground-water quality resulting from the collection, treatment, and disposal of municipal waste water has not been studied in detail.

Untreated sewage is principally composed of domestic wastes. In areas where manufacturing is also served by the community system, the waste products of industry can add important potential contaminants. Storm runoff from streets, parking lots, and roofs contributes salts, inorganic chemicals, and organic matter which have been deposited on exposed surfaces.

According to the 1970 U. S. Census of Housing, the domestic waste from 71 percent of housing units is collected by public sewer lines and piped to central treatment facilities. About 160 million people are served by 500,000 mi (800,000 km) of sewer lines. The total volume of sewage is approximately 15 bgd (57 million cu m/day). More than 5,000 of the almost 22,000 treatment plants in the nation have waste stabilization ponds, which are seldom lined and almost never monitored with wells. Of the more than 2 bgd (7.6 million cu m/day) of sewage treatment plant effluent discharged to the land, a large proportion does not meet secondary treatment standards.

About the only control of potential ground-water contamination related to leaky sewers is the specification by many states of minimum distances between a proposed public supply well and a sewer line. Conformance with pressure test requirements on new sewer line installations in many areas aids in minimizing exfiltration problems. Municipal lagoons and ponds for the retention of waste water are parts of sewage treatment facilities, the construction and operation of which are supervised by state health or environmental departments. In addition, where Federal grants are involved, the design of the impoundment comes under the scrutiny of the EPA. Thus, in the construction of new lagoons and ponds, potential effects on ground-water quality are given consideration. A number of states also require permits for municipal sewage impoundments. Spraying of sewage effluent and other forms of land disposal of sewage wastes are specifically regulated in only a few states. Most states review such practices on a case-by-case basis.

DESCRIPTION OF THE PRACTICE

A major problem in urbanized areas is the collection, treatment, and disposal of domestic waste water. Because a large volume is generated in a small area, urban domestic waste cannot be adequately disposed of by conventional septic tanks and cesspools. Therefore, special facilities are used to collect, treat, and dispose of such wastes in densely populated locales.

Because the area of permeable land surface in urban areas is considerably reduced by the presence of roofs, sidewalks, parking areas, and streets, storm-water runoff also poses a problem. Precipitation, street sweepings, litter, leaves, and salts deposited on exposed surfaces all may be contained in runoff. Storm water may be collected and treated along with the previously mentioned domestic wastes or may be handled in separate collection, treatment, and disposal facilities.

Sanitary Sewer Systems

A municipality constructs a sanitary sewer system in order to collect the community's waste water or sewage and to transport it to a central point for treatment and/or disposal. In the early 1960's more than 120 million persons in the United States, or 65 percent of the then total population, were served by public sewers. 1) Today's population served by sewers stands at an estimated 158 million persons. 2)

The principal role of sewers is to provide a watertight passage through which waste water can be transported quickly, with a minimum of odor production, stoppage, or overflow. The most common causes of sewer problems are grease, grit, and trash accumulation; infiltration; decomposition of organic matter; root penetration; inadequate sewer gradients;

and potential danger from flammable liquids. [3)]

The design of sanitary sewers varies with needs and local conditions. Gravity sewers allow waste water to flow naturally to the lowest point in the system. Because they contain no moving parts, operation is largely a matter of cleaning and maintenance. However, sewer systems can rarely be designed to operate solely by gravity because of topography. It is rare that the topographic low in a small municipality can serve as the discharge point for the community's sewers.

Force mains are often integrated into gravity systems, or in some cases they may operate independently. Force mains are most often used to transport sewage upgrade or to move it faster in one section of the system than would naturally occur.

It has been established that about 60 to 80 percent of the per capita consumption of water will become sanitary sewage. Actual sewage flow rates vary with location, water usage, and the type and condition of the sewers. Sewage flows are cyclic, and the ratio of peak flow to average flow, called the peaking factor, will range from less than 1.3 to more than 2.0. [4)]

The major cause of ground-water contamination from sanitary sewer systems (if above the water table) is through outflow leakage (exfiltration) from gravity sewers. Common factors causing leakage in gravity sewers are:

1. Poor workmanship, especially in the past when mortar was applied by hand as a joining material.

2. Cracked or defective pipe sections.

3. Breakage by tree roots penetrating or heaving the sewer lines.

4. Pipeline rupture by superimposed loads, heavy equipment, or earthfill on pipe laid on a poor foundation.

5. Rupture by downhill creep of soil in hilly terrain.

6. Fracture and displacement of pipe by seismic activity; e.g., a sewerage system in California still suffers from fractures caused by an earthquake in 1909.

7. Loss of foundation support due to underground washout.

8. Poorly constructed manholes, or shearing of pipe at man-

holes due to differential settlement.

Where a pressure sewer develops a leak, exfiltration can occur regardless of whether the sewer is above or below the water table because the sewer operates at a greater hydrostatic pressure than the ground-water system.

Storm Sewer Systems

Storm sewers conduct surface runoff from locations where it poses problems of safety, public health, or inconvenience to the nearest discharge location. A typical storm-sewer system consists of a network of underground piping, at a depth sufficient to receive water from the ground surface by gravity. Storm-water inlets, openings in the street curb, gutter or pavement, allow the entry of storm runoff to the pipe network. Frequently, a catch basin beneath the inlet serves to retain the heavy grit, sand, and debris that passes through the grating of the storm-water inlet.

Storm-sewer sizes vary depending upon the quantity of storm water that must be conveyed through the pipe. In general, storm sewers are not less than 12 in. (30.5 cm) in diameter, and can range in size to 96 in. (244 cm) or greater. The majority of present-day storm sewers are constructed of reinforced concrete and galvanized steel; however, in earlier days, the use of brick and masonry was prevalent. Storm sewers for residential areas are most commonly designed to convey runoff from storms with a five-year recurrence interval. 5) (The sewer has sufficient hydraulic capacity to accommodate the quantity of runoff from a rainfall event that has a statistical probability of occurring only once every five years.)

As with sanitary sewer systems, the potential for ground-water contamination from storm-sewer systems lies with physical failure in the pipe network, which allows exfiltration of storm flows to the surrounding soil and ground water.

Lagoons and Ponds for Sanitary Waste-Water Treatment

Lagoons and ponds used for waste-water treatment are essentially biological waste treatment units. They have a wide variation in function depending on their basic design. These units may operate under aerobic, anaerobic, or facultative (capable of being either at any given time) conditions. They use microorganisms to break down the wastes. Design features vary with waste-water characteristics, location (geology, soil conditions) and the requirements of the controlling regulatory agency.

Ponds are also classified according to their usage. If used to treat raw sewage prior to other treatment, they are referred to as primary waste stabilization ponds. However, since these ponds are generally capable of achieving the effluent limitations required for secondary treatment, they may effectively provide both the primary and secondary treatment functions. When ponds are used to treat effluents from primary settling tanks or secondary biological treatment units, they are called secondary waste stabilization lagoons or polishing ponds, respectively.

That portion of municipal sewage which is being treated by stabilization ponds in the United States is small as compared with the total volume. Data collected during this study indicate that of the 21,787 sewage treatment plants in the United States, 5,132 plants had lagoons as a part of their treatment facilities. 2) Leakage from unlined lagoons may be a significant threat to ground water, if the pond bottom and walls are not properly sealed.

Lagoons and Ponds for Storm-Water Storage and Treatment

Storm-water lagoons can function as storage sites to attenuate storm-water flows and reduce the shock effects of discharges. When lagoons are utilized in this manner, feedback of the stored waters to the sanitary sewer is practiced, with treatment and ultimate discharge of the storm water occurring at other locations.

A storm-water lagoon acts as a sedimentation chamber for grit, sand, and other suspended solids in the storm runoff, thereby providing a degree of treatment. In certain instances, biological treatment is provided by a storm-water lagoon similar to that provided by a sanitary waste-water lagoon. Biological treatment in lagoons is limited by various factors such as temperature, dissolved oxygen, toxicity of waste, etc.

As is true for sanitary waste-water lagoons, the primary threat to ground-water quality from storm-water lagoons is leakage. The prevalent practice of constructing lagoon systems without adequate seals, either intentionally or unintentionally, can contribute to ground-water contamination.

Land Spreading and Basin Recharge (Municipal Waste Water)

Land disposal dates back at least four centuries, and some systems presently in use began operation before the twentieth century. Historically, the purpose of land treatment of sanitary waste water has emphasized disposal, whereas the current trend is toward the concepts of treatment and/or reuse.[6] The increased use of land for sanitary waste-water treatment and disposal in the United States is shown in Table 25. There are three general methods of applying waste water to land areas -- namely, irrigation, overland flow, and infiltration-percolation.[7]

Reclaimed sanitary waste water has been used to irrigate certain field crops such as cotton, sugar beets, and vegetables for seed production. It cannot be used for field crops that are normally consumed in a raw state.

Land application of waste water by infiltration-percolation is often referred to as ground-water recharge because the major portion of the water applied percolates to the water table. Depending upon its final quality, the recharged water may be recovered and used for irrigation, recreation, or municipal or industrial supply.[7] Although it has been employed to raise the level of the water table -- e.g., to maintain baseflow in nearby streams -- the principal use of recharge from lagoons has been to halt salt-water intrusion. This practice has been used extensively in California and is known as basin recharge.

The physical design of a land-treatment system is governed by the type of application involved. Each system has specific characteristics which make it applicable to certain situations (see Table 26).

Flows to land spreading and basin-recharge systems will vary depending primarily on the type of system employed. Typical liquid loading rates for various systems are presented in Table 26. The actual application rates employed are a function of the soil type, character of waste water and degree of pretreatment, and the desired waste removal efficiency.

Land spreading of waste water can pose a significant threat to ground-water quality. A summary of the effectiveness of removal of the more common constituents appears in Table 27. Since the overland-flow method functions more as a land-treatment system than a land-disposal system because a substantial portion of the waste water applied is designed to run off, it has not been included in the tabulation.

Land Spreading and Basin Recharge (Storm Water)

Besides those lagoon facilities specifically constructed for storm-water storage and/or treatment, urban storm-water disposal is often accomplished by land spreading in areas of high infiltrative capacity, particularly in the southwestern

Table 25. MUNICIPALITIES USING LAND APPLICATIONS AND THE POPULATIONS SERVED.[6)]

Year	Number of systems	Population served (millions)
1940	304	0.9
1945	422	1.3
1957	461	2.0
1962	401	2.7
1968	512	4.2
1972	571	6.6

Table 26. CHARACTERISTICS OF THE IRRIGATION, OVERLAND FLOW, AND INFILTRATION-PERCOLATION SYSTEMS.[8]

Factor	Type of Approach		
	Irrigation	Overland flow	Infiltration-percolation
Liquid loading rate	0.5 to 4 in./week (12.7 to 101.6 mm)	2 to 5.5 in./week (50.8 to 139.7 mm)	4 to 120 in./week (101.6 to 3,048 mm)
Annual application	2 to 8 ft/yr (0.6 to 2.4 m)	8 to 24 ft/yr (2.4 to 7.3 m)	18 to 500 ft/yr (5.4 to 152 m)
Land required for 1-mgd flow	140 to 560 acres (566.5 thousand sq m to 2.2 million sq m) plus buffer zones	46 to 140 acres (266 thousand sq m to 566.5 thousand sq m) plus buffer zones	2 to 62 acres (8 thousand sq m to 251 thousand sq m) plus buffer zones
Application techniques	Spray or surface	Usually spray	Usually surface
Soils	Moderately permeable soils, with good productivity when irrigated	Slowly permeable soils, such as clay loams and clay	Rapidly permeable soils, such as sands, loamy sands, and sandy loams
Probability of influencing ground-water quality	Moderate	Slight	Certain
Needed depth to ground water	About 5 ft (1.5 m)	Undetermined	About 15 ft (4.5 m)
Waste water lost to:	Predominantly evaporation or deep percolation	Surface discharge dominates over evaporation and percolation	Percolation to ground water

Table 27. SUMMARY OF CONTAMINATION LIKELY FROM LAND DISPOSAL OF DOMESTIC WASTE WATER. compiled by E. Clark from [7]

Parameter	Irrigation method	Infiltration-percolation method
Nitrogen	Nutrients that are not used by plants or fixed in the soil can leach to ground water	Significant quantities passed to ground water at most sites
Phosphorus	Leaching of excess phosphorus is rare occurrence. Organic and clay soils absorb practically all of the phosphorous	Removal may be limited because granular soils are used
Organics	Usually broken down by microorganisms and used by plants. Can appear in ground water when application rate is highest or when in open soil, such as sand or gravel, with a high percolation rate	Evidence is that little organic matter reaches ground water
Trace Element	Toxic compounds can be changed by the chemical reaction of cation exchange and can be rendered non-toxic by bacteria. Chemical precipitates formed can be leached out, however	Retention may be limited due to granular nature of soils
Total Dissolved Solids	Leaching can occur and build-up is possible	Build-up is possible
Enteric Organisms	Usually are removed or die out and do not reach ground water especially if water table is kept low	Spread of bacteria and viruses by insects or percolating water is of concern but unlikely under proper soil conditions

and western states [9] and on Long Island, New York. In California, seepage ponds are used to dispose of storm water from beneath highway underpasses, where it would otherwise collect. On Long Island, more than 2,000 recharge basins dispose of storm runoff from urban and suburban areas.

Storm-water recharge facilities usually consist of simple excavations or pits in unconsolidated material. Deep excavated basins are popular in the northeastern states. In the western states, dry stream beds and shallow, broad man-made depressions in alluvium adjacent to perennial streams are often employed. Overland flow and spray irrigation methods are not useful for spreading storm water because of the erratic nature of precipitation.

The mechanism for ground-water contamination by basin recharge of storm water is the same as that of sanitary waste water. While no specific reported instances of contamination have been reported, a potential for ground-water contamination does exist. [9]

CHARACTERISTICS OF CONTAMINANTS

Untreated Municipal Sewage

Municipal waste-water treatment plants handle wastes which vary in composition corresponding to certain patterns of everyday life. Typical hourly variation in flow and domestic sewage is shown in Figure 59.

Although waste water is primarily liquid (99.9 percent water), the composition can probably best be studied by first considering the total solids content. The sources of domestic solids in waste water include toilets, sinks, baths, laundries, garbage grinders, and water softeners. In addition to domestic sewage, municipal waste water contains storm water and commercial and industrial discharges. The contributors of total solids in waste water expressed as grams/capita/day (gpcd) are shown in Table 28.

Compositions of strong, medium, and weak domestic sewage are presented in Table 29. The physical, chemical, and biological constituents make up what is referred to as the sewage composition. In addition to the chemical and physical constituents, there are also pathogenic agents present in waste water.

Waste Water After Treatment in Sewage Treatment Plants

Treated waste water can range from almost raw sewage to pota-

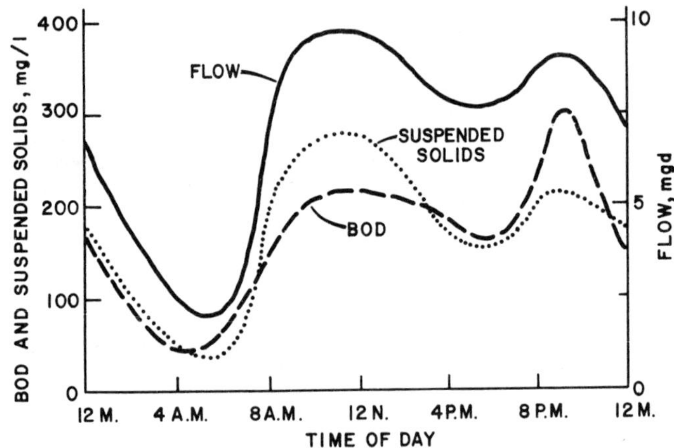

Figure 59. Typical hourly variation in flow and strength of domestic sewage.[4)]

Table 28. ESTIMATE OF THE COMPONENTS OF TOTAL SOLIDS IN WASTE WATER.[4)]

Component	Dry weight (gpcd)
Water supplies and ground water, assumed to have little hardness	12.7
Feces (solids, 23 percent)	20.5
Urine (solids, 3.7 percent)	43.3
Toilet (including paper)	20.0
Sinks, baths, laundries, and other sources of domestic wash waters	86.5
Ground garbage	30.0
Water softeners	- [a)]
Total for domestic sewage from separate sewerage systems, excluding contribution from water softeners:	213.0
Industrial wastes	200.0 [b)]
Total for industrial and domestic wastes from separate sewerage system:	413.0
Storm water	25.0 [c)]
Total for industrial and domestic wastes from combined sewerage system	438.0

a) Variable
b) WIll vary with the type and size of industries
c) Will vary with the season

Table 29. TYPICAL COMPOSITION OF DOMESTIC SEWAGE.[4)]
(All values except settleable solids are expressed in ppm)

Constituent	Concentration		
	Strong	Medium	Weak
Solids, total	1,200	700	350
Dissolved, total	850	500	250
Fixed	525	300	145
Volatile	325	200	105
Suspended, total	350	200	100
Fixed	75	50	30
Volatile	275	150	70
Settleable solids, (ml/l)	20	10	5
Biochemical oxygen demand, 5-day, 20°C (BOD_5 20°)	300	200	100
Total organic carbon (TOC)	300	200	100
Chemical oxygen demand (COD)	1,000	500	250
Nitrogen, (total as N)	85	40	20
Organic	35	15	8
Free ammonia	50	25	12
Nitrites	0	0	0
Nitrates	0	0	0
Phosphorus (total as P)	20	10	6
Organic	5	3	2
Inorganic	15	7	4
Chlorides [a)]	100	50	30
Alkalinity (as $CaCO_3$) [a)]	200	100	50
Grease	150	100	50

a) Values should be increased by amount in carriage water

ble water, depending on the type of treatment received. Although there have long been laws specifying minimum treatment efficiencies (usually secondary treatment or 85 percent removal of suspended solids and biochemical oxygen demand, and most recently even more stringent levels as a result of PL 92-500), many plants still treat inadequately. Many areas only use screening, or at best primary treatment, to remove the more obvious solids. Other plants designed for secondary treatment do not meet their design requirements due to plant overloadings or operational difficulties.

The total biochemical oxygen demand contributed by the sewered population in the United States was 7.3 billion lb (3.3 billion kg) in 1963.[10] Even if a 90 percent removal efficiency is assumed for secondary treatment, about 730 million lb (331 million kg) of biochemical oxygen demand would have been discharged. This points out the fact that although treatment plants remove most contaminants from waste waters, the remaining contaminants might still be significant at some locations.

Bacterial reduction is also obtained by secondary treatment. The usual indicator of fecal contamination of receiving waters is the presence of the fecal coliform group of bacteria. Reductions for fecal coliform bacteria of 90 to 99 percent are realized from biological treatment. However, these numbers are misleading. When one considers that raw waste water may contain a fecal coliform count of up to 500,000/100 ml, then 99 percent removal would leave up to 5,000/100 ml, a still sizeable number. Disinfection before discharge of the treated waste water can remove close to 100 percent of the bacteria.

Significant by-products of waste-water treatment plants after solids removal are grit, screenings, and sludge, of which sludge constitutes the largest volume. These also must be treated and disposed of in an environmentally acceptable manner.

Waste Water Treated by Lagoons and Ponds

It is difficult to characterize waste water that has been treated by lagoons and ponds. If the system serves as a polishing pond (following other treatment), effluent will be of considerably higher quality than if the system serves as the sole treatment process. Multiple ponds, operating in series, also provide more detention time, which reduces the organic load by sedimentation and further oxidation by bacteria. The number of disease-causing bacteria can also be reduced due to natural death.

Lagoons and ponds do not always function effectively; thus the threat of contamination is a variable. In a nationwide survey on lagoon performance in 1973, all 50 states reported problems with odor, 21 with algae in the effluent, 23 with short-circuiting, 6 with organic overload, and 20 with poor effluent. 11)

Waste Water Treated by Land Spreading and Basin Recharge

The resulting water quality after land spreading of waste water will depend on the initial waste-water characteristics, site topography, hydrologic and geologic conditions, type of vegetation, and application method.

Of prime importance in land spreading and basin recharge is the survival of pathogenic bacteria and viruses in the soil, in sprayed aerosol droplets, and on vegetables. It has been found that the survival of pathogenic organisms in the soil can vary from days to months depending on the soil moisture, soil temperature, and type of organism. The travel distance of bacteria in air is limited to the distance of travel of the mist from sprinklers. It was also found that as the relative humidity decreased and air temperature increased, the death rate of bacteria increased. 7) Pathogens, in general, will not enter healthy, unbroken vegetables but may be harbored in broken, bruised, or unhealthy plants and vegetables.

Chemical compounds found in waste water such as nitrate, mineral salts, and toxic trace organics may reach the ground water as a result of land spreading. Nitrate is of concern because it is reported to be a cause of methemoglobinemia in infants. High salt content can be harmful to people with cardiac, renal, or circulatory diseases.

Storm Water

In most studies of urban runoff, it has been observed that higher concentrations of contaminants may be expected (a) during the early stages of a storm; (b) in densely populated, highly paved or industrialized areas; (c) in response to intense rainfall periods; (d) after prolonged dry periods; and (e) in areas where construction activities are underway. Contaminant concentrations tend to decrease as storms progress, and as storm frequency increases.

The phenomenon of higher contaminant concentrations discovered during the earlier stages of a storm has been coined the "first flush" effect. The contaminants occurring in higher concentrations are those which have remained in the sewer from a previous storm. Storm-water runoff is compared to sanitary sewage in Table 30.

EXTENT OF THE PROBLEM

According to the 1970 U. S. Census of Housing,[13)] the domestic waste from 71 percent of the housing units in the United States is collected by public sewer lines and piped to central treatment facilities. The total volume of this sewage is approximately 15 bgd (57 million cu m/day), a portion of which alters the quality of ground water. Municipal waste water may follow one of three direct routes to reach ground water: leakage from the collecting sewers, leakage from the treatment plant during processing, and land disposal of the treatment-plant effluent containing constituents either not present in or in greater concentrations than the natural ground water at the site. In addition, there are two indirect routes: effluent disposal to surface-water bodies which recharge aquifers; and land disposal of the residual sludge, which is subject to leaching.

Storm-water runoff may enter the ground-water system from leaks in storm sewers, from sewer overflows, or by flowing directly from city streets onto unpaved areas and percolating to the water table.

Sanitary Sewer Systems

Infiltration of ground water into typical existing sanitary sewers may range in volume from 1,000 to over 40,000 gpd/mi (2,352 to 94,096 l/day/km) of sewer.[4)] Table 31 presents regional data for the total length of residential sewer pipe in service from 1940 to 1980 (estimated), with the corresponding estimates of sewage flow rates, not including infiltration. The principal variables that control the volumes of infiltration are: the quality of the materials and workmanship in the sewers, the type of joints used to connect individual lengths of pipe, and the elevation of the water table with respect to the sewer line.

Infiltration of ground water into sewers has been the subject of much investigation because the excess flow can overstress the sewage treatment plant. On the other hand, little attention has been paid to sewage leakage into the ground, or exfiltration, because the resulting loss of flow is frequently ignored or is considered an asset by the treatment plant operator. From a ground-water contamination standpoint, however, exfiltration is known to be a serious problem in some areas, and undoubtedly contributes to problems in many others. The dearth of data regarding this problem precludes, at present, estimations of the volume of ex-

Table 30. GENERALIZED WATER QUALITY COMPARISON OF VARIOUS WASTES. [12]

Type	BOD_5, ppm	SS, ppm	Total coliforms, MPN/100ml	Total nitrogen, as N ppm	Total phosphorus, as P ppm
Untreated municipal	200	200	5×10^7	40	10
Treated municipal					
Primary effluent	135	80	2×10^7	35	8
Secondary effluent	25	15	1×10^3	30	5
Combined sewage	115	410	5×10^6	11	4
Surface runoff	30	630	4×10^5	3	1

Table 31. SEWER PIPE LENGTH AND ESTIMATED FLOWS. after 14)

Year	Northeast		North Central		South		West		Total c)	
	Pipe length a) (miles)	Estimated flow b) (mgd)	Pipe length a) (miles)	Estimated flow b) (mgd)	Pipe length a) (miles)	Estimated flow b) (mgd)	Pipe length a) (miles)	Estimated flow b) (mgd)	Pipe length a) (miles)	Estimated flow b) (mgd)
1940	40,115	2,310	30,474	2,020	40,740	1,930	23,599	810	134,918	7,070
1950	53,619	2,580	41,687	2,390	54,812	2,350	41,062	1,240	191,180	8,560
1960	81,025	3,060	67,540	3,050	89,510	3,040	69,472	1,920	307,547	11,070
1970 (est)	116,121	3,670	107,636	3,910	135,478	3,820	112,826	2,600	472,061	14,000
1980 (est)	173,718	4,460	161,577	4,880	202,923	4,870	168,774	3,580	706,992	17,790

a) Total residential sewer pipe in service regionally
b) Based on 100 gpd per capita
c) Conterminous United States

Note: Miles times 1.609 gives kilometers.
Million gallons times 3,785 gives cubic meters.

filtration or the severity of its impact on ground-water quality, either locally or regionally.

The available data regarding the sewered population throughout the United States make possible the definition of areas with a potential sewer leakage problem. Figure 60 shows three ranges of sewered population by county: less than 50,000; 50,000 to 500,000; and greater than 500,000. While ground-water degradation from sewer leakage can occur wherever a sewer line is present, the magnitude of such problems is probably proportional to the ranges in density of population served as shown on Figure 60. The impact on ground-water quality in these areas is dependent on local geology and hydrology.

Sanitary Waste Stabilization Lagoons and Ponds

The waste-water stabilization pond is the most popular form of municipal secondary treatment in the United States, especially among the smaller sewage plants. The EPA "Municipal Waste Facilities Inventory" lists 5,132 treatment plants, serving 7,800,191 people, currently using this method of treatment. As previously noted, the total number of facilities listed, including those which are currently providing no treatment, is 21,787. [2)]

The volume of waste water that seeps into the ground from unlined stabilization ponds is quite large. A rough estimate of the volume of pond seepage in all areas of the United States has been calculated, based on an approximate median infiltration rate for sealed sewage stabilization ponds, of 0.008 ft/day (0.2 cm/day). [15,16,17,18)] Rule of thumb engineering estimates for lagoon leakage are: new, unlined lagoons, 0.05 to 6.2 ft/day (1.5 to 189 cm/day) and older, self-sealed lagoons, 0 to 0.1 ft/day (0 to 3.0 cm/day). The total estimated leakage for the United States equals 50 million gpd (157,000 cu m/day). Figure 61 provides a county breakdown of the number of people per square mile being served by treatment facilities using stabilization ponds.

The impact of leakage from waste-water stabilization ponds depends on several factors which cannot be generalized for this type of survey. Among the more important of these are the nature of the waste water contained in the pond; nature of the soils through which the leakage must migrate; depth to the water table; and quality of the natural ground water.

Land Spreading and Basin Recharge

Disposal practices that are covered in this section include:

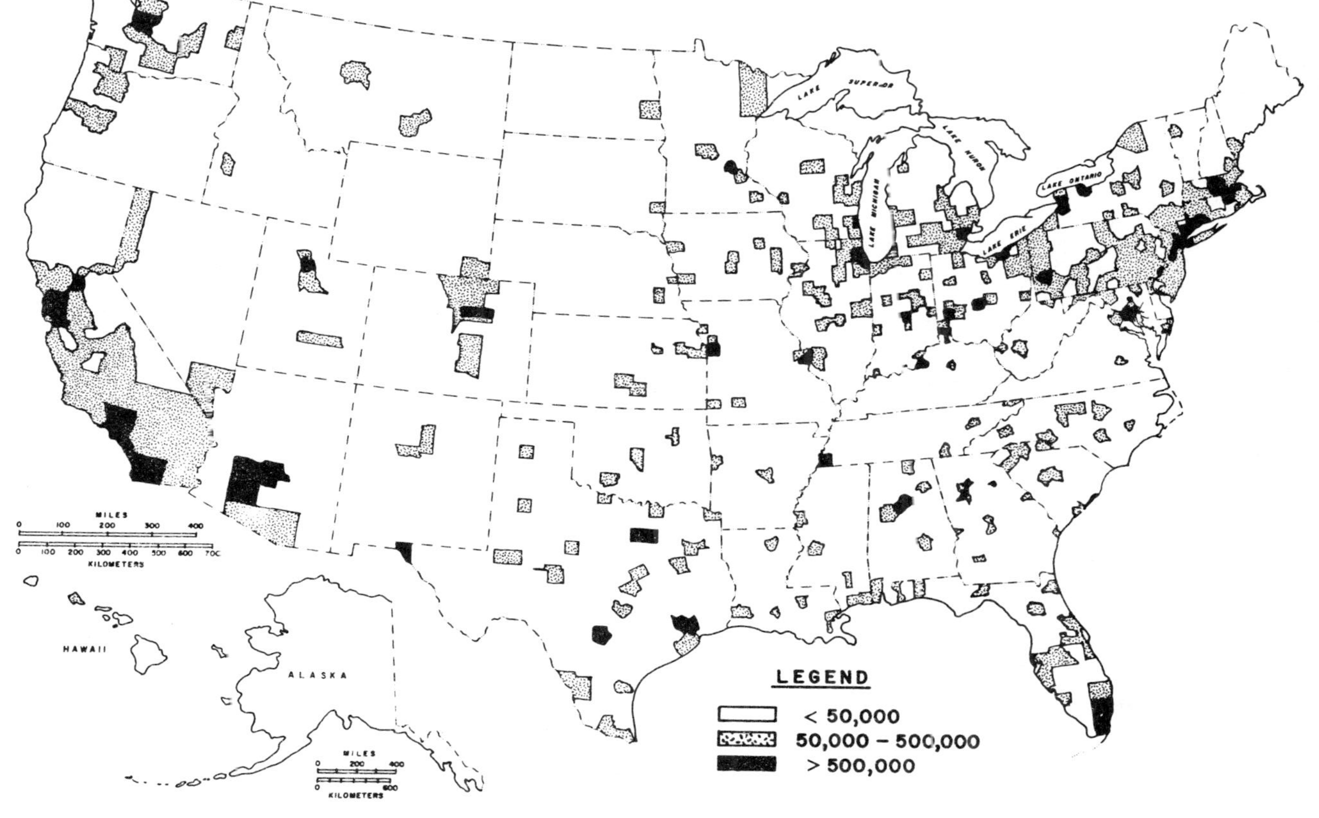

Figure 60. Number of people using public sewer systems for disposal of domestic waste (by county).

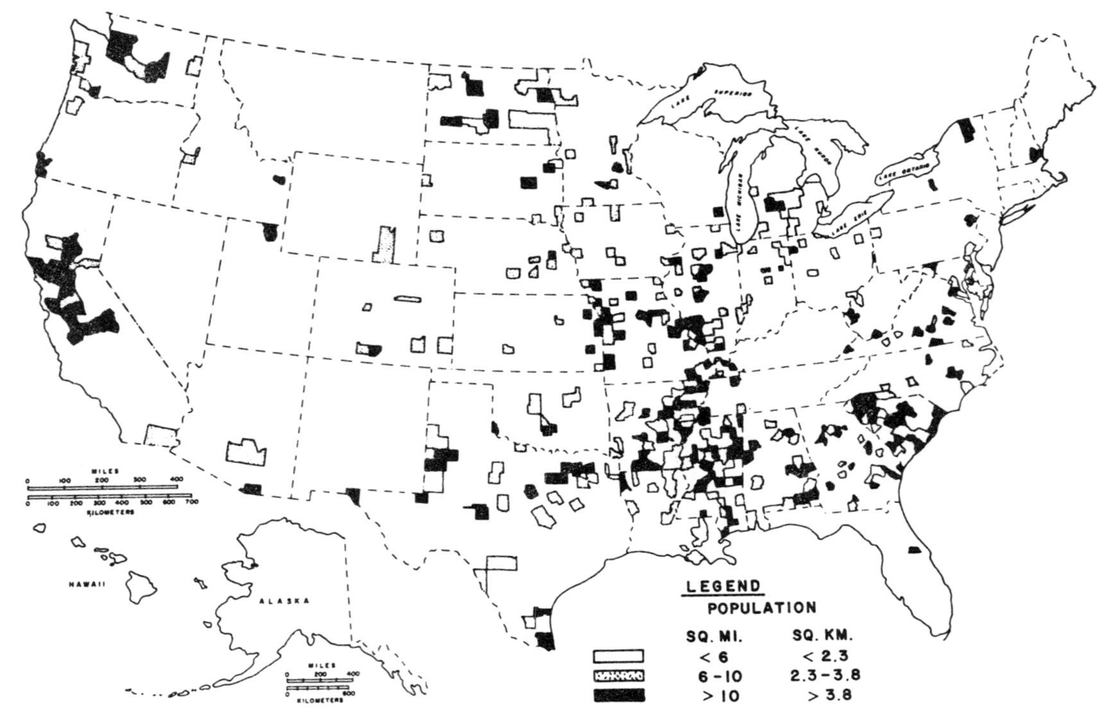

Figure 61. Density of population served by municipal sewage treatment facilities using stabilization ponds (by county).

irrigation of agricultural land, rapid infiltration ponds, overland runoff, and discharge to dry stream beds and ditches. Also discussed in the following section is discharge to intermittent or perennial streams where a portion of the surface water enters the ground-water system as recharge.

The lack of available data makes it impossible to specifically define the impact on ground-water quality of municipal waste disposal on land. Two important factors can be estimated, however, providing a good indication of the regional potential for ground-water degradation. The first is the quality of municipal effluent as it is discharged to land and the second is the degree to which this disposal practice is used in a particular region.

Effluent which meets the standards for secondary treatment, as established by the Federal Water Pollution Control Act Amendments of 1972, is not likely to seriously degrade ground-water quality. Thus, facilities providing effective secondary treatment can generally be disregarded as ground-water quality threats except for nitrate, in some cases, and also where the geology allows for rapid migration of contaminants. On the other hand, effluent receiving only primary treatment is limited in acceptability for land application.

Table 32 shows the degree of treatment for existing plants. While the bulk of effluent discharged has received secondary treatment, about 19 percent has received only primary treatment or less. In addition, there are numerous reported cases where plants claim secondary treatment but are in fact discharging effluent that does not meet the Federally established standards. In October 1975, the EPA promulgated the Best Practicable Waste Treatment Technology Standards which require that effluent discharged to land not degrade ground water to a non-potable condition. This action plus the establishment of monitoring programs called for in the regulations should help minimize the threat of future ground-water contamination from land disposal of sewage effluent.

The second factor is related to those regions in the United States where municipal waste treatment plant effluent is disposed of directly on land. In order to obtain a picture of the importance of this practice on a national basis, a list was compiled of the 2,665 facilities which discharge effluent to sources other than intermittent and perennial streams, lakes, or the ocean. Based on these data, Figure 62 was prepared to show by county, three ranges of population density served by facilities discharging effluent to land.

Table 32. DEGREE OF TREATMENT AT MUNICIPAL WASTE FACILITIES.[2]

Treatment degree	Number of plants	Population served (millions)	Percent of total population served
None	1,118	2.92	1.8
Minor	68	0.76	0.5
Primary	2,777	37.85	23.9
Intermediate	68	6.15	3.9
Secondary	16,809	107.74	68.0
Tertiary	947	2.98	1.9
Total:	21,787	158.40	100.0

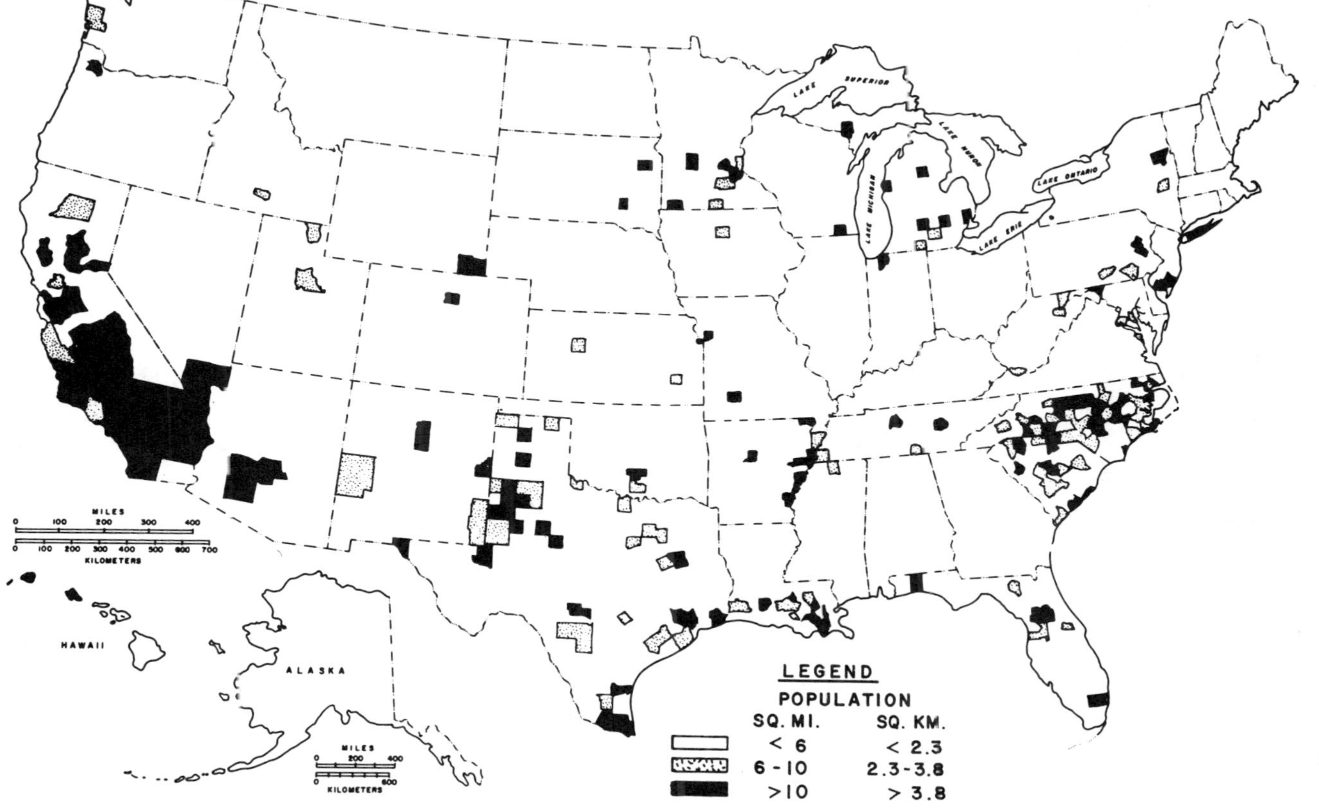

Figure 62. Density of population served by municipal sewage treatment facilities discharging effluent to land (by county).

It has been estimated that over 26 bgd (98.4 million cu m/day) of sewage effluent from domestic, industrial, commercial, storm water and all other sources, are discharged by municipal treatment plants.[19] Taking the proportion of land disposal to total disposal, approximately 2.3 bgd (8.7 million cu m/day) of effluent, some of which has received only primary treatment, are discharged onto the land.

Figure 62 indicates that the principal region for land disposal of effluent is the arid southwest. Several reasons for this are: (1) there is a lack of surface water in which to discharge effluent; (2) ground water in such areas is "mined" (taken from storage), and water recycling is practiced; and (3) arid conditions make the effluent valuable for crop irrigation. There is, however, substantial land disposal practiced in other regions of the United States, for example, North and South Carolina.

Stream Discharge

About 90 percent of municipal waste treatment plant effluent is discharged to surface-water bodies, particularly flowing streams. Upon entering the stream, the effluent is diluted to some degree. All of the water in the stream may discharge to the ocean and have no effect on ground water. However, ground-water quality may be significantly degraded where stream water recharges the ground-water system and the effluent-flow to stream-flow ratio is large enough to significantly degrade the quality of the stream. Effluent discharged to such streams may at times make up a substantial part of the total stream flow which enters the ground. Because of the many variables involved, it is not possible to estimate, on a broad basis, the impact on ground-water quality of effluent discharged to these streams from municipal waste treatment facilities. The extent of this problem should decrease significantly because of the general implementation of more stringent discharge requirements.

Storm Water

As previously mentioned, ground-water contamination from urban runoff may result from leaks in, and overflowing of, storm sewers; intentional ground-water recharge of untreated runoff; and runoff flowing from streets directly into unpaved areas and percolating to the water table. The volume and quality of contaminated runoff entering ground water from leaks in the storm sewers cannot be estimated, due to the lack of published data on this problem.

Storm-sewer overflow is another potential ground-water quality threat. During rainstorms, the capacity of the sewer lines is easily exceeded; as a consequence, regulators or overflow weirs are used to discharge excess volumes of mixed street runoff and, in the case of combined sewers, sanitary wastes. There may be hundreds of discharge points involved in a single storm-sewer system. Combined sewer overflow pollutant loads are normally higher than the sum of street runoff loads and "steady-state" sanitary loads, due to the between-rainfall accumulation of wastes on sewer floors and in the various traps throughout a sewer system.

Although there now is virtually no construction of combined sewer systems, many cities have extensive old combined sewer systems. 4) A total of 1,010 separate facilities exist which are using combined sewers, and an additional 600 use both combined and separate sewer systems. Also, there are 3,459 facilities for which information regarding the type of sewer system used is unknown. 2) Figure 63 shows the relative use of combined sewers compared to total sewered population.

As noted previously, ground-water recharge by urban stormwater runoff is practiced in many areas, including Los Angeles and Fresno, California; Long Island, New York; and Orlando, Florida. Recharge is accomplished by directing runoff via storm sewers to infiltration basins or pits.

Case Histories

Sewer Exfiltration (Kings County, New York) -

A study conducted in 1972 in Kings County, the Brooklyn borough of New York City, found that leaky sewers are a significant source of artificial recharge in the county and a major contributor to the high nitrogen content of the ground water. 21) The report notes that while the nitrogen content of typical sewage ranges from 16 to 73 ppm, the total nitrogen content of ground water from 17 widely scattered wells in Kings County is 16 ppm or greater. This finding, together with the total absence of agricultural activities and domestic waste disposal systems, and the observed existence of old and damaged sewer lines, indicates that leaky sewers are the principal source of the high nitrogen levels.

In addition, the investigation found that the rapid recovery of the water table after a marked reduction in ground-water pumpage in 1948 probably could not have been caused by recharge from precipitation alone, and attributed it largely to a combination of leakage from sewer lines and water mains. The report concludes that the value of sanitary sewer sys-

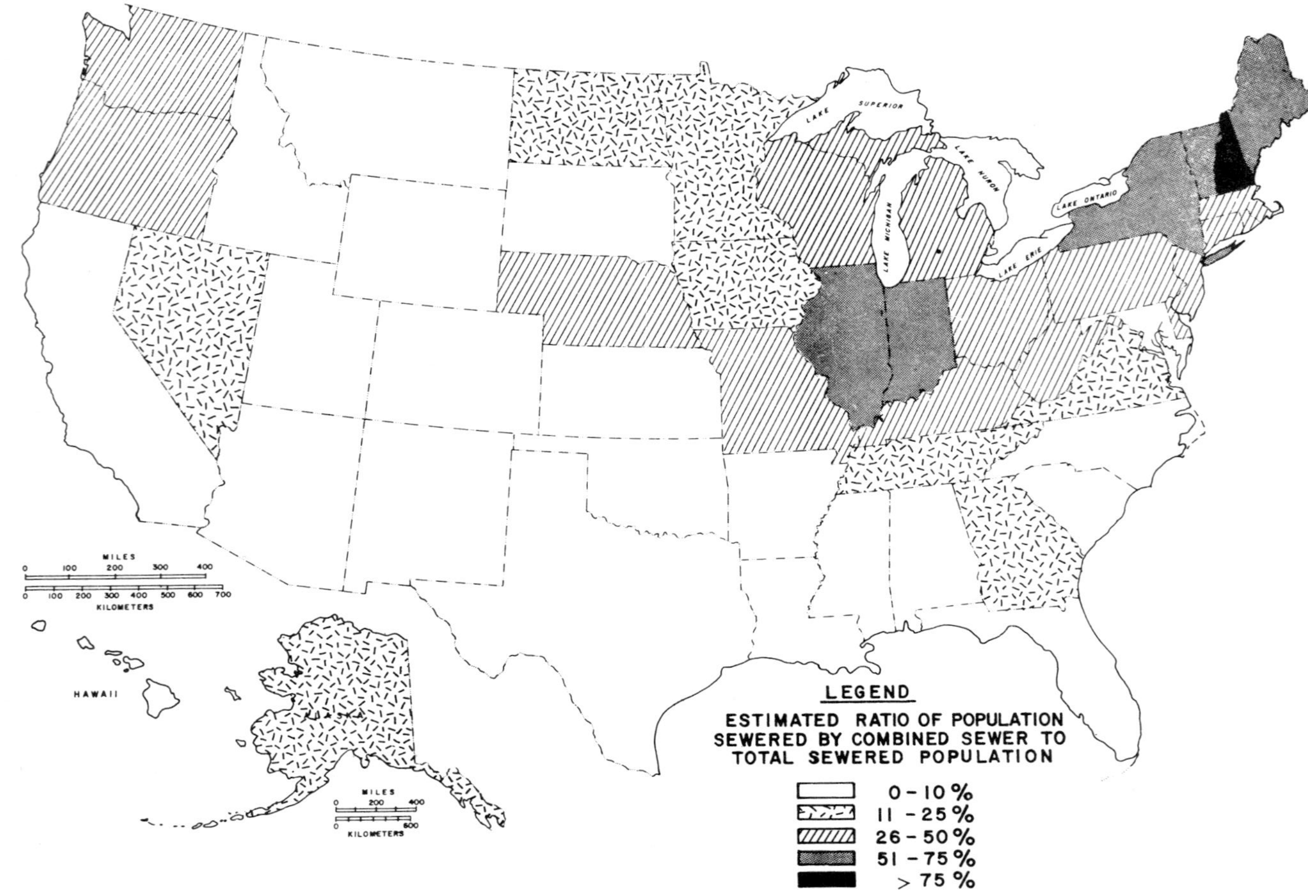

Figure 63. Relative use of combined sewers compared to total sewered [illegible] 20)

tems in preventing contamination of ground water in any particular area may be partially offset if the sewers are not carefully built and maintained.

Discharge to Lagoon (Tieton, Washington) -

Ground-water contamination from sewage lagoons has occurred at Tieton, Washington. 22) Wastes of the community were discharged to a lagoon located in a narrow valley underlain by permeable sands and gravel. Average daily waste flow was 130,000 gal. (490 cu m), of which 50,000 gal. (190 cu m) were domestic, with the balance constituting industrial waste waters. Within a few months after discharge to the pond began, a well located 250 ft (76 m) south of the lagoon became contaminated. Investigation showed that coliform bacteria had traveled from the lagoon to the well.

Anionic synthetic detergent had also entered the aquifer. A tracer study using chloride showed that sewage infiltrating through the bottom of the lagoon formed a shallow, elongated mound of fluid resting on top of the water table. Flow velocities of ground water below the lagoon were in excess of 300 ft/day (90 m/day). About 1,000 ft (300 m) down the valley, the velocity decreased to 200 ft/day (60 m/day). Within 6 days, water from the lagoon reached wells located as far as 1,500 to 2,000 ft (450 to 600 m) down the valley.

Little consideration in the design of the system had been given to potential contamination of ground water. Infiltration rates at the lagoon ranged from 3 to 15 in./day (7.6 to 38 cm/day) indicating highly permeable soil conditions with little or no filtration or treatment capability.

Disposal to Dry Stream Bed and Oxidation Ponds (Barstow, California) -

For the past 60 years, part of the alluvial aquifer along the Mojave River near Barstow has been contaminated by the percolation of wastes and sewage from industrial and municipal sources. 23) The contamination has forced the abandonment of several domestic wells due to taste, odor, and foaming, and threatens the well field serving a U. S. Marine Corps supply center.

The City of Barstow has operated three sewage treatment plants since 1938. Effluent from the 1938 plant was discharged directly into the Mojave River bed (dry except during periods of flooding). In 1953, a new sewage treatment plant was constructed 0.5 mi (0.8 km) downstream from the first plant, which was then abandoned. Disposal of effluent from the new facility was by direct percolation from oxidation ponds. In 1968, a larger plant was built several miles downstream from the 1953 plant, and just upstream of the supply center. Treatment in the new plant was performed by seven oxidation ponds.

In addition to the municipal treatment facilities, a railroad yard has been disposing of emulsified oil, grease, and synthetic detergents in the Mojave River bed since 1949. Dissolved solids are also added to the ground water by irrigation return flow from farms in the area. The concentrations of dissolved solids, detergents, and dissolved organic carbon were used to identify the extent of the degraded ground water in the area. Results of the study indicated that several plumes of degraded ground water were moving downgradient toward wells at the supply center. One plume near the base of the aquifer is probably the result of percolation from the abandoned, upstream, waste-disposal sites. This plume has moved about 4 mi (6 km) since 1910. A more recent overlying plume occurs near the downstream edge of the deeper plume, and has been produced by percolation from the facilities installed in 1968.

A third plume in the vicinity of the U. S. Marine Corps golf course is moving toward the U. S. Marine Corps well field in response to pumpage from these wells. Higher concentrations of dissolved constituents in this third plume are the result of the use of reclaimed industrial and domestic effluent for irrigation on the golf course.

Stream Disposal (Denver, Colorado) -

In 1965, 23 plants discharged about 105 mgd (400,000 cu m/day) to the South Platte River basin. By 1974, 140 mgd (530,000 cu m/day) were being discharged. In studies made in 1965 and 1967, significant quantities of ABS detergents and nitrate were found in the valley-fill aquifer, principally between Denver and Kersey.[24,25] This aquifer is the principal source of water for the majority of public water supply systems in the area.

TECHNOLOGICAL CONSIDERATIONS

Sewer Systems

Control of leaky pipe joints and pipe failures starts with proper installation techniques, good quality construction materials, and testing before the system becomes operative. Of prime importance in sewer pipe connections is joint flexibility since differential soil settlement often shears joints and connections, causing the pipe to leak. A number of joining methods are available to allow flexibility between sewer-pipe sections, and between sewer pipes and manholes. Later, if it is determined that a sewer pipe is defective, possible rehabilitation methods include digging up and replacing defective pipe, inserting plastic liners, or chemical grouting. The most universally used method is grouting. 26)

Lagoons and Ponds

Of primary concern in the design and operation of waste-stabilization ponds is the infiltration of contaminants to the underlying ground water. To minimize leakage, the pond bottom can be sealed by compaction, often with the addition of clay, or by the use of artificial liners. The types of liners used and some of their limitations are:

PVC - destroyed by aromatic solvents
soil-cement blanket - not completely impermeable
concrete - high cost
reinforced chlorinated polyethylene - high cost
asphalt - deteriorated by sunlight
hypalon or butyl - requires reinforcing to prevent rupture and puncture

Land Spreading and Basin Recharge

Contamination of ground water can occur from design failures or misuse of the system. The suitability of any particular site for land spreading is of key importance and depends upon climate, soil characteristics, soil depth, topography, hydrology, and geology. 7) Two major factors in the design of sprinklers for spray-irrigation systems are wind and soil infiltration rates. Wind can create a health hazard by carrying spray containing bacteria and viruses into populated areas. Odor and mosquito problems and water flooding may occur when the infiltration rate of the soil is exceeded.

Criteria to be evaluated in the design of a basin system include degree of waste-water pretreatment, loading rates, geologic conditions, and surface topography. For infiltration-percolation systems to function properly, soil infiltration rates of 4 to 12 in./day (10 to 30 cm/day) are needed. The permeability of the lower soil layers is also important. The vertical percolation of these layers must equal or exceed infiltration rates. A minimum depth of 15 ft (4.5 m) from the basin bottom to the water table is usually required to prevent the recharge mound from intersecting the basin bottom with a resultant reduction in infiltration rates.

Storm Water

Certain efforts may be made to limit contamination from storm water. Examples of controls that might be implemented include the following:

1. Improvements in street cleaning practices and equipment. Current evaluations indicate that the contamination impact from this source is largely associated with the fine solids fraction (less than 43 micrometers), which existing broom-type street sweepers are inefficient in collecting.

2. Enforcement of anti-litter laws to reduce the quantities of wastes.

3. Restrictions on the indiscriminate use of chemicals such as fertilizers and pesticides and deicing and anti-skid agents used in northern climates.

4. Institution of more stringent regulations on erosion and sediment control during construction activity. It has been observed that sediment, generated when ground covers are stripped away to allow construction activity, is a major contaminant in storm water. 27)

5. Establishing requirements that will reduce the quantity of contaminants washed through the storm sewerage system. These preventative measures can include the requirement of storm runoff detention ponds for new developments, which can be incorporated into development layouts in an aesthetically pleasing manner.

6. Prohibiting the filling of lowlands, open channels, and unprotected areas which are subject to accumulations of storm water, and exercising controls over what substances can be stored openly outdoors without contamination controls.

To further reduce the contributions to ground-water contamination made by storm sewer systems, the following types of collection system controls are available:

1. A systematic program of cleaning sewers of deposited solids and debris to help reduce the flushing effect of contaminants in the early portions of a storm.

2. Improved catch basin designs to eliminate the flushing of accumulated solids into the system during storms.

End-of-pipe controls are the most sophisticated means of removing contaminants in storm water, and include actual treatment processes similar to those employed in the waste-water treatment field. The treatment alternatives are separated into those types that provide physical treatment (adsorption, sedimentation, filtration, screening, concentration), those that provide treatment by means of biological organisms (lagoons of various types, modifications of the activated sludge process, trickling filters, rotating biological contractors), those that enhance treatment by the addition of chemicals which act as coagulants, flocculants, and conditioners, and finally, those that employ a combination of physical-chemical processes.

Specific preventative measures that can be employed to reduce ground-water contamination from storm-water storage basins include the following:

1. Lining of lagoons and basins with impervious barriers.

2. Use of barrier wells to intercept plumes of contaminated ground water where leakage has occurred.

3. Banning the use of seepage lagoons per se, but allowing construction of properly designed retention lagoons.

4. Determination of suitable detention pond locations by means of subsurface soil and ground-water investigations.

5. Installation of observation wells around the perimeter of storm-water storage lagoons for the purpose of detecting the presence of contaminants and defects in the basin seal.

In considering all of these control techniques, it must be noted that cost estimates in the range of $300 billion have been developed by EPA and the National Commission on Water Quality for adequate treatment and disposal of excess discharges from storm and combined sewers, on a national basis.

INSTITUTIONAL ARRANGEMENTS

Sewers

The possibility of ground-water contamination caused by leaky sewers has been given little attention in regulations. Where a state agency has the authority to approve sewer construction, it may, as a matter of policy, prohibit construction of a sewer line within a specified distance, such as 100 ft (30 m), of a well used for drinking-water purposes.

One guideline used by a number of states, requires sewer joints to be designed to minimize infiltration and to prevent entrance of roots; test requirements are that leakage outward or inward not exceed 500 gal./in. (745 l/cm) of pipe diameter per mile per day. [28]

Delaware is unusual in its requirement that one obtain a permit to construct any sewer or pipeline which conveys liquid waste. [29]

Waste-Water Lagoons and Ponds

Municipal lagoons and ponds for the retention of waste water are parts of sewage treatment facilities, the construction and operation of which are supervised by state health departments or environmental protection agencies. Virtually all municipal waste treatment facilities currently are built with Federal grant contributions, subjecting them to requirements established by the EPA pursuant to the Federal Water Pollution Control Act Amendments of 1972. That law requires that contaminants not migrate to cause water or other environmental contamination. [30]

An example of a recent adoption of design criteria for sewage stabilization ponds is that of the Minnesota Pollution Control Agency. A number of these criteria are directed at prevention of ground-water contamination. [31] One requires that permeability of a pond seal be as low as possible, and in no case should seepage loss through the seal exceed 500 gal./acre/day (4.7 cu m/ha/day). A testing program is required; specifications for construction and placement are to be based on test results. A minimum of 4 ft (1.2 m) between the top of the pond seal and the maximum high water table is to be maintained. An approved system of ground-water monitoring wells or lysimeters is required around the perimeter of the pond site, well locations to be determined on a case-by-case basis depending on proximity of private water supply wells and maximum ground-water levels. Information required to be filed prior to construction includes a log of each well within one mi (1.6 km) of the proposed pond. The Agency also requires submission of information as required in Recommended Standards for Sewage Works [28] and Federal Guidelines, Design, Operation and Maintenance of Waste Water Treatment Facilities. [32]

A number of states, including Michigan, Pennsylvania, Delaware, and California, require permits or approvals for municipal sewage impoundments similar to those for industrial waste impoundments.

Land Spreading and Basin Recharge

Spraying of sewage effluent on land as a disposal method is specifically regulated in only a few states. Most states review such disposal on a case-by-case basis with due consideration to effect on water quality, the same as any other discharge. Maryland specifically requires a discharge permit for waste-water effluents disposed of by means of spray or other land irrigation systems; this permit program is one of those used to enforce the water-quality standards Maryland has established for its 3 classifications of aquifers. 33) New York presents detailed requirements for a design report for any facility employed for ground disposal of waste waters.

Detailed requirements specifically directed at spray irrigation for the purpose of protecting ground water are exemplified by the Spray Irrigation Manual published by the Bureau of Water Quality Management of the Pennsylvania Department of Environmental Resources. 34) An introduction to the manual states:

> "Since roughly 50 percent of waste water discharged to the land surface in Pennsylvania will infiltrate and recharge ground water, all spray irrigation installations are considered discharges to the waters of the Commonwealth. As such, each installation will require a Department of Environmental Resources permit under the Clean Streams Law."

The manual provides guidelines for locating and evaluating sites, and in designing spray irrigation systems. Factors include soils, geology, hydrology, weather, the agricultural practice involved, and adjacent land use. The guidelines set standards for treatment, storage, screening, controls, piping, sprinklers, distribution diameter (not in excess of 140 ft or 43 m), spacing and application rate.

Florida's Department of Environmental Regulation specifies type of treatment for low rate (between 2 to 3 in./week or 5 to 8 cm/week) and high rate (maximum 4 in./week or 10 cm/week) of application for irrigation and crop harvesting, and spraying for purposes of recharge (1/40 of initial percolation rate). For each category the guidelines specify detention time for holding basin, depth to ground water, and buffer zone for adjoining property. 35) The introduction states:

> "It is significant to note that provision has been made in these guidelines for use of soils as treat-

ment media as compared to their conventional use as sinks for treated waste waters."

Idaho's regulation states that land disposal of liquid waste "shall not create a ground-water mound or result in a salt build-up on another person's property." It requires that the waste water used be biologically degradable, but allows use of other waste water if it can be shown that it will have no adverse effect on ground water. An applicant for a permit must "provide reasonable assurance that the earth material underlying the proposed disposal site will not allow direct rapid movement of pollutants into the underlying ground water." 36)

Regulations typically require at least secondary treatment of wastes that are to be sprayed, and contain requirements for monitoring ground water. In Wisconsin, according to the "Wisconsin Administrative Code, Chapter NR-214, land disposal of liquid waste - discharge limitations and monitoring requirements," spraying of untreated dairy, canning, and meatpacking waste is allowed. Washington requires only primary treatment of sprayed wastes, but allows land disposal only for 5 years in one location. 37)

The Center for the Study of Federalism of Temple University reported that in 1972, only 14 states regulated land treatment (land disposal) of wastes. General patterns reported from a survey of all states' regulations were "a disappointing lack of information about land treatment, a great deal of misinformation, and an even greater lack of interest." 38)

California's Water Reclamation Law applies broadly to any use of treated waste water, whether for land disposal, injection for recharge, or other use. It requires each regional water quality board to prescribe water reclamation requirements for water used or proposed to be used as reclaimed water. These must be in conformance with statewide reclamation criteria established by the State Department of Health. Anyone reclaiming or proposing to reclaim water, or using or proposing to use it for any purpose for which reclamation criteria have been established, must file with the regional board a report containing such information as the board requires. Reclaimed water may not be injected directly into an aquifer that is suitable for domestic water supply until a finding by the State Department of Health, after a public hearing, that the proposed recharge will not impair the quality of water in the receiving aquifer. 39)

REFERENCES CITED

1. Water Pollution Control Federation. 1966. Sewer maintenance, manual of practice No. 7. Washington, D. C.

2. U. S. Environmental Protection Agency. 1975. Municipal waste facilities inventory. Computer retrieval of Storet system, June 6.

3. Symons, G. E. 1967. Water and wastes engineering, manual of practice No. 3. Reuben H. Donnelly Corp., New York.

4. Metcalf & Eddy, Inc. 1972. Wastewater engineering: collection, treatment, disposal. McGraw-Hill Book Co., New York. 782 pp.

5. Water Pollution Control Federation. 1970. Design and construction of sanitary and storm sewers, manual of practice No. 9. Washington, D. C.

6. Thomas, R. E. 1973. An overview of land treatment methods. Journal Water Pollution Control Federation. July:1476-1484.

7. Pound, E. E., and R. W. Crites. 1973. Waste water treatment and reuse by land application, Vol. II. U. S. Environmental Protection Agency.

8. Godfrey, K. A. 1973. Land treatment of municipal sewage. Journal American Society of Civil Engineers 43(9).

9. Fuhriman-Barton and Associates. 1971. Ground-water pollution--Arizona, California, Nevada, Utah. U. S. Environmental Protection Agency Report 16060 ERU.

10. Sopper, W. E., and L. T. Kardos. 1973. Recycling treated municipal waste water and sludge through forest and cropland. Pennsylvania State University Press, University Park, Pennsylvania.

11. Barsom, G. 1973. Lagoon performance and the state of lagoon technology. U. S. Environmental Protection Agency Report EPA-R2-73-144.

12. Lager, J. A., and W. G. Smith. 1974. Urban stormwater management and technology: an assessment. U. S. Environmental Protection Agency Report EPA-670/2-74-040.

13. U. S. Census of Housing. 1970. Detailed housing characteristics. U. S. Department of Commerce HC (1)-B. 53 vols.

14. Kollar, K. L. 1966. Regional requirements for sewer pipe in sewerage utilities. U. S. Department of Commerce. 20 pp.

15. Change, A. C., and others. 1974. The sealing mechanism of wastewater ponds. Journal Water Pollution Control Federation 46(7):1715-1721.

16. Wilson, L. G., W. L. Clark, III, and G. G. Small. 1973. Subsurface quality transformations during the initiation of a new stabilization lagoon. Water Resources Bulletin 9(2):243-257.

17. Fossom, G. O. 1971. Water balance in sewage stabilization lagoons. Civil Engineering Department, North Dakota University, Grand Forks, North Dakota. 38 pp.

18. Preul, H. C. 1968. Contaminants in ground water near waste stabilization ponds. Journal Water Pollution Control Federation 40(4):659-669.

19. Sopper, W. 1973. Review of symposium on recycling treated municipal waste water and sludge. Irrigation Journal 23(4):16-19.

20. U. S. Department of Interior, Federal Water Pollution Control Administration, American Public Works Association. 1967. Report on problems of combined sewer facilities and overflows. WP-20. Storm and Combined Sewer Pollution Control Branch, Division of Engineering Development, Research and Development, Federal Water Pollution Control Administration. Washington, D. C. Page 21.

21. Kimmel, G. E. 1972. Nitrogen content of ground water in Kings County, Long Island, New York, U. S. Geological Survey Professional Paper 800-D. Pages D199-D203.

22. van der Leeden, F., L. A. Cerrillo, and D. W. Miller. 1975. Ground-water pollution problems in the northwestern United States. U. S. Environmental Protection Agency Report 660/3-75-018.

23. Hughes, V. L., and S. G. Robson. 1973. Effects of waste percolation of ground water in alluvium near Barstow, California. Underground Waste Management and Artificial Recharge.

24. U. S. Department of Health, Education and Welfare, Water Supply and Pollution Control Division. 1965. Ground-water pollution in the South Platte River valley between Denver and Brighton, Colorado. Published report 4.

25. U. S. Department of the Interior, Federal Water Pollution Control Administration. 1967. Ground-water pollution in the Middle and Lower South Platte River Basin of Colorado. Published report 9.

26. Fairweather, Virginia. 1974. Sewer pipe: infiltration is the issue. Journal American Society of Civil Engineers 44(7).

27. Economic Systems Corporation. 1970. Storm-water pollution from urban land activity. U. S. Environmental Protection Agency Report 11034 FKL.

28. Great Lakes-Upper Mississippi River Board of State Sanitary Engineers. Recommended standards for sewage works.

29. Delaware Water Pollution Control Regulation. Section 4.

30. Federal Water Pollution Control Act Amendments of 1972. (PL 92-500).

31. Minnesota Pollution Control Agency, Division of Water Quality. 1975. Recommended design criteria for sewage stabilization ponds.

32. U. S. Environmental Protection Agency. Federal guidelines, design, operation and maintenance of waste water treatment facilities.

33. Maryland Water Pollution Control Regulation 08.05.04.04. Ground water quality standards.

34. Pennsylvania Department of Environmental Resources, Bureau of Water Quality Management. 1972. Spray irrigation manual. 49 pp.

35. Florida Department of Pollution Control, Division of Operations. 1973. Guidelines for treatment and/or disposal of wastewaters by irrigation on land. Memo No. 149.

36. Idaho Board of Environmental and Community Services. Rules and regulations for the establishment of standards of water quality and for wastewater treatment requirements for waters of the State of Idaho.

37. Washington Department of Social and Health Services, Health Services Division. General criteria for land treatment sites.

38. Stevens, R. M. 1972. Green land - clean streams. Center for the Study of Federalism, Temple University, Philadelphia, Pennsylvania. 330 pp.

39. Water Code of the State of California. Chapter 7. Water Reclamation Law. Section 13500.

SECTION X

LAND SPREADING OF SLUDGE

SUMMARY

Municipal and industrial sludge is the residue remaining after treatment of waste water. The impact of diffuse land spreading of municipal and industrial sludge on ground water is not documented even though the potential for contamination exists. Less than one percent of the present municipal sludge disposal facilities are monitored for effects on water quality. Even fewer industrial sludge sites are monitored because this potential source of ground-water contamination has received less attention than municipal sources.

Sludge may be a product of physical, biological, or chemical treatment or a combination thereof. Ground-water quality degradation can be caused by land spreading of sludge because organisms (such as viruses) and chemical ions and compounds can be leached by precipitation and carried in percolate to ground water.

Land and air (through incineration) remain the depositional areas for an ever increasing volume of sludge from a growing population and from higher degrees of waste-water treatment, the latter brought on by more stringent environmental protection of rivers, lakes, the ocean, and the atmosphere. Most municipal and industrial sludge now goes to landfills and impoundments. As controls over these two methods of disposal become more restrictive with respect to type of waste accepted, the amount of sludge diverted to land-spreading sites will increase rapidly.

In the United States, municipal sludge production amounts to about 5,000,000 dry tons/yr (4,540,000 dry tonnes/yr). Accurate data on quantities of industrial sludge are not available. However, the total volume certainly exceeds municipal sludge production by many times. The organic and inorganic chemicals industries and coal-fired utilities are the largest contributors of residues and account for over half of the total production. Industrial expansion and growing pollution control activities should increase the volume of industrial sludges dramatically over the next 10 years.

The key to correct management combines proper site selection with sludge composition, application rates, and land use (crops). Of major importance to ground water is the availability of soil, such as a loam or silt loam, that is the most efficient for attenuating contaminants.

In most states, the basic provision of law applicable to land spreading of municipal and industrial sludges is the all inclusive prohibition against polluting waters of the state. Before action can be taken, the presence of a contamination problem must be established. In a few instances, control over sludge disposal can be asserted where states have enacted "potential pollution" statutes which include sludge spreading in the same provisions as those that apply to waste lagoons and landfills. Other states have developed special laws that apply to disposal of hazardous or general industrial process wastes including sludges.

DESCRIPTION OF THE PRACTICE

Sludge is the residue remaining after treatment of either municipal or industrial waste water. Sludge is a product of physical, biological, or chemical treatment, or any combination thereof (Figure 64). As it leaves the treatment plant, it is composed mostly of water, with biological matter and small amounts of metals and other chemicals. The two most common land disposal methods for this product are land spreading and landfilling, the latter being described in the SOLID WASTE section of this report. Disposal methods other than land include incineration; high-temperature, wet-air oxidation (Zimpro); and ocean dumping (soon to be prohibited).

Surface spreading of raw sewage on agricultural land is a practice dating back to antiquity. However, raw sewage is a vector for disease and parasites, both of which have been associated with the land disposal practice. The use of raw sludge has produced a prejudice against land spreading of any form of human waste, whether treated or untreated. As a result, land spreading of sludge has not enjoyed great popularity in the United States. In Europe and Great Britain, where population densities are greater and agricultural utilization of land more intense, sludge farms have been included as an integral part of many standard sewage-treatment systems for over a century.

Unlike municipal sludge which has a relatively uniform sewage base, industrial sludges range widely in composition. (As used here, industry refers to manufacturing, not mining, forest products, or agricultural crop production. These industries are discussed elsewhere, or do not produce sludges.) Extremes in industrial sludge composition are exemplified by the food processing industry and the electroplating industry. Food wastes, with treatment to decompose the organic matter,

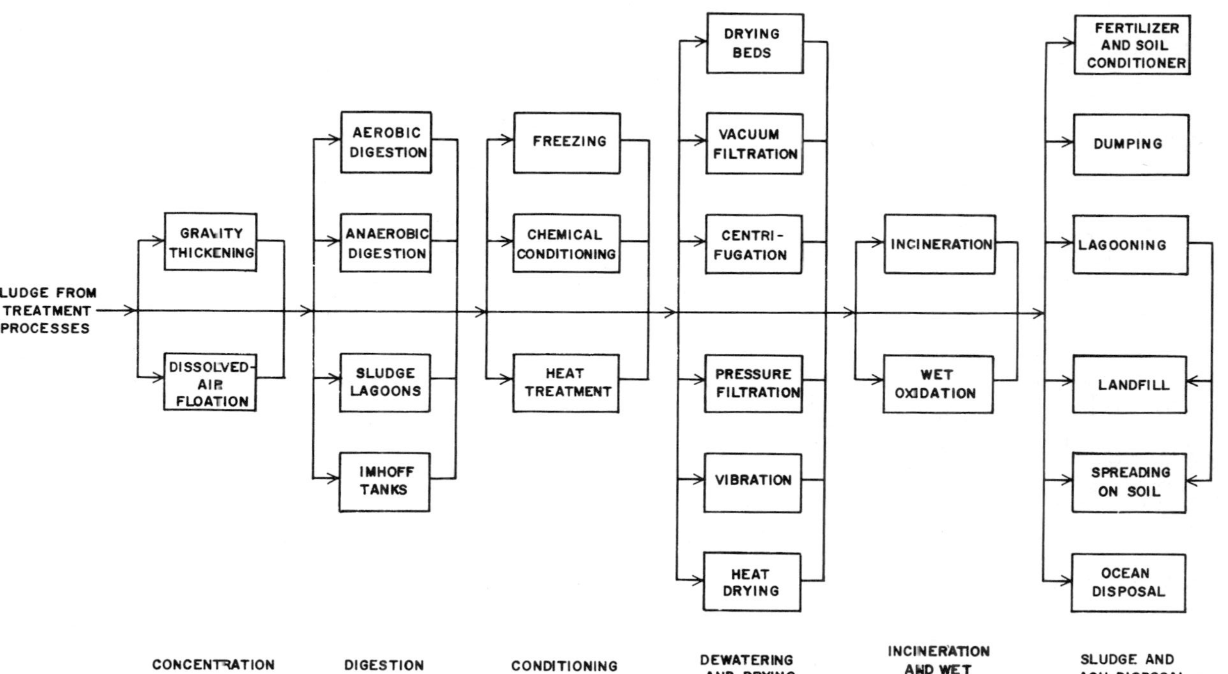

Figure 64. Generalized sludge processing and disposal flowsheet. [1)]

are reduced to a source of plant nutrients. On the other hand, electroplating sludge contains precipitated salts of heavy metals, frequently including some of the more toxic ones -- cadmium, chromium, nickel, and cobalt. There is no way to decompose these metals and land disposal methods run the risk of recycling the metals back into the environment. Between these extremes are a spectrum of organic, inorganic, and mixed sludges varying from highly degradable to highly refractory.

Before application to land, sludge must be properly stabilized. It may be heat dried, composted, digested, incinerated, or chemically stabilized; and then dewatered by filtration, centrifugation, or sand beds; or it may be applied as a liquid slurry. Dried and dewatered sludges are solid products although the mechanically dewatered sludge usually contains 20 to 30 percent moisture. Liquid sludge of about 5 percent solids content can be pumped and applied as a liquid. Normally land spreading is accomplished by tank trucks or wagons traversing the receiving area and releasing sludge by gravity feed.

Ground-water contamination from land spreading of sludge is most likely to be in the form of chemical contamination. Some constituents are soluble and are likely to be leached more readily than others. These include sodium, potassium, sulfate, chloride, and nitrate ions. Other constituents are held more strongly in the sludge matrix or are attenuated more strongly in soil. These include calcium, magnesium, and the suite of heavy metals. Constituents which are more strongly held in sludge or move relatively slowly into or through the soil profile pose less of a threat to ground water, but may affect the quality of crops to a greater extent.

The rate and extent to which chemical constituents of sludge are leached depend upon the amount of precipitation and the relationship between precipitation and evapotranspiration. Only where precipitation occurs in excess of evapotranspiration to the degree necessary to effect recharge, can sludge constituents be carried to ground water.

Soil type influences the potential for ground-water contamination. Soils of light (sandy) texture have a low ion-exchange capacity which results in little retention of the sludge-applied ions. Sludge often contains a suite of heavy metal ions which can be toxic at higher concentrations. The ion-exchange capacity of a specific soil can make a portion of the heavy metals contained in the sludge inaccessible to percolation and to plant absorption. The more clay and organic colloids in the soil the greater its ion-exchange capacity and the lower its permeability.

The type of crop grown on a sludge-treated soil can materially affect the quantity and quality of percolate. Nutrients and water absorbed in plant tissue are removed from the soil-water system unless the plants are reincorporated into the soil. Surface incorporation is an attractive method of sludge disposal because the water and nutrients are beneficial to plant growth. Heavy metals and phosphate are immobilized (relative to landfilling) in soil and pose less of a threat to ground water. In addition, soil harbors a large population of microorganisms which have the ability to decompose almost any kind of organic molecule.

CHARACTERISTICS OF CONTAMINANTS

Municipal sludge is a mixture of substances whose sources are metabolic wastes, industrial waste, street runoff, and other household waste water. The residues comprising sludge include partially decomposed organic compounds, inorganic salts, and heavy metals (Table 33). Included in this mixture is a large bacterial population originating from the sewage. Viruses introduced in raw sewage may or may not be viable depending upon the degree and type of treatment. Potential contaminants from sludge may be categorized as nutrients, heavy metals, and pathogenic organisms.

Information concerning the composition of industrial sludges and residuals is far more limited than it is for municipal sludges. Waste products from industry are predominantly handled by the industry through the treatment process, and ultimate disposal. Because chemical compounds or elements in the sludges reflect proprietary products or processes, industrial management is not amenable to outsiders analyzing the waste. Thus, until recently, under pressure of more stringent environmental restrictions, little information about industrial wastes was available.

Because industrial sludge characteristics may differ substantially from municipal sludge, a section describing the sludge characteristics of several major industries is included following the general description below.

Chemical Components

Nutrients -

Many of the components of sludge could be categorized as plant nutrients. However, in usual agricultural practice, the primary plant nutrients are limited to nitrogen, phos-

Table 33. TYPICAL CHEMICAL COMPOSITION OF RAW AND DIGESTED SLUDGE. [1)]

Item	Raw primary sludge		Digested sludge	
	Range	Typical	Range	Typical
Total dry solids (TS), percent	2.0 - 7.0	4.0	6.0 - 12.0	10.0
Volatile solids (percent of TS)	60 - 80	65	30 - 60	40.0
Grease and fats (ether soluble, percent of TS)	6.0 - 30.0	-	5.0 - 20.0	-
Protein (percent of TS)	20 - 30	25	15 - 20	18
Nitrogen (N, percent of TS)	1.5 - 4.0	2.5	1.6 - 6.0	3.0
Phosphorus (P_2O_5, percent of TS)	0.8 - 2.8	1.6	1.5 - 4.0	2.5
Potash (K_2O, percent of TS)	0 - 1.0	0.4	0.0 - 3.0	1.0
Cellulose (percent of TS)	8.0 - 15.0	10.0	8.0 - 15.0	10.0
Iron (not as sulfide)	2.0 - 4.0	2.5	3.0 - 8.0	4.0
Silica (SiO_2, percent of TS)	15.0 - 20.0	-	10.0 - 20.0	-
pH	5.0 - 8.0	6.0	6.5 - 7.5	7.0
Alkalinity (ppm as $CaCO_3$)	500 - 1,500	600	2,500 - 3,500	3,000
Organic acids (ppm as HAc)	200 - 2,000	500	100 - 600	200
Thermal content (Btu/lb)	6,800 - 10,000	7,600 [a)]	2,700 - 6,800	4,000 [b)]

a) Based on 65 percent volatile matter.
b) Based on 40 percent volatile matter.

phorus, and potassium. Nitrogen usually is present in the largest concentration.

The nitrogen content of digested sludge shown as "typical" in Table 33 provides 676 lb (307 kg) of nitrogen per acre-inch liquid application (11.25 dry tons of solids per acre) to the land. Generally, 40 percent of the total nitrogen in digested sludge is ammonium and the other 60 percent is organic nitrogen.[2)] Thus, about 270 lb (123 kg) of ammonium-nitrogen would be added in an acre-inch application. This constitutes a nitrogen source immediately available to plants.

Because of the cost of transportation and the relationships between sludge production and available land, maximizing the sludge application rate is a typical occurrence. The result is that several inches of sludge or equivalent dry sludge may be applied during a growing season. The addition of 270 lb (123 kg) of "available" nitrogen in one inch of sludge is slightly above the usual agricultural recommendation for even the most nitrogen demanding crops. Multiplying that application several times over provides more nitrogen than a crop can remove. Consequently, the remaining nitrogen is available for leaching to and contaminating of the ground-water system. The normal nitrogen content of sludge places it in an awkward position. It is too low to make sludge competitive with commercial inorganic fertilizers, and too high to allow heavy soil loadings without a nitrate contamination hazard. Proper management of sludge irrigation can minimize such hazards.

Phosphorus, another major plant nutrient, is found in relatively large quantities in municipal sludges, primarily as phosphate. Usually it is slightly lower than nitrogen in concentration, equivalent to an average of 2,500 ppm wet weight. An acre-inch sludge application would supply about 560 lb (255 kg) of phosphate expressed as P_2O_5. However, only a portion of that amount is available to plants.

Phosphorus reactions in soil generally lead to fixation of the element. Under normal conditions of fertilization and irrigation, phosphorus is not a threat to ground-water quality. Although phosphorus is attenuated by movement through the soil when irrigation of waste-water effluents or sludge farming is practiced, heavy loadings of phosphorus can occur and, depending upon the specific soil type, phosphorus may exceed the fixation capacity. For most medium to fine textured soils the fixation capacity is on the order of several thousand pounds per acre. If the fixation capacity is exceeded, phosphate is then free to move through the soil profile and enter the ground-water system. Phosphate in ground water may not be detrimental *per se*, but if the ground water discharges to surface water, eutrophication may result.

Potassium in plant tissue typically ranges from one to 3 percent on a dry weight basis. Therefore, a nitrogen-phosphorus-potassium balance is important in supplying plant nutrients. Generally, sludge must be supplemented with a potash fertilizer to provide that balance. Potassium is not considered a threat to soil or water quality.

Heavy Metals -

In chemical terms, the heavy metals group includes vanadium, chromium, manganese, iron, cobalt, nickel, copper, zinc, lead, molybdenum, silver, cadmium, platinum, gold, and mercury, and a few more not mentioned here because of their rarity. Several elements, which differ chemically from the heavy metals, are frequently listed with them. Because of their common occurrence in sludge, they will be included here. They are: boron, arsenic, selenium, tin, and antimony.

The suite of heavy metals found in a given sample of sludge is dependent upon the source of waste water. Heavy metals present in sludge from domestic sewage include those required for human nutrition -- chromium, cobalt, copper, iron, manganese, molybdenum, selenium, and zinc; metals contributed through dissolution of plumbing -- lead, copper, tin, and zinc; and, in systems of combined sewers, metals from storm runoff -- cadmium, lead, and zinc. Although humans excrete iron in only small quantities, it is often found in relatively high concentrations in sewage sludge because ferric chloride is frequently used in the treatment process. Industrial wastes may contribute any or all of the heavy metals in concentrations which are sometimes high enough to be toxic to the biota effecting treatment.

Data on heavy metal concentrations in sewage sludge from selected locations are shown in Table 34. The ranges frequently extend over two orders of magnitude. These large fluctuations reflect mixing in some sewerage systems of industrial effluents in various ratios with domestic sewage.

Table 35 compares ranges of reported concentrations of heavy metals from various locations to concentrations of the metals typical of sludge from treatment plants with no industrial input. The contribution by industrial wastes can amount to several thousand parts per million. The extent to which ground water may become contaminated with heavy metals

Table 34. CONCENTRATIONS OR CONCENTRATION RANGES OF TRACE ELEMENTS IN SEWAGE SLUDGES FROM VARIOUS LOCATIONS IN THE UNITED STATES. [3]
(in parts per million)

Location	Boron	Cadmium	Chromium	Copper	Nickel
Athens, Georgia	9- 18	-	-	350- 530	11- 37
Columbus, Ohio	99-126	5.6- 10.5	-	282- 728	17- 23
Dayton, Ohio	-	830	-	6,020	200
Cincinnati, Ohio	-	40	-	4,200	600
Chicago, Illinois	6- 67	495	4,200	385- 1,225	425
Milwaukee, Wisconsin	8	-	-	435	-
Des Moines, Iowa	7	-	-	315	-
Houston, Texas	8	-	-	1,035	-
Rochester, New York	12	-	-	1,980	-
Maryland	4- 12	-	-	100- 490	-
Connecticut	160-360	-	-	465- 1,025	-
Southern California	360-680	1 -140	40- 600	135- 800	10-2,140
Oklahoma	70-100	-	Tr - 600	800- 6,000	100-3,000
Indiana	-	3 -810	50-19,600	300-11,700	70-3,500

Table 34 (Continued). CONCENTRATIONS OR CONCENTRATION RANGES OF TRACE ELEMENTS IN SEWAGE SLUDGES FROM VARIOUS LOCATIONS IN THE UNITED STATES.[3)]
(in parts per million)

Location	Manganese	Molybdenum	Lead	Zinc
Athens, Georgia	123- 268	-	-	1,850- 2,850
Columbus, Ohio	148- 232	-	72- 88	1,605- 1,764
Dayton, Ohio	-	-	-	8,390
Cincinnati, Ohio	-	-	-	9,000
Chicago, Illinois	135- 250	6.5- 6.7	1,500	3,050- 7,450
Milwaukee, Wisconsin	130	13.5	-	1,550
Des Moines, Iowa	420	4.9	-	1,350
Houston, Texas	65	6.7	-	950
Rochester, New York	60	5.1	-	3,400
Maryland	60- 790	2.1- 118	-	610- 3,100
Connecticut	105- 280	-	-	2,200- 3,500
Southern California	-	2 - 25	15-1,120	373- 3,400
Oklahoma	-	Tr -1,000	-	3,000- 7,000
Indiana	200-1,470	-	450-1,900	870-28,400

Table 35. RANGE OF METAL CONTENTS IN DIGESTED SEWAGE SLUDGES.[4]
(dry weight basis)

Analysis	Observed range	"Domestic" sludge [a]
Zn	500 - 50,000 ppm	< 2000 ppm
Cu	250 - 17,000 ppm	< 1000 ppm
Ni	25 - 8,000 ppm	< 200 ppm
Cd	5 - 2,000 ppm	< 15 ppm
Cd	0.1 - 40% of Zn	< 1.0% of Zn
B	15 - 1,000 ppm	< 100 ppm
Pb	100 - 10,000 ppm	< 1000 ppm
Hg	< 1 - 100 ppm	< 10 ppm
Cr	50 - 30,000 ppm	< 1000 ppm

a) Typical sludge from communities without excessive industrial waste inputs, or with adequate abatement.

from sludge applied to soil is dependent upon: heavy metal content of applied sludge, loading rate, physical and chemical soil properties, and distance to the water table. Shallow ground-water tables, coarse textured soils, and high rates of recharge combine to present the highest risk to the ground-water system. Conversely, fine textured soils with organic matter, deep water tables, and moderate recharge rates create the most favorable conditions for diffuse land disposal of sludge.

Other Hazardous Chemical Constituents -

Various micro-organic constituents are also present in sludge, some of which are ubiquitous and others of which are typical only of industrial waste water. These waste waters can contain materials such as dyes, inks, oils, pesticides, polychlorinated biphenyls (PCB), detergents, polynuclear aromatic hydrocarbons (PAH), and organic solvents. These constituents of waste water usually become constituents of sludge because they are resistant to microbial decomposition. PCB's and pesticides have been detected in sludge in small quantities. [5)]

The contamination potential posed by these materials toward ground water is difficult to assess at present. These materials are exposed to adsorption, microbial degradation, and plant, microbial, or protist absorption after they are applied to soil. Alternatively, no significant interactions may occur between micro-organic constituents and soil components.

Biological

A variety of organisms are associated with the biological treatment process in waste-water treatment plants. These include bacteria, viruses, fungi, algae, protozoa, rotifers, and macro-organisms including worms and flukes. Of these, bacteria and viruses are of concern because of possible harmful effects if pathogenic species should contaminate ground water. Pathogenic organisms are generally associated with sanitary waste. Thus, industrial waste water which does not include sanitary waste would not be likely to include pathogens.

The public health aspect of spreading a material in the environment which harbors pathogenic organisms is of great importance. The questions of what and how many pathogens may have survived the treatment process are immediately raised. Further, assurances must be given that any residual potentially harmful organisms will quickly die off for land spreading of sludge to be accepted.

Stabilization of sludge prior to land spreading is necessary in order to reduce public health hazards and to prevent nuisance odor conditions. A common stabilization method is anaerobic digestion which, however, does not result in a complete elimination of coliform organisms. Standard practice in monitoring for pathogenic organisms has been to measure fecal coliform (Escherichia coli) and/or fecal streptococcus (Streptococcus faecalis). These bacteria are not in themselves pathogenic, but their presence indicates the possible presence of other enteric bacteria and viruses. Pathogenic enteric bacteria include those causing typhoid, cholera, and dysentery. The absence of fecal coliforms is indicative of the disappearance of all enteric bacteria.

It has been generally agreed by investigators that bacterial migration through soils under most circumstances seldom exceeds 10 ft (3 m). In one report, bacteria were found to have traveled through a porous medium under percolation beds a distance of 400 ft (122 m). 6) Percolation beds consist of coarse sands and gravels and are designed for maximum percolation rates. These materials allow relatively free flow of water over a medium with little or no surface-active materials. Thus, the extensive movement of bacteria is not surprising. In circumstances involving fractured or cavernous rock, the distance traveled by live bacteria would be a function of their die-off rate as well as the rate of percolation and ground-water flow.

Far more research has been done on bacteria in sludges and their routes in the environment after disposal than has been done on viruses. Facilities for virological testing are more costly, complicated, and not nearly as common as those for bacteriological testing. Until recently, responses of bacterial populations to treatment were used as indicators for virus.

Laboratory investigations of virus migration through soils have concluded that adsorption on soil particles was not rapid enough to prevent breakthrough. 7) T7 and polio-I viruses were loaded on a 7.5-in. (19-cm) long soil column and leached with water. Intermittent leaching resulted in fewer T7 viruses passing through the soil, but had no effect on the polio-I. In a field irrigation experiment, viruses were detected in wells screened at 10 and 20 ft (3 and 6 m) below a sandy soil surface. 8) Not only did the viruses migrate that distance, but they also survived aeration and sunlight prior to doing so. The authors concluded that reevaluation of the ultimate danger of aquifer contamination from spray irrigation of waste water or land spreading of sludge may be necessary.

Industrial Sludge/Residuals

The use of the term sludge/residuals is necessitated because in most instances, no distinction is made between sludge and such wastes as fly ash, bottom ash, or slag. For example, in the iron and steel industry, the sludge/non-sludge waste ratio is lower than it is for the coal-fired electric power industry. The latter produces little residue which is classed as sludge, but a large quantity of fly and bottom ash residue. Land spreading of residues other than sludge (fly ash, pulp, whey) is practiced, thus reducing the difference between total waste production and that accommodated by land spreading.

A list which includes substances commonly found in industrial wastes illustrates the wide range of chemical elements and compounds possible (Table 36). No single sludge would be likely to contain all of the components in the list. The known organic compounds, for example, exceed 2 million, most of which are synthetic. The organic synthesis industry produces sludges which include some of these.

The available information on specific components of industrial residues varies from being nearly complete to being only scanty. Moreover, only a few industries have well established practices of land spreading sludges. A brief description of the volume and characteristics of the waste water produced by these industries follows.

Industries Which Employ Land Spreading of Sludges

Canned Fruit and Vegetable Industry - 9,10)

Wastes generated by this industry are simple carbohydrate, starch, and cellulosic substances. These are biodegradable, and part of their environmental impact comes from the high oxygen demand resulting from their easy biodegradability. These wastes place a severe stress on receiving streams unless high dilution or extensive treatment is possible.

For these reasons, land spreading or irrigating are the preferred methods of disposal where land is available. In a recent survey, 41 percent of the vegetable processing plants and 37 percent of the fruit processing plants used land disposal.[10)] An application rate of 10 to 20 tons/acre/yr (22 to 45 tonnes/ha/yr) for residuals was reported. These wastes do not contain elements considered as contaminants to

Table 36. EXAMPLE OF POLLUTANTS WHICH MAY BE PRESENT IN INDUSTRIAL WASTE STREAMS AND RESIDUES. [9]

Alkalinity	Aluminum
BOD	Boron
COD	Cadmium
TS	Chromium
TDS	Cobalt
TSS	Copper
Ammonia	Iron
Nitrate	Lead
Phosphorus	Magnesium
Turbidity	Manganese
Fecal Coliform	Mercury
Acidity	Nickel
Hardness, Total	Selenium
Sulfate	Sodium
Sulfite	Vanadium
Bromide	Zinc
Chloride	Oil and Grease
Fluoride	Phenols
	Polychlorinated biphenyls
	Surfactants
	Algicides
	Chlorine
	Organics specific to organic synthesis

either soils or ground water. As long as loading rates do not create aesthetic problems, they can be considered safe.

Textile Industry - [9,11]

Waste production in the textile industry includes organic and inorganic wet chemicals and purely dry products. There are four types of textile products -- animal, vegetable, regenerated, and synthetic.

Animal fibers create wastes of high biochemical oxygen demand. These are amenable to biological treatment, and the residues can be handled like municipal sludge. The most difficult problem with these wastes is the high grease content, and in some cases, residual hair which resists microbial decomposition.

Of the vegetable fibers, cotton is predominant. An example of a composite waste from an integrated cotton textile plant consists of: starches, dextrins, gums, glucose, waxes, pectins, alcohols, fatty acids, acetic acid, soaps, detergents, sodium hydroxide, carbonate, sulfide, sulfate, chloride, dyes and pigments, carboxymethyl cellulose, gelatine, dye carriers (phenols and benzoic acid), peroxide, and chlorine bleach compounds.

Waste treatment usually terminates with lagoons. The sludges have been used for soil conditioning, but costs of transport and application exceed the benefits derived. When land spreading is of economic advantage to the industry, it will probably increase in popularity.

Petroleum Refining Industry - [9,12]

Materials typically found in refinery residuals and their sources are listed in Table 37. As can be surmised by noting materials on the list, the sludges pose a problem from both physical and chemical points of view. There are four general types of sludges: oily, oil-free, chemical, and biotreatment.

Oily sludges as they are removed from oil-water separators, tank bottoms, etc., consist of about 97 percent water, one percent oil, and 2 percent solids. Thickening is required before final disposal. Landfilling and incineration are popular disposal methods, and some refineries also use land spreading. [13] The aerobic soil environment with a diverse microbial population is capable of assimilating wastes with these characteristics.

Table 37. TYPICAL REFINERY SOLID WASTE INVENTORY ASSOCIATED WITH WASTE-WATER TREATMENT.[12)]

Type	Major Constituents	Source
Slop Oils, Oily Solids, Emulsions	Oils, Sand, Catalyst Fines, Coke Fines	Gravity Separators Skimmings and Sludges, Contaminated Storm Water Skimmings, Dissolved Air Flotation Unit Skimmings and Sludges
Water Treating Sludges	Lime, Alum, Clays, Sand, Polyelectrolytes	Water Treatment Plant
Biological Sludges	Microorganisms, Organics	Activated Sludge Plant, Aerated Lagoon Systems, Tertiary Filter Backwash
Absorbents	Oil, Solids, Excelsior	Absorbed Materials from Fibrous Media Coalescers
Sulfur	-	Sour Water Stripping, Sulfur Plant

Oil-free sludges result from water conditioning, and consist of silt, calcium carbonate, magnesium hydroxide, and traces of organic matter. These residues must be segregated from oily residuals because the water conditioning wastes encourage the formation of emulsions. These wastes may be beneficial to land, especially if the soil has a low pH. Because it is cheaper, landfilling is the usual disposal method.

Chemical sludges are produced by processes such as aluminum chloride, sulfuric acid, or hydrofluoric acid alkylation. The sludges are not suitable for land spreading, and are usually disposed of in pits with alkaline wastes.

Biotreatment sludges, consisting mainly of microorganisms, are produced from trickling filters or the activated sludge process. They may be anaerobically digested if low in oil. Like municipal sludges, these may be used in sludge farming. These sludges are more acceptable to regulatory authorities for land spreading than oily sludges.

Heavy metals may be present in refinery sludges as a result of catalyst contamination, metals being released from petroleum feedstock compounds, and refined product and water treatment additives. The impact from land spreading of these metals must be estimated on the same basis as that described for municipal sludges.

<u>Steam Electric Power Industry</u> - 9,14)

The major contaminants produced by the coal-fired steam electric power industry are chemical wastes from fuel, fuel residues, and water treatment additives.

Fly ash is the solid residue usually associated with coal-fired electric generators. The fly ash may be simply piled near the generating facility, or it may be utilized as an asphalt additive. Fly ash has been used as a soil conditioner because of its calcium, magnesium, potassium, and trace element components. A factor limiting soil application rates is the boron content. Many crops have a low tolerance to boron.

Fly ash and bottom ash transport water contains contaminants which are removed to a large extent before the water is discharged as effluent. These substances, therefore, become associated with sludge (Table 38). Many heavy metals are included among the contaminants (Table 39). Land spreading of the sludge must be managed accordingly.

Table 38. TYPICAL CHEMICAL CONSTITUENTS IN ASH TRANSPORT WATER (COAL PLANT). [14)]

Constituent	Fly Ash Transport water (ppm)	Bottom Ash Transport water (ppm)
Silica (SiO_2)	4,824 - 11,040	4,656 - 11,736
Aluminum Oxide (Al_2O_3)	4,176 - 9,768	4,536 - 8,688
Iron Oxide (Fe_2O_3)	1,824 - 7,896	2,808 - 9,600
Calcium Oxide (CaO)	24 - 1,464	2.4 - 1,008
Sulfur Trioxide (SO_3)	2.4 - 1,080	2.4 - 240
Potassium Oxide (K_2O)	288 - 576	408 - 672
Titanium Oxide (TiO_2)	312 - 480	312 - 432
Magnesium Oxide (MgO)	96 - 288	120 - 216
Sodium Oxide (Na_2O)	72 - 192	48 - 192
Phosphorus Pentoxide (P_2O_5)	2.4 - 120	2.4 - 96
Arsenic (As)	< 0.25 - 1.44	< 0.25
Boron (B)	< 2.4	< 2.4 - 7.2

Table 39. TYPICAL HEAVY METAL CONCENTRATIONS IN ASH TRANSPORT WATER (COAL PLANT).[14)]

Constituent	Fly Ash Transport water (ppm)	Bottom Ash Transport water (ppm)
Magnesium	28.8 - 1,200	240 - 720
Titanium	216 - 480	120 - 360
Sodium	28.8 - 288	48 - 192
Cesium	2.4 - 192	0.36 - 19.2
Vanadium	2.76 - 3.6	2.4 - 7.2
Lead	2.64 - 3.6	3.6 - 6
Nickel	2.64 - 3.6	3.6 - 6
Manganese	2.64 - 3.6	3.6 - 4.8
Copper	2.16 - 3.6	0.00024 - 7.2
Chromium	2.16 - 2.88	1.92 - 3.6
Arsenic	0.192 - 2.88	0.048 - 6
Cobalt	0.168 - 2.16	0.48 - 1.92
Selenium	0.6 - 1.8	0.072 - 0.24
Tin	0.00024 - 0.36	0.048 - 0.36
Cadmium	0.00024 - 0.192	0.00024 - 0.36
Mercury	0.0168 - 0.036	0.024 - 0.048

Pulp and Paper Industry - [9]

The most common types of paper produced are kraft, sulfite, neutral sulfite, semichemical, and groundwood. Some textile fibers are grown for use in specialty papers. These include flax, cotton, and jute.

Paper and pulp wastes exert a high biochemical oxygen demand load because they contain organic compounds such as sugars, resins, tannins, and lignins. Inorganic compounds which are in a reduced state such as sulfite also utilize oxygen as they oxidize after being discharged.

Underflow from clarifiers treating pulp and paper mill effluents carries 2 to 12 percent solids. Dewatering is accomplished by filtration, use of polyelectrolytes, or drying beds. Land disposal of the sludge is almost universal. The materials in the sludge are generally beneficial to soils and crops, and the aerobic soil environment minimizes odors.

The volume of waste production and a brief description of its characteristics will also be given for industries which do not apparently employ land spreading as a disposal method. This will allow the reader to assess the future possibility of utilizing land spreading for the disposal of these wastes.

Industries Which Presently Do Not Employ Significant Land Spreading of Sludges

Inorganic Chemicals Manufacturing - [9,15]

The most recent (1972) data for 27 manufactured chemicals reported an annual production of 77 million tons (70 million tonnes). Chromium, zinc, copper, nickel, cadmium, and other heavy metals are frequently discharged. Because these have a low threshold of toxicity, they must be removed from effluent and thus constitute or partially constitute the sludge fraction. Diffuse land disposal is not practiced because the sludges have little value, or may be quite harmful to growing plants.

Plastics and Synthetics Industry - [9,16]

A large amount, almost 17 million lb/yr (7.7 million kg/yr), of plastics and synthetic resins are produced. The water demand is also quite large, amounting to 580 mgd (25 cu m/sec) or 12.6 gal./lb (105 litre/kg) of product. Wastes present in the waste-water stream may be harmful and pass relatively unchanged through municipal treatment plants. (It was reported that 27 percent of the manufacturing plants discharged to municipal treatment plants.)[16]

Sludges from some plastics manufacture can be difficult to stabilize biologically because of the presence of aromatic hydrocarbons or their derivatives. In addition, heavy metals may be present from inorganic salts used in the process or from process catalysts.

Depending upon sludge composition and physical characteristics, these sludges may be disposed of by land spreading. The aerobic soil environment with its variety of microorganisms acts as a medium for biological decomposition of refractory compounds. One plant in Illinois is utilizing land spreading for process sludge. There is no information on the extent to which means of disposal is utilized nationwide.

Iron and Steel Industry - [9,17]

Industrial residues from iron and steel manufacture are significant by their magnitude alone. In 1973, the industry produced 150 million tons (136 million tonnes) of crude steel with an average water use of 21,000 gal./ton (72,000 litre/tonne). It is estimated that the industry uses over 7 billion gpd (26.5 million cu m/day) of cooling water and 3.5 billion gpd (13 million cu m/day) of process water.

Residuals from the steel making processes are low in organic matter and high in inorganic substances. In general, they are not suitable for land spreading.

Organic Chemical Industry - [9]

Over 500,000 products are associated with the organic chemicals industry. Many hundreds of these are produced on a large scale for commercial use. Many thousands more are produced in smaller amounts for organic research, pharmaceuticals, additives for a variety of products, and cosmetics. There is no typical organic industrial sludge. Some sludges are no doubt suitable for land spreading as a disposal method. Others, because of physical or chemical hazards are unsuited. No information is available concerning the extent to which sludge farming with these wastes is practiced.

Metal Finishing Industry - [9,18,19]

The metal finishing industry has three major categories: cleaning and conversion coating, organic coating, and plating and anodizing. The industry consists of thousands of small shops, a lesser number of large independent plants, and many facilities oriented toward specific end products such as the automobile.

Cleaning and conversion coating includes processes with solutions ranging in pH from one to almost 14. Some solutions contain biochemical-oxygen-demand producing compounds, primarily organic solvents. Toxic substances are also included such as chromate and metal cations (Table 40).

Organic finishing wastes are primarily organic solvents with smaller amounts of oil and pigment (Table 40).

Plating and anodizing wastes contain heavy metal ions and cyanide (Table 40). These wastes are difficult to handle because the usual treatment is to precipitate the metals with hydroxide. The resulting sludge is slimy and does not filter or dewater easily. Because the metals are toxic at relatively low concentration, these sludges are not amenable to land spreading.

EXTENT OF THE PROBLEM

In the United States, domestic sludge production amounts to about 5,000,000 dry tons/yr (4,540,000 dry tonnes/yr) as shown on Table 41. Table 42 lists the number of treatment plants by type and the population served. Because sewering is a function of population density, the greatest sludge production comes from the most densely populated regions. These are located along the East, Gulf, and West Coasts, and along the Great Lakes' shores.

Faced with ever-increasing quantities of sludge, rising treatment costs, and decreasing space for storage (lagoons), cities with large populations have been forced to investigate other options for ultimate disposal. One option common to all is land disposal. Chicago and Milwaukee have the largest land disposal programs, but they are dissimilar in approach. Chicago has contracted for the removal of its digested liquid and lagooned sludge for application to agricultural fields and strip-mined areas. Milwaukee heat dries, supplements with nitrogen, bags, and distributes its sludge as fertilizer and soil conditioner under the trade name of Milorganite.

Total municipal sludge production represents 2.3 percent of the total national production of fertilizer nitrogen and 3.1 percent of the production of fertilizer phosphorus. Were all of the sludge spread on land, less than one percent of the agricultural land in the United States would be utilized. At present, about 0.3 percent of the agricultural land is utilized for land spreading.[20)]

Table 40. CLASSIFICATION OF METAL FINISHING WASTES. [9]

Process	Impurities	Origin
Cleaning and conversion coatings	Oil and grease	Cleaning
	Chlorinated solvents	Degreasing
	Hydrocarbon solvents	Diphase and other cleaning
	Alkalies: caustics, carbonates, silicates, and phosphates [a]	Cleaning and phosphating
	Acids: HCl, H_2SO_4, HNO_3, HF, H_3PO_4, and HOAc	Acid dips
	Sludge	Metal hydroxides, metal particles, and buffing compound residues
	Chromates	Chromating solutions
Organic finishing	Solvents	Lacquering
	Oils	Painting
	Sludge	Pigments
Plating and anodizing	Metal ions: Cu, Ni, Zn, Cd, precious metals, etc.	Plating and anodizing
	Simple and complex cyanides	Plating solutions
	Chromates	Plating and anodizing

a) Phosphates of iron, manganese, and zinc.

Table 41. ANNUAL SLUDGE QUANTITIES FOR DISPOSAL.[20]

Physical characteristics	1974 Sludge production (tons)	1985 Sludge production (tons)
All sludge generated on a 100% dry solids basis after approximately 65% of the sludge has been digested.[1]	5,000,000	9,000,000
Assuming 50% of the sludge is dewatered (i.e. 20% solids) and 50% is in the liquid state (i.e. 5% solids) and that 65% of both the liquid and dewatered sludge is digested.[2]	62,500,000	112,500,000

1) Sewage sludge is seldom if ever produced as a totally dry solid. Such an expression is only a theoretical means of presenting sludge data for comparative purposes.

2) Sludge is normally produced as a liquid or semi-solid and as such must be disposed of along with its inherent water content. Based on average values it is assumed that half of the current U.S. sludge is dewatered and half is disposed of as a liquid. Dewatered sludge is more suitable for land disposal in a sanitary landfill and is generally required before incineration. Liquid sludge on the other hand is more conducive to land spreading by tank truck or conventional irrigation equipment.

Table 42. MUNICIPAL WASTE TREATMENT FACILITIES IN THE U.S. [1,21]

Type of treatment	Generation rate a) (lb/cap/day)	1974 Population b) (million)	1974 Sludge quantity (dry tons)	1985 Population c) (million)	1985 Sludge quantity (dry tons)	Percent increase or decrease
I. No treatment (collection only)	-	2.9	-	-	-	-
II. Minor treatment	-	0.8	-	-	-	-
III. Primary treatment						
Primary settling basins	0.125	2.9	66,156	-	-	
Septic tanks	0.081	1.1	16,260	-	-	
Imhoff tanks	0.069	1.9	23,925	-	-	
Mechanically cleaned tanks	0.125	21.6	492,750	-	-	
Hopper bottom tanks	0.125	1.6	36,500	-	-	
Miscellaneous	0.125	8.9	203,031	-	-	
Total (before digestion)			838,622			
Raw 35% d)			293,531			
Digested 65% e)			354,310			
Total primary sludge			647,841			-100
IV. Secondary treatment						
Activated sludge	0.225	47.6	1,954,575	83.5	3,428,718	
Extended aeration	0.090	4.0	65,700	7.1	116,617	
High rate trickling filter	0.200	19.4	708,100	34.2	1,248,300	
Std. rate trickling filter	0.170	10.3	319,557	18.2	564,655	
Sand filter	0.234	0.5	21,352	0.9	38,434	
Effluent to land	-	0.4	-	0.7	-	
Oxidation ponds	-	7.7	-	13.5	-	

Table 42 (continued). MUNICIPAL WASTE TREATMENT FACILITIES IN THE U.S. [1,21]

Type of treatment	Generation rate [a] (lbs/cap/day)	1974 Population [b] (million)	1974 Sludge quantity (dry tons)	1985 Population [c]	1985 Sludge quantity (dry tons)	Percent increase or decrease
IV. Secondary treatment (cont'd.)						
Miscellaneous	0.200	14.4	525,600	25.5	930,750	
Total (before digestion)			3,594,884		5,227,474	
Raw 35% [d]			1,258,209		1,829,615	
Digested 65% [e]			1,518,839		2,208,609	
Total secondary sludge			2,777,048		4,038,224	+45
V. Intermediate treatment				-	-	
Total (before digestion)	0.330	6.8	409,530			
Raw 35% [d]			143,335			
Digested 65% [g]			199,647			
Total intermediate sludge			342,982			-100
VI. Tertiary treatment						
Total (before digestion)	0.687 [f]	2.4	300,906	20.4	2,557,701	
Raw 35% [d]			105,317		895,195	
Digested 65% [g]			146,692		1,246,880	
Total tertiary sludge			252,009		2,142,075	+850
Total U.S. sludge production						
before digestion		155.2	5,143,942	204.0	7,785,175	+ 52
after digestion (65% digestion)		155.2	4,019,880	204.0	6,180,299	+ 52

Table 42 (continued). MUNICIPAL WASTE TREATMENT FACILITIES IN THE U.S. [1,21]

Notes:

a) Hecht, N.L., A.S. Rashidi, D.S. Duvall. Characterization and Utilization of Municipal and Utility Sludges and Ashes. University of Dayton, Research Institute. Dayton, Ohio, Oct. 1973. p. 6. Data based on 100 gal/cap/day and 0.25 lb/cap/day of suspended solids.

b) U.S. Environmental Protection Agency, Office of Water Programs. Storet Municipal Waste Facilities Directory, July 1974. (Unpublished Data).

c) The 1985 Sludge Quantities have been estimated on the assumption that 80% of the total projected U.S. population (i.e., 204.0 million) will be served by municipal sewage treatment facilities by 1985 and further that tertiary treatment will comprise 10% of the total population served with the remaining 90% receiving secondary treatment. U.S. Department of Commerce, Social and Economic Statistics Administration. Population and Economic Activity in the United States and Standard Metropolitan Statistical Areas -- Historical and Projected 1950-2020. July 1972. p. 14.

d) The general assumption is made that 65% of the sewage solids generated will at a minimum be digested and that the remaining 35% will be stabilized by some other means or disposed of as raw sludge. Memorandum. R.E. Fuhrman, To Director, Municipal Waste Water Systems Division, March 15, 1973.

e) Assuming a 50% reduction in volatile solids after digestion (the volatile solids content of the sludge is assumed to be 70%), the net result will be a 35% reduction in the total solids digested.

f) Battelle Northwest Laboratories. Evaluation of Municipal Sewage Treatment Alternatives. Final Report Prepared for: Council on Environmental Quality, Contract EQC 316, Feb. 1974. p. 126-138.

g) Assuming a 45% increase in total solids due to the addition of lime, iron, and aluminum, and a 50% reduction in volatile solids after digestion, the net result will be a 25% reduction in the total solids digested.

Many cities have undertaken land spreading research and demonstration projects of varying size. None of these programs are typical of nationwide sludge disposal practices. Sludge handling and disposal accounts for nearly half of the budget for a municipal sewage treatment plant. From that point of view alone, it can be considered a national problem. The water pollution potential of directly discharging sludge into surface waters is enormous. Therefore, fresh waters are completely restricted for sludge disposal. Ocean disposal is currently being debated, and may be prohibited within the next few years.

Total solid industrial waste residuals produced annually in the United States calculated from data covering the period 1970-74 amounted to 260 million tons/yr (236 million tonnes/yr). A breakdown of industrial residuals production is given in Table 43. Industrial organic chemicals, coal-fired utilities, and industrial inorganic chemicals are the largest contributors of residues and account for over half of the total production. The environmental impact of these residues has not received the attention which the volume of production would warrant. Municipal sludges have attracted far more attention from environmentalists and the public. The major reason is probably that industrial residues are primarily handled "in-house" rather than by municipalities or other tax-supported institutions.

Industrial sludges/residuals production from total processing contrasted with production from pollution control for nine manufacturing categories is shown in Figure 65. The nonferrous metal industry is the only single industry in which pollution-control wastes constitute more than 50 percent of the total production. It is also the industry which is projected to have the greatest increase in pollution-control residuals by 1980 (Figure 66). [23)]

Nonradioactive hazardous waste (pesticides, carcinogens, toxic materials) produced by industry amounts to about 5 percent of the total. The mid-Atlantic, Great Lakes, and Gulf Coast states comprise the area in which 70 percent of the hazardous waste is generated. Table 44 shows some hazardous waste components in selected waste-water streams. Many of these components are heavy metals which are relatively valuable in the form of raw materials. Because metals such as chromium must be imported, economics may soon dictate that recoveries from waste streams be made. Some potential materials recovery from selected industries is shown in Table 45. Removal of metals from waste water and sludge would lead not only to resource conservation, but also would make the treated waste more amenable to land spreading.

Table 43. BREAKDOWN OF TOTAL INDUSTRIAL RESIDUALS.[22]

Industry	Quantity of Sludge/Residual Dry Weight [a)]	
	10^6 tons/yr	10^6 tonnes/yr
Meat and dairy products	1.6	1.4
Food processing	9.5	8.6
Grain mill products	0.2	0.2
Textile mill products	1.2	1.1
Paper and allied products	16.8	15.2
Industrial inorganic chemicals	41	37
Plastics and synthetics	0.4	0.4
Drugs	10.2	9.3
Soaps and detergents	?	-
Paints and allied products	0.4	0.4
Industrial organic chemicals	55	50
Agricultural chemicals	27.5	24.9
Miscellaneous chemical products	0.1	0.1
Petroleum refining	13	12
Rubber and miscellaneous plastics	1.5	1.4
Leather and leather products	0.4	0.4
Glass products	?	-
Cement/clay/pottery products	11.6	10.5
Blast furnaces and steel works	9.2	8.3
Iron and steel foundries	0.7	0.6
Primary smelting/refining nonferrous metals	7.7	7.0
Fabricated metal products	4.0	3.6
Machinery, except electrical	4.0	3.6
Electrical equipment	2.0	1.8
Transportation equipment	1.8	1.6
Coal-fired utilities	43	39
Total:	260	

a) Data values from references dated 1965-1974.

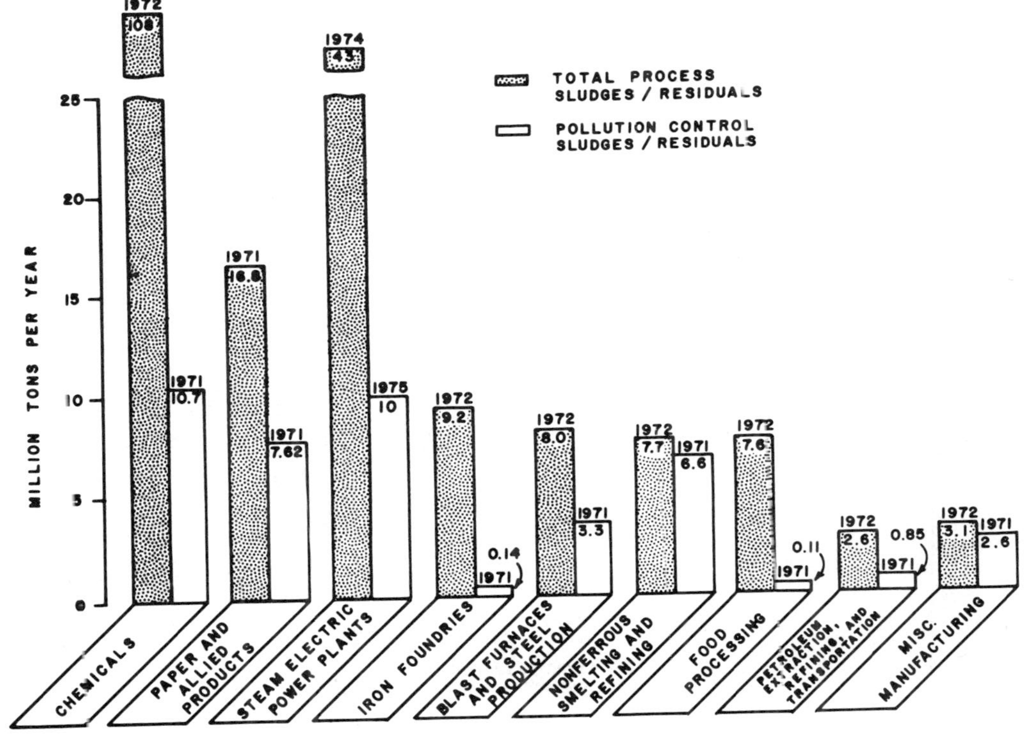

Figure 65. Industrial sludges/residuals, 1968–1972. 23)

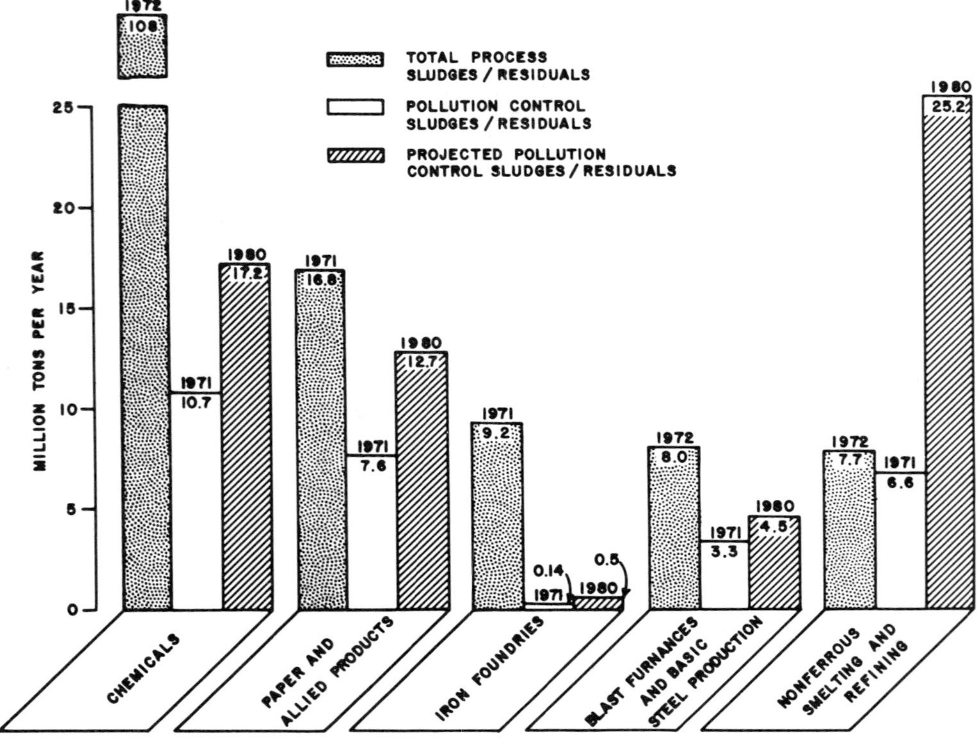

Figure 66. Industrial sludges/residuals. 23)

Table 44. HAZARDOUS COMPONENTS OF WASTE STREAMS.[22)]

Industry category	Component	Percent of waste stream	Percent of National total
Blast furnaces, steel works & iron foundries	Chromium	1.28	86
	Cadmium	0.24	65
	Fluorine	0.95	34
	Zinc	2.62	18
	Phenols	0.007	10
	Cyanide	0.004	8.6
	Berylium	Neg.	0.48
Primary smelting/refining of nonferrous metals	Arsenic	1.0	82
	Lead	0.64	72
	Mercury	0.008	62
	Cadmium	0.59	35
	Fluorine	2.6	21
	Selenium	0.02	20
	Zinc	13.2	20
	Chromium	0.16	2.4
Industrial inorganic chemicals	Mercury	Neg.	36
	Fluorine	0.41	17
	Lead	0.005	2.8
	Zinc	0.002	1.61
Petroleum refining	Phenols	0.33	90
	Lead	0.17	17
	Vanadium	0.03	11
Fabricated metal products	Copper	1.17	100
	Cyanide	5.22	91
	Chromium	2.7	1.5
	Zinc	1.97	0.1

Table 45. POTENTIAL MATERIALS RECOVERY FROM SELECTED INDUSTRIES. [22]

Industry category	Substance	(short ton x 10^3) U.S. production [a]	(short tons x 10^3) Imports	(short ton x 10^3) Tons potentially available
Blast furnaces, steel works, and iron foundries	Chromium	0	1,450	510
	Cadmium	3.65	0.65	96
	Zinc	500	345 [b]	1,050
Primary smelting/refining of nonferrous metals	Lead	552	60 [b]	55
	Arsenic	?	17.3	86
	Zinc	500	345 [b]	1,135
Industrial inorganic chemicals	Mercury	0.684	1.03	0.37
	Lead	552	60 [b]	2.05

a) 1971 data, multiply by 0.908 for tonnes
b) In ore and concentrate

Land and air remain as the sinks for ultimate disposal. Land spreading is a viable option only if it is used and managed wisely. Sufficient land area must also be accessible and readily available, which is not always the case as is commonly found in many of the locations of greatest sludge production. Air can be used if incineration facilities have proper air pollution control devices. There are several major disadvantages with incineration. Incineration requires expensive air pollution monitoring and abatement equipment, the necessity of adding energy for complete combustion, and high capital and operating costs. With energy costs rising, one promising alternative is to utilize a refuse-sludge incineration/electrical generation scheme.

Sludge disposal throughout the country as previously indicated, is a serious problem, and one for which there are no simple, cheap, and easy solutions. Most cities of over 100,000 population have some difficulties with adequate disposal. Land spreading can be an effective solution, but only if the proper ingredients of climate, land availability, topography, and geology are present.

Although it is not a ground-water problem associated with land spreading, the effect of this mode of sludge disposal on crops should be noted here. Excess uptake of nitrogen and phosphorus may occur, but of even more concern is the uptake of heavy metals. Some (chromium, manganese, iron, copper, zinc, molybdenum) are required for proper plant growth. The others may be absorbed into plant tissue, but perform no known metabolic functions. Animal nutrition requires a few more heavy metals -- tin, vanadium, cobalt, zinc, and selenium. All of these metals become toxic to plants or animals at some level specific to that metal. Often, as with selenium, there is very little difference between levels in the diet that cause deficiency and toxicity. Therefore, increasing heavy metals concentrations in plants could have detrimental as well as useful (ameliorating trace element deficiencies) effects.

Limits for heavy metal applications have been suggested. One expresses heavy metal additions as zinc equivalents. The total amount of sludge that may be applied is limited to the addition of zinc equivalents up to 10 percent of the soil's CEC. 24)

$$\text{Total sludge (dry wt. tons/acre)} = \frac{32{,}700 \times \text{CEC}}{\text{ppm Zn} + 2\ (\text{ppm Cu}) + 4\ (\text{ppm Ni})}$$

This limit is formulated to protect against plant toxicity. Additional controls based on cadmium/zinc ratio and maintenance of soil pH above 6.5 have been proposed to reduce plant uptake of heavy metals.

These controls are only designed to reduce metals uptake. More research is needed to determine the relationship between plant uptakes and the impact on animals by injestion. It must be insured that not only are cropland resources protected, but that harmful contaminants are not accumulated in the human food chain.

The effect of diffuse land disposal of sludge on ground water is not documented. A few monitoring wells have been installed at sites for agricultural and strip-mine utilization of sludge in Illinois. To date, no data have been published from these sites.

Over the past few years, EPA has sponsored a survey of municipal waste treatment plants to obtain information concerning liquid sludge land spreading operations. Of the 987 respondents to the survey, 225 are currently land spreading liquid sludge on a routine basis. Applying the survey results on a region-by-region basis to the total population surveyed in Regions II, III, IV, V, and IX, it would appear that about 400 plants are currently land spreading by this technique. Over 68 percent of the plants responding indicated they have been land spreading liquid sludge for less than 10 years. It is estimated 25 percent of the total municipal sludge production is utilized in land application. [20)]

Industrial residuals create the greatest potential for ground-water contamination in areas where net recharge to ground water from precipitation is greatest. The soluble substances in the waste solids are transported to ground water percolating through soil. The Great Lakes and mid-Atlantic industrial regions receive enough precipitation to virtually assure that soluble waste components will be carried to ground water barring geochemical attenuation or geological barriers. Along the Gulf Coast, evapotranspiration rates are higher and reduce the net recharge rate. However, the shallow ground-water table in the region increases the vulnerability of the ground-water system to contamination. In southern California and other southwestern states, recharge from precipitation is only associated with unusually intense storms, or storms of long duration.

Ground-water contamination from land spreading of industrial residuals, unlike municipal sludges, can be classified as a regional problem. Because industries dispose of their residuals primarily with their own methods and on their own land, there are no reliable figures to indicate the extent to which land spreading of residuals is practiced. Land spreading, in contrast to ocean dumping, presumably would be practiced near the site of production, hence the heavily industrialized regions will have the greatest potential for ground-water contamination.

A larger population, industrial growth, and higher degree of waste-water treatment in the future will cause an increase in sludge production. The EPA projects an increase of about 50 percent (Table 42) in annual dry weight production of municipal sludge in the United States in 1985. Some projections for industrial residuals increases are listed in Table 46 for 1977 and 1983.

TECHNOLOGICAL CONSIDERATIONS

The choice of land disposal of the various alternatives for ultimate disposal of sludge is made on the basis of several types of criteria. These may be described as (1) efficacy of the method, (2) environmental impacts, (3) availability of agricultural land or land for reclamation, and (4) economics. Land has the greatest assimilative capacity for sludge. The limit on loading rates is not well established, but the limit is higher than the limits on direct disposal to water or air. Given careful management, the efficacy of land spreading is probably best.

Land spreading of sludge is frequently the method of choice in reducing environmental stresses associated with sludge disposal. Where sludge production occurs in metropolitan areas with little access to useful agricultural land or strip-mined land, land disposal may not be the most viable solution. Under such circumstances, alternative methods such as incineration (with adequate air pollution controls) or heat drying and shipping to more suitable sites may be desirable.

Perhaps the factor which most controls decisions on the ultimate fate of sludge is economics. From the point of view of the municipal waste-water treatment plant operator, land spreading is economically attractive. This is illustrated in Table 47 in which disposal costs for various methods are listed. Lagooning and ocean disposal are the least expensive methods. A more recent site-specific study was reported by Troemper.[26] The costs for land spreading of sludge for corn and soybeean production from 1965 through 1971 were discussed. The net cost per dry weight ton of sludge ranged from $0.97 to $17.05 ($1.07 to $18.78/dry tonne), and averaged $2.46 ($2.71). Most of the fluctuation was caused by the variation in crop yield which affected in-

Table 46. PREDICTIONS OF FUTURE RESIDUALS PRODUCTION BY INDUSTRY.[22]

	Residuals Production, Dry Weight			
	1977		1983	
Industry Category	10^6 tons/yr	10^6 tonnes/yr	10^6 tons/yr	10^6 tonnes/yr
Blast furnaces, steel works, foundries	60	54	62	56
Hazardous fraction	3.3	3.0	3.5	3.2
Industrial inorganic chemicals	66-77	60-70	83-104	75-94
Hazardous fraction	0.4	0.36	0.5	0.45
Coal-fired utilities	117	106	155	141
Oil-fired			9.6	8.7
Hazardous fraction	0.49	0.44	0.53	0.48
Primary smelting and refining of nonferrous metals	13.4	12.2	14.2	12.9
Hazardous fraction	2.5	2.3	2.7	2.4
Paper and allied products	20.5	18.6	22.7	20.6
Hazardous fraction	0		0	
Petroleum refining	-		15	
Hazardous fraction	0.06	0.054	0.08	0.07
Agricultural chemicals without phosphatic fertilizers	2.5	2.3	3.2	2.9
Hazardous fraction	0.08	0.073	0.1	0.09

Table 47. ULTIMATE DISPOSAL COSTS OF SLUDGE INCLUDING TREATMENT COSTS (1968).[20,25]

Method	Capital and operating costs ($/dry ton) Average	Range
Heat drying	50	40 - 55
Incineration		
Wet combustion	42	-
Multiple hearth and fluidized bed	30	10 - 50
Landfilling dewatered sludge	25	10 - 50
Dewatered, for use as soil conditioner, gross cost	25	10 - 50
Landspreading as liquid	15	8 - 50
Lagooning	12	6 - 25
Ocean disposal by barge	12	5 - 25
Ocean disposal by pipeline	11	-

	Total disposal costs, including operating and construction costs (1975 $/dry ton)		
	1 mgd	10 mgd	100 mgd
Land application	127 - 168	53 - 71	57 - 84
Landfill	171 - 208	77 - 116	63 - 98
Incineration	250 - 320	111 - 174	72 - 120
Ocean dumping	376 - 417	93 - 134	56 - 93

Note: Dollars/ton divided by 0.9078 equals dollars/tonne.

come from crop sales. Table 48 lists factors influencing the cost of land disposal of sludge.

Land spreading of industrial sludges/residuals has been essentially limited to biological sludges from the food, petroleum, and paper and pulp industries. More stringent standards for effluents, however, are making land spreading more attractive. First, more sludge is produced as solids are removed from effluent, and second, land application removes the chemical elements from the water route.

Alternatives to land spreading of industrial sludge are landfilling, lagooning, incineration, burial in pits, disposal to marine waters, or deep well injection. There are also contractors who accept industrial sludges for treatment, material recovery, and ultimate disposal by the means listed above. Landfilling and lagooning are the most popular methods in use because of their convenience and low cost. These disposal methods and their potential for ground-water contamination have been discussed in appropriate sections of this report.

Land spreading of most sludges on agricultural land usually requires a minimum of land preparation. For example, berms may be required to contain runoff, but leveling is seldom required. Land spreading on strip-mine spoils frequently requires leveling, terracing, or even bulldozing of large amounts of overburden to fill gullies and trenches. Costs for site preparations of this scale can amount to several thousands of dollars per acre. If leveling is a requirement regardless of revegetation method, sludge irrigation may be economically advantageous. Transportation of sludge to either type of site and equipment for irrigation incur equal expenditures per unit distance and area.

From the point of view of the recipient, the nutrient content of sludge is not high enough to cover even the cost of transportation and distribution. Sludge does have properties which make it an effective soil conditioner. It adds stabilized organic matter which increases the bulk density of soil, increases its moisture holding capacity, cation exchange capacity, and builds soil structure. Home gardeners frequently utilize the entire sludge supply from small municipal treatment plants. Municipalities usually provide distribution services for large-scale commercial farmers.

Siting is a major technological consideration because the impact of sludge on the environment can be maximized or minimized by the choice of spreading site. The physical aspects of siting include some political considerations as well as

Table 48. FACTORS AFFECTING COST OF MUNICIPAL SLUDGE DISPOSAL ON LAND.

Sludge type	Liquid Dewatered Heat dried
Transportation method	Pipeline Unit train Truck
Distribution method	Ridge and furrow Trench Sprinkler irrigation Tanker Spreader
Climate	Year round access Requires storage facility
Site	Acquisition cost Distance from source Access to transport system Agricultural, no land preparation Agricultural, minor land preparation Reclamation, minimum land preparation Reclamation, major land preparation

those more tangible. The proximity to the waste source to a degree determines the transportation cost. Transportation accounts for most of the cost associated with land spreading of sludge. Tank trucks and wagons are the most common means of transport. Only for long distance handling (more than 40 miles) does dewatering pay.

Zoning of the site and adjacent properties may enhance or greatly complicate its utilization. Problems such as the transmission of odors and disease are more imagined than real with the utilization of properly stabilized sludge. However, what people imagine can make real problems for the administrators of the program. Locating a sludge farm in a residential neighborhood can be done (Hanover Park, Illinois), but it requires a good public-relations program ahead of time.

Related to the zoning considerations is the recognition that expansion will likely take place, so land should either be acquired or be available for future growth. Space for storage, roads, and buildings should also be provided.

Siting should be done with concern for minimizing environmental contamination. Thus, the distance to surface water and depth to ground water should be reasonable. Runoff directly entering a stream or lake, and percolate entering ground water without passing through an aerobic, unsaturated zone can degrade the quality of the receiving waters.

Physical aspects of siting can be used to help protect the ground-water supply. The deeper the water table, the greater the chance for renovation of percolate before it enters the ground-water system. A water table several tens of feet below ground surface allows chemical and biological reactions to occur which remove some of the components in percolate.

An adequate area for the volume of waste should be acquired. The greatest protection of ground water is achieved when the application rate is commensurate with crop requirements and soil capacities rather than being determined solely on the amount of sludge requiring disposal.

Geology and ground-water hydraulics should be investigated as part of the procedure for ground-water protection. Subsoil formations can react beneficially with percolate, or they may be essentially inert. Sediments or rock forming the aquifer also may interact beneficially with percolate-enriched ground water. Knowing the direction and rate of ground-water movement, one can predict the path substances entering ground water will take. Qualitative predictions about attenuation may also be possible. A knowledge of existing ground-water quality will help in future interpretation of water-sampling results. Thus, a prior knowledge of the geohydrology can reduce the hazard of ground-water contamination.

Sludge application methods must be chosen with regard to crops as well as the previously described physical characteristics of the site. Spray application after a crop is growing may either damage it, restrict photosynthesis (by caking of solids on leaf surfaces), or contaminate it directly with undesirable constituents.

Use of liquid or dry sludge brings different application techniques and moisture relationships. There are conditions favoring each, and the compatible method should be used.

Application of sludge at a rate commensurate with the crop and soil system, at times compatible with crop demands and favorable climatic conditions, and with suitable means as determined by the overall system, will result in a minimal environmental insult and a maximum benefit.

INSTITUTIONAL ARRANGEMENTS

Formal regulations governing land application of waste-water treatment sludges exist in 21 of the 54 states and territories. In other states, the basic provision of state law which may be applied to land spreading of sludges is the prohibition against polluting waters of the state. A state with only this provision generally has the burden of proof that surface- or ground-water pollution results from sludge disposal.

Pennsylvania has changed this burden with its "potential pollution" statute, which allows the state, where storage, disposal, etc., of materials creates a danger of water pollution, or where regulation of the activity is necessary to avoid such pollution, to require by rule that the activity be conducted only pursuant to a permit issued by the Department of Environmental Resources, or it may make an order regulating the activity.[27] The Department has by rule required that a person or municipality engaged in an activity which includes the impoundment, production, processing, transportation, storage, use, application or disposal of polluting substances take all necessary measures to prevent such substances from reaching waters of the Commonwealth, directly or indirectly.[28] The Department may require a report or plan setting forth the nature of the activity and the nature of the preventative measures taken to comply with the requirement that pollution be prevented. A permit is required for land spreading of sludges. The Department is currently preparing guidelines and regulations on industrial sludges and sewage sludge.

Michigan also uses a general provision, which requires the same procedures in the case of spreading of sludge as for lagoon storage of wastes. A person who wants to dispose of wastes on the ground must file a "new use statement," drill three initial observation wells, and file an application for a ground-water discharge permit, which is reviewed by the Water Resources Commission. The permit allows disposal of specified wastes under a specified monitoring program. The permittee must sample and report each month, and the agency also checks monthly. The Commission establishes requirements industry by industry. 29)

Various types of special laws may apply to spreading of industrial sludges, such as Massachusetts' Hazardous Waste Regulations, requiring approval of the site by the Division of Water Pollution Control; 30) New York's "Industrial Waste Scavenger" Law which requires a license for anyone engaged in the business (among other things) of scavenging or disposing of industrial process waste products including sludges, by which the Department of Environmental Conservation may control place and manner of disposal; 31) and Virginia's law regulating industrial establishments, which requires anyone constructing or operating an establishment from which there is a potential or actual discharge of wastes to state waters, to provide approved facilities for treatment or control. 32)

In some state regulations, only the provisions relating specifically to municipal sludges anticipate that the sludge will be spread on land. For example, Minnesota's regulation prohibiting various sources of pollution states that it is not to be construed as prohibiting land disposal of acceptable organic wastes. 33) Oregon is another state with a specific regulation for spreading of sewage sludge. It requires either that sewage sludge disposal be adequately covered by specific conditions of a Waste Discharge Permit, or that a special permit be obtained based upon detailed plans and specifications. Spreading of septic-tank pumpings and raw sewage sludge is prohibited unless it is specifically determined by the Department of Environmental Quality or state or local health agency that such disposal can be conducted with assured, adequate protection of public health and safety and the environment. If non-digested sludge is spread on land within 1/4 mi (0.4 km) of a residence, community, or public use area, it must be plowed into the ground, buried, or otherwise incorporated into the soil within 5 days after application. Where disposed of in a lagoon and there is a potential for ground-water contamination, monitoring wells are required. [34]

Provisions of state specifications for the operation of municipal waste treatment plants may contain specific provisions affecting land spreading of sludge. The criteria for review of waste-water treatment facilities of the Colorado Department of Health, for instance, allows land spreading of stabilized sludge only. Plans must be submitted containing a detailed description of the process and design data. [35]

Illinois' Design Criteria for Waste Treatment Plants requires that ultimate disposal of sludge wastes not cause air, land, or water pollution, including ground and surface waters. A permit must be obtained from the Division of Land Pollution to dispose of non-liquid sludges, or from the Division of Water Pollution Control to dispose of liquid sludges. Basic feasibility study information is required to be submitted for review prior to submitting the detailed project design. [36] The Illinois regulation by reference incorporated requirements of the Great Lakes-Upper Mississippi River Board of State Sanitary Engineers recommended standards for feasibility studies and design proposals for ground disposal of waste waters. [37]

REFERENCES CITED

1. Metcalf & Eddy, Inc. 1972. Wastewater engineering. McGraw-Hill Book Co., New York.

2. Kardos, L. T. 1965. Soil fixation of plant nutrients. Pages 369-394 *in* F. E. Bear, editor. Chemistry of the soil. Reinhold Publishing Corp., New York.

3. Page, A. L. 1974. Fate and effects of trace elements in sewage sludge when applied to agricultural lands. U. S. Environmental Protection Agency Report EPA-670/2-74-005. 97 pp.

4. National Program Staff, Soil, Water, and Air Sciences. 1974. Factors involved in land application of agricultural and municipal wastes. U. S. Department of Agriculture, Beltsville, Maryland.

5. Chawla, V. K., J. P. Stephenson, and D. Liu. 1974. Biochemical characteristics of digested chemical sewage sludges. Pages 63-94 *in* Proceedings sludge handling and disposal seminar. Conference proceedings No. 2, Environment Canada, Ottawa, Canada.

6. U. S. Environmental Protection Agency. 1973. Wastewater treatment and reuse by land application - Vol. II. U. S. Environmental Protection Agency Report EPA-660/2-73-006b. 249 pp.

7. Duboise, S. M., and others. 1974. Virus migration through soils. Pages 233-240 *in* J. F. Malina, Jr. and B. P. Sagik, editors. Virus survival in water and wastewater systems. Water resources symposium No. 7, University of Texas, Austin, Texas.

8. Wellings, F. M., A. L. Lewis, and C. W. Mountain. 1974. Virus survival following wastewater spray irrigation of sandy soils. Pages 253-260 *in* J. F. Malina, Jr., and B. P. Sagik, editors. Virus survival in water and wastewater systems. Water resources symposium No. 7, University of Texas, Austin, Texas.

9. Gurnham, C. F., editor. 1965. Industrial wastewater control. Academic Press, New York. 476 pp.

10. Environmental Associates, Inc. 1975. Capabilities and costs of technology (study area III) canned and preserved fruits and vegetables industry. Unpublished draft. Corvallis, Oregon.

11. Lockwood Greene Engineers, Inc. 1975. Textile industry technology and costs of wastewater control. Unpublished draft. Spartanburg, South Carolina.

12. Engineering-Science, Inc. 1975. Petroleum refining industry technology and costs of wastewater control. Unpublished draft. Houston, Texas.

13. Weisberg, G. 1975. Assessment of industrial hazardous waste practices; petroleum refining. Unpublished draft. Jacobs Engineering Co., Pasadena, California.

14. Teknekron, Inc. 1975. Capabilities and costs of technology for water pollution control for the steam electric power industry. Unpublished draft. Berkeley, California.

15. Catalytic, Inc. 1975. Capabilities and costs of technology for the inorganic chemicals industry to achieve the requirements and goals of the Federal Water Pollution Control Act Amendments of 1972. Unpublished draft. Philadelphia, Pennsylvania.

16. Procon, Inc. 1975. Technological assessment for the plastics and synthetics industry. Unpublished draft.

17. Arthur G. McKee & Company. 1975. Analysis of effluent control technology and associated costs for the iron and steel industry. Unpublished draft. Cleveland, Ohio.

18. Roulier, M. H. 1975. Personal communication. U. S. Environmental Protection Agency, Cincinnati, Ohio.

19. Lancy Laboratories. 1975. The capabilities and costs of technology associated with the achievement of the requirements and goals of the Federal Water Pollution Control Act as amended, for the metal finishing industry. Unpublished draft. Zelienople, Pennsylvania.

20. Bastran, R. K. and W. A. Whittington. 1976. Municipal sludge management: EPA construction grants program. U. S. Environmental Protection Agency, Office of Water Program Operations, Washington, D. C.

21. Prior, L. A. 1975. Office of Solid Waste Management Programs, U. S. Environmental Protection Agency, Washington, D. C.

22. Lehman, J. P. 1975. Industrial waste disposal overview. Preprint of presentation given at National Conference on Management and Disposal of Residues from the Treatment of Industrial Wastewaters. U. S. Environmental Protection Agency, Washington, D. C.

23. U. S. Environmental Protection Agency, Office of Solid Waste Management. 1975. Unpublished data.

24. Chaney, R. L. 1973. Crop and food chain effects of toxic elements in sludges and effluents. Proc. Joint Conference on Recycling Municipal Sludges and Effluents on Land. Champaign, Illinois. EPA, USDA, NASULGC, Washington, D. C.

25. Burd, R. S. 1968. A study of sludge handling and disposal. U. S. Department of Interior, Water Pollution Control Research Series Publication WP-20-4. 326 pp.

26. Troemper, A. P. 1974. The economics of sludge irrigation. Pages 115-122 in Information Transfer, Inc. Municipal sludge management, proceedings of the national conference on municipal sludge management. Washington, D. C.

27. Pennsylvania Regulations. 1937. Clean streams law. Section 402.

28. Pennsylvania Department of Environmental Resources. Rules and regulations. Chapter 101.

29. Michigan Department of Natural Resources, Water Resources Commission. General rules. Part 21.

30. Massachusetts Division of Water Pollution Control. Hazardous waste regulations.

31. New York Environmental Conservation Law. Section 27-0301, Subpart 75-5.

32. Virginia, Code of. Chapter 3.1. Virginia state water control law.

33. Minnesota Pollution Control Agency. Rules and regulations. WPC 22. Classification of underground waters of the state and standards for waste disposal.

34. Oregon Department of Environmental Quality. Regulations pertaining to solid waste management. K. Special rules pertaining to sludge disposal sites.

35. Colorado Department of Health. 1973. Criteria used in the review of waste water treatment facilities.

36. Illinois Environmental Protection Agency. 1971 revision. Design criteria for waste treatment plants and treatment of sewer overflow. Technical policy 20-24, revised July.

37. Great Lakes-Upper Mississippi River Board of State Sanitary Engineers. 1970. Recommended standards for sewage works. Addendum No. 2, ground disposal of waste waters.

SECTION XI

BRINE DISPOSAL FROM PETROLEUM EXPLORATION AND DEVELOPMENT

SUMMARY

Disposal of brine from oil and gas production activities has been a major cause of ground-water contamination in areas of intense petroleum exploration and development. The principal problem has been related to the long-term practice of discharging to unlined pits, which is now almost universally prohibited. The large number of instances of ground-water contamination from brine disposal stem mainly from days when there was very little regulation of oil exploration and development. Today, the major problem is discharge of saline water from abandoned oil and gas wells rather than disposal of waste brine through injection or secondary recovery wells at active petroleum recovery fields.

The first method of brine disposal was uncontrolled discharge to streams and ditches, and later to evaporation pits. These pits were unlined shallow excavations which could leak salts and minerals into shallow fresh-water zones. Evaporation pits range in area from tens of square feet to a few acres. It is impossible to even roughly estimate the total number, areal extent, and brine input to such sources of contamination, especially since so many have been abandoned over the past decade.

Most oil-field brines today are returned to oil-producing zones or deep saline aquifers through old production wells or brine injection wells for the purpose of water flooding, or just as a disposal method. However, many of these wells are poorly designed for injection, and they offer the opportunity for the salt water to enter fresh-water formations through ruptured or corroded casings.

A tremendous volume of oil-field brine is produced every day. Some states keep detailed records, others none at all. In 1963, the Interstate Oil Compact Commission made a study to determine the production and ultimate fate of brine. Of the 24 states for which data were obtained, almost 24 million bbl (3.84 million cu m) were produced daily that year. About 8 million bbl/day (1.28 million cu m/day) were reinjected for water flood and 9 million bbl/day (1.44 million cu m/day) were reinjected for disposal only. Unlined pits received about 3 million bbl/day (480,000 cu m/day). Brine production in some states has increased significantly since 1963. For example, brine production in California had increased by about 2 million bbl/day (320,000 cu m/day) by 1974.

Enactment of state oil and gas laws has been primarily motivated by recognition of the need for orderly development of oil fields in order to prevent waste of this resource and to stop losses that result from unregulated competition. Although such laws reveal an awareness of the close relationship of petroleum activities to ground-water resources, they are principally concerned with economics of petroleum production and not environmental considerations. In almost every state, disposal of brines to streams, rivers, ditches, and unlined pits is prohibited. Many states allow use of lined evaporation pits and most allow the use of brine injection wells.

DESCRIPTION OF THE PRACTICE

In 1859, the first oil well was drilled in Titusville, Pennsylvania. Since then, oil and gas exploration and production activities have caused countless numbers of ground- and surface-water contamination incidents. Surprisingly to some, the principal contaminant is natural brine rather than oil or gas. The term brine, as used in this section, refers to saline water which is usually associated with oil below ground and is pumped out with the oil.

Thus far, all oil samples analyzed from petroleum reservoirs have been found to contain some water, presumably of indigenous origin.[1)] The amount of water in such reservoirs can be considerable, occupying up to more than 50 percent of the pore space. It is this water or brine and, if the reservoir is not developed correctly, basal salt water beneath the oil that create disposal problems after being brought to the surface with the oil.

The amount of brine produced from a given well depends upon the geologic formation tapped and the well's location, construction, and age. Many wells yield very little brine when first pumped but produce more with time; others yield large quantities of brine initially. In older well fields, as many as 100 bbl (16 cu m) of brine may be pumped for each barrel of oil.

The first method of brine disposal was uncontrolled discharge to streams and ditches; later "evaporation" pits came into use. These pits were shallow excavations into which brine was pumped so that it would evaporate. Such reasoning was faulty as only fresh-water vapor is lost to the atmosphere and the salts and minerals remain. Where the soil was permeable, brine seepage from the pits contaminated ground water. If the soil was impermeable, too much brine would often be added to the pit causing overflow, or in effect, uncontrolled discharge.

With few exceptions, evaporation exceeds precipitation in areas west of, roughly, the eastern borders of the Dakotas, Nebraska, Kansas, Oklahoma, and Texas. In such areas, the liquids in brine pits are concentrated because more water evaporates than is replenished by precipitation. Such highly mineralized fluids have an even greater contamination potential than ordinary brines. In more humid areas, brines might be somewhat diluted by precipitation, but the degree of dilution is not significant.

Evaporation pits range in area from tens of square feet to a few acres. Their design is simple, with no more than a bulldozer needed in most cases, to excavate and build up a berm around the pit. Brine from the pit percolates downward until it reaches the water table, where it moves under the influence of hydraulic gradient to a point of discharge. The volume of brine reaching the water table can be considerable, depending upon conditions. The point of discharge -- a well, spring, or surface-water body -- might be situated miles from the pit.

With a typical ground-water movement rate of one to two ft/day (0.30 to 0.61 m/day), brine seeping from an abandoned pit may eventually contaminate a fresh-water well miles away. The brine moves as a body, undergoing little dilution. In addition, ground water in the area will remain contaminated for years to come. In such a situation, the property owner has little or no recourse.

Because of the lack of data on the use of evaporation pits, it is impossible to estimate in any detail their economic and environmental impact. The High Plains region, a broad alluvial terrace east of the Rocky Mountains extending northward from Texas to South Dakota, may be one of the most vulnerable areas in the country to ground-water contamination. In the southern part of the Plains, ground-water pumpage for irrigation alone amounts to more than 10 percent of the total national pumpage for all uses. [2] Withdrawals have greatly exceeded replenishment, lowering the water table to a level from which it may never recover. With a decline of water in storage, in an area where evaporation exceeds precipitation and brine disposed of in pits becomes more concentrated, the brine contamination potential increases drastically.

Kansas was one of the first states to recognize the problems associated with evaporation pits, and in 1934, its legislature passed several acts. [2] One allowed disposal of brine to subsurface formations containing highly mineralized water. Another act encouraged the oil-field operator to inject brine into the oil-producing zone to increase reservoir yields, a procedure termed water flooding, secondary recovery, or repressuring.

Water flooding is used extensively in most of the oil producing states today, employing brine for the most part but also fresh water in certain locales. Of interest is the fact that in California, there are about 12,000 wells used for secondary recovery; of that total, more than 9,500 use water converted to steam. [3] No mention of the use of steam for injection was found for other states.

Brines may also be disposed of in injection wells with no intention of secondary recovery. Depending upon the regulations of a given state, the wells may have to penetrate only to the shallowest saline-water zone. Such is the case in Florida where oil producing formations are 11,000 to more than 15,000 ft (3,353 to 4,572 m) deep but saline water occurs at relatively shallow depths. Assuming that the injection well is constructed and equipped in the same manner as an oil well, considerable savings are realized by injection to a shallower depth rather than to the oil-producing zone. An 1,800-ft (549-m) brine disposal well would cost about $22,000 as opposed to nearly $450,000 for one 11,100 ft (3,383 m) deep at 1973 prices. [4]

The use of wells to dispose of brine is the best method, considering cost and the environment, that has been developed to date. However, numerous problems are inherent with this method. Successful disposal requires the availability of permeable formations sufficiently thick to accomodate the amount of brine produced. These formations must lie below fresh-water zones and be separated from them by impermeable layers that will prevent the upward migration of the mineralized water, a condition that is not found everywhere.

A disposal well can be an expensive investment that pays no direct dividends (although in Arkansas there are tax incentives applied to disposal wells). Because of the expenses involved, operators of neighboring leases may join together in constructing a disposal system. In addition, where possible, dry holes or abandoned wells are converted to disposal use, thereby eliminating the cost of drilling a well specifically for this purpose. [5)]

The character of brine presents problems in that it is highly corrosive, and the well casing often has to be lined with plastic or an inert material to prevent rapid deterioration, which could result in brines contaminating fresh water. The chemistry of the brine must be compatible with that of the water in the injection zone, for if it is not, chemical precipitates might be formed which could plug the receiving formation, thereby greatly reducing its permeability. In certain instances, enough plugging can take place to render the disposal well useless. Exposure to air often alters the chemistry of a brine to the extent that it can no longer be injected back into the zone from which it was produced without the occurrence of precipitation and concomitant plugging. Unless the brine is properly treated, difficulties with both secondary recovery and disposal are common.

The mechanics by which brine disposal or secondary recovery wells (using brine) can contaminate a fresh-water aquifer are more complex than those for evaporation pits. Improper disposal well construction may be intentional to save money. For example, an operator might drill to only a shallow depth and inject brine directly into a fresh-water formation, even though most states discourage disposal into fresh-water aquifers. A common occurrence is improper sealing or cementing of casing, allowing brine to travel upward in the annular space between the casing and bore hole to contaminate shallow zones. Use of casing that easily corrodes and leaks may similarly contaminate.

Where injection to the shallowest saline water-bearing formation is permitted, care must be taken to insure that there is no hydraulic connection between that formation and overlying ones that contain fresh water. If injection pressures are high and large quantities of brine are disposed of, the poor quality water can be forced upward into fresh-water zones. Detection of such contamination is particularly difficult if there are no regularly monitored wells in the vicinity penetrating the deepest aquifer containing fresh water. The lack of such monitoring may result in the loss of large portions of usable aquifers without anyone's knowledge.

Improperly plugged, abandoned wells and test holes are excellent conduits for the migration of brine. Where either hydrostatic pressure or pressure caused by secondary recovery operations is sufficient, brines are pushed upward in these holes to escape to fresh-water zones or the land surface. Even if the well is plugged at land surface, considerable leakage of brine into fresh-water zones below the plug can occur. Resealing poorly plugged wells after leakage occurs is difficult from a physical as well as monetary standpoint because in many cases the party responsible is not known or cannot be found.

In California, abandoned gas wells which discharge large volumes of salt water into the Tuolumne River are targets of a study to determine if the wells can be capped, sealing off a major source of contamination.[6] The wells, which were drilled in the 1930's without the approval or knowledge of the Division of Oil and Gas, contribute an estimated 110,000 tons/yr (99,792 tonnes/yr) to the dissolved solids loading of the river. The tops of the wells were destroyed when the river gravels were dredged for gold recovery. Proper abandonment procedures were not taken before the wells were destroyed, and it is thought that the wells continue to discharge beneath the gravel tailings.

Dry holes probably seem like an insignificant source of contamination until one examines data on drilling. In 1973, more than 24,000 mi (38,616 km) of hole were drilled onshore, of which nearly 10,000 mi (16,090 km) did not produce oil and were, therefore, abandoned for the most part (some may have been used for brine disposal).[4] Over the years, the total number of miles of dry holes undoubtedly is in the hundred thousands, and although at present dry holes are generally properly plugged, those drilled in the past were commonly left open.

As to abandoned wells, there were over 90,000 fewer producing oil wells in 1973 than in 1965.[4] Since drilling has been going on for more than 100 years, the magnitude of the situation can be understood. Until recently, few states had regulations for proper plugging of abandoned wells. One must conclude that thousands and thousands of miles of vertical conduits for brines exist.

Related Problems

Although brines present the major threat to water quality, other potential contaminants are an inherent part of oil and gas exploration and development activities. These include drilling fluids, chemicals used in treating wells, corrosion inhibitors and other additives, and of course both oil and gas.

Most oil and gas wells are drilled by the rotary method in which drilling fluid is circulated for removal of drilled cuttings from the bottom of the hole and to keep the bottom of the hole and drill bit clean. The fluids are pumped from ground surface down through the drill pipe and out the bit, then returned to the surface through the annulus outside the drill pipe. Various chemicals are added to drilling fluids to cope with different situations, and they are capable of contaminating water and land if the fluids are spilled during drilling or escape into a fresh-water aquifer.

A well is often treated with acid to increase the permeability of the reservoir rocks in order to increase oil recovery or improve fluid injection in disposal or repressuring wells. The acids used include hydrochloric, nitric, sulfuric, hydrofluoric, formic, and acetic acids. The volume of acid used to treat a single well can be as much as several hundred thousand gallons. Soluble compounds such as calcium chloride, sodium sulfate, sodium fluoride, and others are produced as a result of these treatments and, in addition, partially neutralized acids may be left in solution. Contamination can occur when the salt-enriched solutions and any unneutralized acid are withdrawn from the well and are not properly disposed of. Also, the acids are corrosive and can cause pipe failure with possible resultant contamination.

The best corrosion inhibitors used in acid treatment contain arsenic compounds. Other additives are employed to reduce friction, reduce loss, maintain permeability, prevent emulsion formation and avoid precipitation. The corrosion inhibitors and most of the other additives are potential contaminants.

Finally, oil and gas have the potential to contaminate ground water, either through leaky casing or, in the case of oil, through spills. Some crude oils contain mercury in concentrations in excess of 20 ppm. The U. S. Public Health Service limit for mercury in drinking water is 0.002 ppm. Probably of more significance is the fact that the taste and smell of oil can be detected in water with oil concentrations of only one part in 10 million. [7)] Casing leaks which develop opposite fresh-water zones in natural gas wells can also cause contamination of ground water. Hydrogen sulfide, often present in natural gas, gives water the odor of rotten eggs, and it is possible for a person to detect amounts of only two parts in one billion.

CHARACTERISTICS OF CONTAMINANTS

The quality of waters found in oil and gas reservoirs varies widely, often reflecting characteristics of the geologic stratum in which they are found. The water may be fresh, brackish or saline, but it is usually saline. In general, the water chemistry depends upon the chemistry of the ancient sea within which sediments were deposited. The salinity may have been reduced through dilution with fresh water, concentrated by evaporation, or altered by bacterial action. Table 49 shows the composition of some different brines as compared to sea water. Their salinity may reach about 25 percent by weight of solids, the greatest part of them commonly sodium chloride. Lesser amounts of other constituents are also present, but a few instances have been reported where sodium chloride makes up more than 99 percent of the total dissolved solids present. [1)]

The average chloride content for brines is around 50,000 ppm, more than twice that of sea water. The U. S. Public Health Service considers water containing more than 250 ppm of chloride to be unsatisfactory for human consumption if more suitable supplies are available. It is primarily for this reason that brines pose a significant threat to fresh surface and ground waters. One volume of brine (50,000 ppm chloride) can raise the chloride content of almost 200 volumes of fresh water above the acceptable limit, assuming that the fresh water contains no chloride initially, an extremely rare situation. The contamination potential of brine increases with the chloride content of fresh water, which is naturally high in some sections of the country.

Certain brines may contain toxic elements. One, from a well in southwestern Arkansas, was found to have 5.8 ppm of lead and 87 ppm of barium. [8)] The U. S. Public Health Service limits for these two constituents in drinking water are 0.05 ppm and 1.0 ppm, respectively. The fact that most brines contain high concentrations of chloride, which gives water a salty taste when present in concentrations of about 400 to 500 ppm, means that these brines would have to be highly diluted before one would drink them. The large amount of dilution would probably reduce levels of toxic elements to below the acceptable limits. However, this would not be the case for low-chloride brines which are found in a few areas and are used for irrigation. These brines should be analyzed for a wide range of constituents rather than just a few, to determine whether they contain toxic contaminants.

Brines may contain certain elements in sufficient concentration to make extraction economically attractive. In the Michigan Basin and southern Arkansas, companies recover selected minerals from brines. The W. R. Grace Company has developed a process for manufacturing fertilizer from sea water; certain brines might be substituted for sea water in the process. [9)]

Table 50 shows the approximate concentrations of recoverable

Table 49. ANALYSES OF OIL-FIELD BRINES (in ppm). [1)]

Field	Formation	Chloride	Sulfate	Carbonate	Bicarbonate	Sodium	Calcium	Magnesium	Total
Kawkawlin, Michigan	Dundee limestone	161,200	155	-	60	66,280	25,740	4,670	258,105
Seminole, Oklahoma	Wilcox sand	89,990	515	-	65	44,020	9,460	1,990	146,040
Glenn, Oklahoma	Arbuckle limestone	101,715	120	-	60	50,345	10,160	2,120	164,520
Nikkel, Kansas	Hunton limestone	76,797	207	-	61	40,284	5,440	1,790	124,579
Yates, Texas	San Andres dolomite	2,518	2,135	-	-	1,624	587	288	7,445
Monument, New Mexico	Grayburg limestone	6,630	160	-	1,740	3,735	515	365	13,145
Shelby, Montana	Madison limestone	1,179	659	71	1,270	1,322	143	66	3,388
Frannie Dome, Wyoming	Tensleep sand	27	2,303	0	691	51	760	240	4,022
Grass Creek, Wyoming	Frontier sand	256	6	1,211	-	1,087	5	2	2,565
Edison, California	Upper Duff sand	79	4	29	648	299	17	1	962
Ventura Avenue, California	Pico Repetto sand	14,212	59	-	1,846	8,607	729	242	26,091
Bay City, Michigan	Salina dolomite	403,207	0	-	1,208	21	206,300	7,300	642,798
Ocean waters, (mean)	-	19,410	2,700	70	-	10,710	420	1,300	35,000

Table 50. AMOUNT OF ELEMENT PER 1 MILLION LB. BRINE NECESSARY TO PRODUCE CORRESPONDING CHEMICAL PRODUCT WORTH $250.[9]

Element	Concentration, ppm.	Product
Sodium	50,000	sodium chloride
Lithium	170	lithium chloride
Magnesium	8,000	magnesium chloride
Calcium	11,000	calcium chloride
Strontium	4,000	strontium chloride
Boron	1,400	sodium borate
Bromine	1,700	bromine
Iodine	250	iodine
Sulfur	5,300	sodium sulfate

minerals in one million pounds of brine which would have a market value of $250. [9] The data in this table are not absolute because factors such as product demand, ease of recovery, and proximity to markets affect price.

EXTENT OF THE PROBLEM

A tremendous volume of brine is produced every day. Some states keep detailed records; others none at all. The Interstate Oil Compact Commission made a study to determine how much brine was produced in 1963 and what happened to it after production (see Table 51). Attempts made to secure more recent data were not entirely successful as several states still keep no records. However, it was found that brine production in certain states had increased significantly from 1963 to 1974. For example, California now produces about 2 million bbl/day (320,000 cu m/day) more. Updating and extrapolating of the 1963 data indicate that the present production of brine is about 30 million bbl/day (4.8 million cu m/day). This estimate is made by assuming that brine production is equal to four times the amount of oil production (80 percent of the fluid pumped is brine). Average onshore United States oil production was 7,284,000 bbl/day (1.2 million cu m/day) in 1974. [4]

Accepting a daily brine production of 30 million bbl (4.8 million cu m) as reasonable, brine production on an annual basis totals almost 11 billion bbl (1.8 billion cu m) or 460 billion gal. (1.8 billion cu m). If the brine has a contamination potential of only 200 times, one year's brine production could theoretically contaminate 92 trillion gal. (0.35 trillion cu m) of fresh water. Obviously, such extensive contamination is not happening, but the preceding figures do show that brine disposal is a potentially significant threat to fresh ground-water resources in petroleum areas.

Disposal of oil-field brines is not a nationwide problem; only 30 states had producing oil wells in 1973, and oil production averaged more than 5,000 bbl/day (800 cu m/day) in only 24 states in 1974. [4] These 24 states are shown on Figure 67.

No more than a general correlation can be made between the volume of brine produced and ground-water contamination. Too many factors are involved to make a direct correlation. These include well location, geology, occurrence and quality of native ground water, and regulations and their enforcement.

Table 51. DISPOSAL OF PRODUCED SALT WATER, 1963 (barrels per day). [1)]

State	Current volumes produced	INJECTION	
		For water flood	For disposal only
Alabama	2,493		1,397
Alaska	1,128		1,128
Arizona	100		
Arkansas	539,132	89,082	340,734
California	2,740,850	445,768	208,665
Colorado	202,194	131,500	5,000
Florida (7)	600		600
Georgia (7)			
Idaho (7)			
Illinois	876,712		
Indiana	81,797	50,960	14,724
Kansas	5,011,400	800,000	4,200,000
Kentucky	123,287	73,973	35,616
Louisiana	2,785,000	184,000	1,762,000
Maryland (7)			
Michigan	149,587		147,849
Mississippi	340,079	40,000	203,836
Missouri (7)			
Montana	50,000	10,000	31,400
Nebraska	121,907	17,329	7,567
Nevada (7)			
New Mexico	356,624	55,176	165,423
New York (7)			
North Dakota	31,000	23,500	
Ohio (7)			
Oklahoma	3,751,911	3,160,577	583,280
Oregon (7)			
Pennsylvania	191,780		
South Dakota	68		
Tennessee (7)			
Texas	6,127,671	2,736,755	1,472,954
Utah	81,634	2,981	
Virginia (7)			
Washington (7)			
West Virginia	115,068		
Wyoming (7)			
	23,682,022	7,821,601	9,182,173

Table 51 (continued). DISPOSAL OF PRODUCED SALT WATER, 1963 (barrels per day).[1)]

State	Impervious pits	Unlined pits	Streams and rivers	Other methods
Alabama		493	603	
Alaska				
Arizona		100		
Arkansas		7,444	101,871	
California	3,127	399,933	501	(1) 825,410
				(2) 99,168
				(3) 758,277
Colorado	70	65,624		
Florida (7)				
Georgia (7)				
Idaho (7)				
Illinois				(4) 876,712
Indiana		15,132		(5) 982
Kansas	1,800	9,600		
Kentucky	2,740	5,480	5,480	
Louisiana		698,000		(1) 141,000
Maryland (7)				
Michigan		982		(6) 756
Mississippi	8,219	74,329		(5) 13,699
Missouri (7)				
Montana		8,600		
Nebraska		97,011		
Nevada (7)				
New Mexico		136,025		
New York (7)				
North Dakota		7,500		
Ohio (7)				
Oklahoma	5,370	2,685		
Oregon (7)				
Pennsylvania			191,780	
South Dakota		68		
Tennessee (7)				
Texas		1,262,719	615,566	(5) 39,677
Utah		4,862		(3) 73,790
Virginia (7)				
Washington (7)				
West Virginia			115,068	
Wyoming (7)				
	21,326	2,796,587	1,030,869	2,829,471

(1) To nonpotable water body
(2) Disposal at sites approved by regulatory agency
(3) Fresh water used for irrigation and livestock
(4) Unknown disposition
(5) Unaccounted
(6) Lease operations and dust control on county roads
(7) Either/or no report, no production, no information available

Note: bbl/day equals .16 cu m/day.

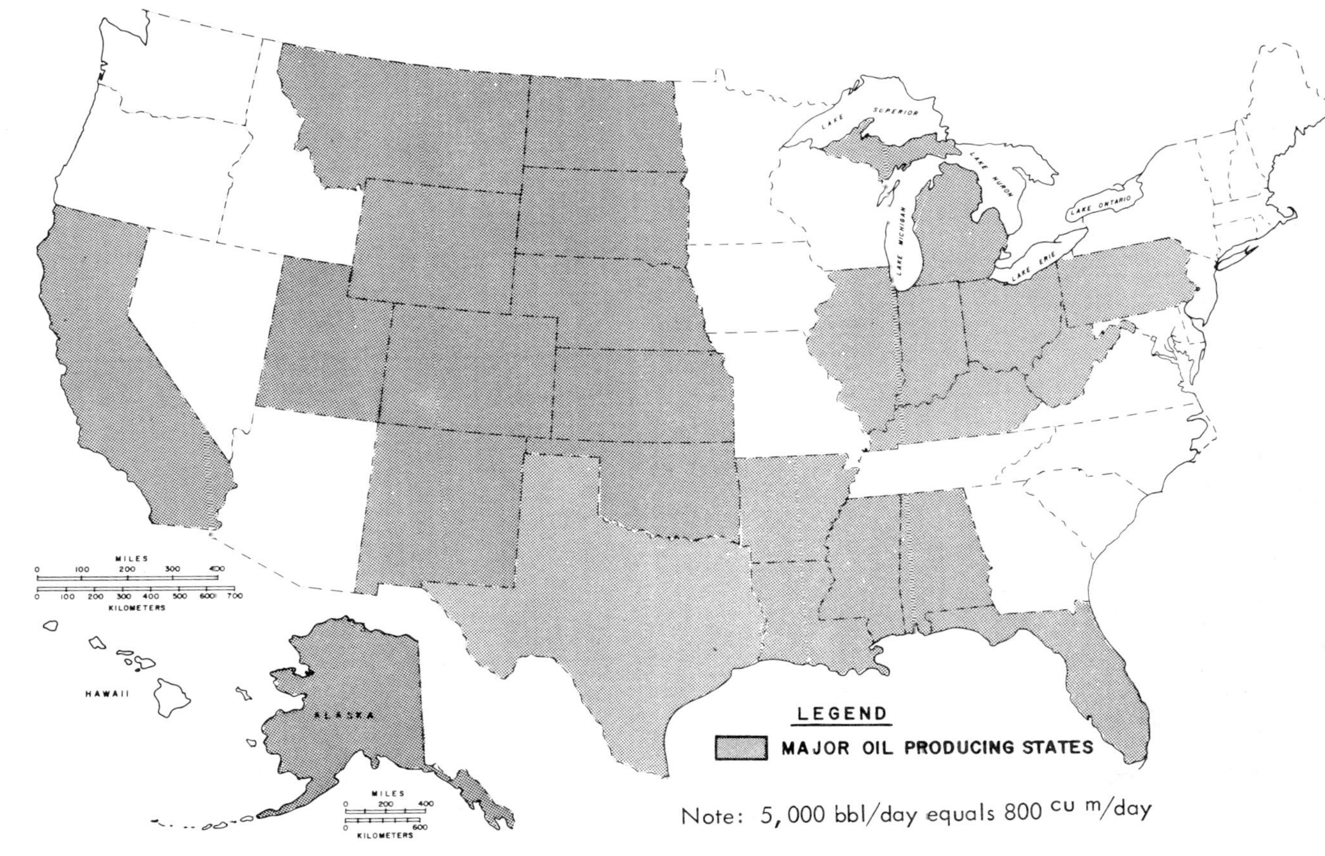

Figure 67. Major oil producing states (more than 5,000 barrels per day in 1974). 4)

Brine-producing wells can be located in uninhabited areas, generating large volumes of brine that contaminate fresh water but affect no one. They can be located in populated areas, produce relatively small amounts of brine, and contaminate an aquifer serving an entire town -- an event which has taken place and is discussed later. Finally, they can be situated where no usable ground water exists. Much brine can be produced in a state that has strictly enforced regulations for proper brine disposal, and no contamination will occur. Conversely, another state could have a small brine output but suffer widespread contamination because disposal is unregulated.

The preceding examples may be more or less the exception rather than the rule. Logic dictates that if more brine is produced, more problems will result. Ground-water contamination by oil-field brines has been documented in at least the following states:

Alabama	Kansas	Ohio
Arkansas	Kentucky	Oklahoma
California	Michigan	Pennsylvania
Colorado	Mississippi	Texas
Georgia	New Mexico	West Virginia
Illinois	New York	

It is most unlikely that contamination has not taken place at one time or another in most or all of the states that have produced or are producing oil.

Case Histories

Case histories of ground-water and surface-water contamination by oil-field brines abound. Four typical ground-water contamination cases are briefly described below.

Arkansas - 8)

In 1967, in southwestern Arkansas, a farmer noted that his irrigation well was producing salty water, and he brought this fact to the attention of state agencies. The well, which tapped an alluvial aquifer, was capable of producing 1,000 gpm (63 litre/sec). It had to be shut down after the chloride concentration reached 1,100 ppm.

Various state agencies jointly conducted an investigation to determine the source of contamination. The investigation consisted of augering holes through the alluvium and sampling the sand and water mixture that was brought to the surface. Results suggested that an unlined brine disposal pit was the source of contamination. It was later found that a leaky disposal well, subsequently repaired, also contributed. At the same time, the U. S. Geological Survey was making a reconnaissance study, obtaining samples from domestic and irrigation wells and other test holes over a 20-sq mi (52-sq km) area. This study found two other contaminated areas where chloride exceeded 500 ppm (the normal range is from 7 to about 50 ppm). All three areas are in or just downgradient of producing oil fields. A later study, authorized by EPA, found that at least four more contaminated areas existed in the county but focused on the contaminated aquifer tapped by the farmer's well and also on whether the aquifer could be rehabilitated.

The extent of the aquifer found to be contaminated by brine was about one sq mi (2.6 sq km). It was calculated that the salty water was moving toward the Red River, 4.5 mi (7.2 km) away and the discharge point for all ground water in the area, and would reach the river in approximately 250 years. Thus, ground water occupying at least one sq mi (2.6 sq km) over the path of travel will remain contaminated for at least 250 years. The opinion was expressed that the entire 4.5 by one mile area will remain contaminated for a much longer period due to dispersion, adsorption, and nonhomogeneous characteristics of the aquifer.

Monetary losses already incurred include the loss of an irrigation well valued at $4,000 and the partial loss of one year's rice crop, worth $36,000 on 120 acres (48 ha) for which irrigation water was a necessity. Assuming that only one sq mi (2.6 sq km) of irrigable land is removed from production, future annual losses would amount to $96,000 for rice, $22,400 for cotton and $12,800 for soybeans. Had a municipal ground-water supply of one mgd (3.8 million l/day) become contaminated and the town been forced to construct a surface supply, the added yearly cost would have been an estimated $73,000.

Obviously, it would be desirable to remedy the situation. Initially four rehabilitation methods appeared feasible: containment, accelerated discharge, use, and deep-well disposal.

Instead of allowing the salty water to spread and move downstream, contaminating more than four times the area already affected, it might be contained by constructing an impermeable underground wall across three sides of the area. The wall would be made by injecting bentonite (a clay) into the ground. The cost of this containment was estimated at about $7,000,000, and there is no documentation on how effective such a system would be.

Two suggested methods of accelerating discharge to the Red River are water drive and direct pumping. Water drive would entail injection of fresh water into the aquifer upgradient of the salty water, causing the salty water to move at a faster rate. This type of system would require the expenditure of nearly $1,300,000. Direct discharge, utilizing four strategically located wells and a pipeline to the river, would be relatively inexpensive (about $180,000) but pollute the river in violation of existing regulations.

Of course, rehabilitation of the aquifer by pumping the water out and putting it to beneficial use would be the ideal solution. Three possible uses for the water were mentioned. Injecting the water for secondary recovery of oil would have been most preferable; unfortunately no operators in the area expressed a need for additional water. This approach has been successful in other areas with threefold benefits in that the aquifer is reclaimed, the contaminated water is used beneficially, and fresh water that might have been used for water flooding becomes available for other uses. Blending the contaminated water with fresh water for irrigation purposes was ruled out, as was desalination, which carried a price tag of $2,000,000.

Deep-well disposal, pumping the shallow salty water out and injecting it into a deeper zone containing saline water, is technically feasible and several different systems were evaluated. Cost estimates for them ranged from $290,000 to $450,000.

In summary, rehabilitation of the contaminated aquifer is technically possible. The most economic and practical methods are pumping into the Red River and deep-well disposal. The report concludes, however, that none of the methods are economically justified at this time.

Oklahoma - [10])

During a recently completed investigation, two areas of widespread salt-water contamination were delineated in the Cimarron Terrace region in northwestern Oklahoma. The terrace extends some 110 mi (177 km) along the north side of the Cimarron River, ranging in width from one to 15 mi (1.6 to 24 km), and is up to 80 ft (24 m) thick. Many towns, an increasing number of irrigated farms, and thousands of households depend upon the terrace aquifers for their water supply.

Two major sources of sodium chloride are found in the study

area. One is natural salt (halite) deposited as lenses in a shale formation which outcrops in part of the area. Although most of the lenses within 600 ft (183 m) of the surface have been dissolved by percolating ground water, salt deposits still exist, e.g., the Little and Big Salt Plains at the northwestern end of the terrace. The Cimarron River acquires most of its salt load while flowing through this area. High sulfate, also common to the river water, is derived from gypsum which overlies the shale.

The second source of sodium chloride is oil and gas exploration and production activities. Many complaints (about 370 by residents of 26 townships) have been filed with the Oklahoma Corporation Commission, the state's oil and gas regulatory agency. The bulk of the complaints originates where oil and gas production takes place and where ground water is more intensively developed.

In the past, the salt water from oil and gas production in the area was disposed of in pits. Some were properly constructed in impervious material or were lined. Others were dug into permeable terrace sands, which permitted ready percolation of discarded brines to the aquifer.

A great number of pits were in use in the study area between 1930 and 1950 when oil production was at its peak and brine disposal was generally uncontrolled. These pits are believed to be the source of most of the salt-water contamination in the terrace. In an area of more than 9 sq mi (23 sq km) southeast of the town of Crescent, ground water, which was once fresh, is now no longer fit for human consumption. Other occurrences are much less extensive. Newer oil and gas fields have caused fewer complaints because more precautions are being taken to prevent contamination. However, it is worth noting that although unlined oil-field brine pits are now prohibited in Oklahoma, several were observed during the course of the investigation.

Salt-water contamination in the terrace deposits was delineated by data review, test drilling, water sampling and analysis, and surface resistivity methods. Sodium/chloride ratios of water samples were used to identify the source of contamination. It was found that residual chloride was being flushed out of the terrace into the Cimarron River by natural ground-water flow. The study concluded that, assuming future ground-water withdrawals do not significantly change the ground-water gradient, approximately 100 years will be required before the salt water is completely flushed out of the aquifer.

Ohio - 11)

In Morrow and Delaware Counties, projects were conducted in two areas to study the effects of ground-water contamination caused by brine disposal in evaporation pits. The investigation focused on determining the source, concentration, areal extent, and probable future movement of the contaminants. Emphasis was placed on detection methods, and electric resistivity was found to be a quite effective monitoring tool in both areas.

In 1961, an oil well was successfully completed in Morrow County, and within the next three years over 2,000 were drilled, more than 600 of which became producers. The amount of crude oil extracted was greater than 25 million bbl (4 million cu m), and it has been estimated that nearly as much brine was also produced and disposed of, for the most part through evaporation pits. In Morrow County there were numerous pits; in Delaware County there were only a few.

Ground-water contamination was first noted in Morrow County in 1964 when the village of Cardington was forced to abandon its municipal well because of an influx of contaminants that most likely originated from an evaporation pit less than 150 ft (46 m) from the well. The total number of wells in the county affected by brines was not precisely determined. Six wells used for potable water supplies were found to have an average chloride concentration of greater than 250 ppm over a 2-year period.

The areal extent of contamination in Morrow County was large, totaling about 13 sq mi (34 sq km), although only four areas were located where contamination levels exceeded U. S. Public Health Service drinking water standards.

In Delaware County, ground water at one site near several disposal pits became extremely contaminated. At times chloride concentrations in the ground water exceeded those of the brines. Because of the geologic and hydrologic conditions of the site, contamination was confined to some 20 acres (8.1 ha) along the Olentangy River. Stream pollution was significant, and it was established that several tons of chloride daily entered the Olentangy River from ground-water discharge. The chloride became appreciably diluted downstream, and no public health problem had occurred or was expected to occur.

The potential for contamination of ground and surface waters by oil-field brines was finally recognized by authorities.

At present, under new laws and regulations enforced by the Ohio Division of Oil and Gas, the only legal method of disposal is through injection wells. Unfortunately, it will be tens of years before brine-polluted aquifers in the two areas are flushed out.

Texas - 12)

Complaints of ground-water contamination have been reported in northwestern Garza County since 1956. In 1962, the Texas Water Commission (now the Texas Water Development Board) investigated the situation.

About 140 sq mi (363 sq km) in the western part of Garza County are within the southern High Plains of Texas. In the county, oil production and agriculture provide the principal sources of income. Most of the cultivated land is found in the Plains' area where about 37,000 acres (15,000 ha) produce principally cotton and grain sorghums. Some 20,000 of those acres are irrigated with ground water.

The Ogallala Formation is the principal aquifer in western Garza County and furnishes practically all of the water used for irrigation. However, it contains only a moderate quantity of water and withdrawals of ground water exceed replenishment. It was estimated in 1963 that only 200,000 to 250,000 acre-ft (0.24 to 0.30 cu km) of water remain in storage in the aquifer in the county, of which only a portion is economically recoverable for irrigation. The water table has been lowered as much as 35 ft (11 m) locally, and virtually all natural discharge has ceased.

During the Texas Water Commission's investigation, water samples from 41 wells were collected and analyzed. It was found that 18 wells (12 irrigation and 6 domestic) were contaminated. Brine disposal in open pits was primarily responsible. Because of the almost total lack of natural discharge in the area, the contaminant cannot be removed from the aquifer except by pumpage, and additional supply wells will probably be affected.

Precise calculation of the environmental impact of brine disposal on this area is not possible. More than 400 volumes of fresh water are required to dilute one volume of brine so that the resultant chloride concentration is 250 ppm. During a one-year period, 100,000 bbl (16,000 cu m) of brine may have entered the aquifer; more than 5,000 acre-ft (0.01 cu km) of fresh water would be required to dilute that amount. A large portion of the available fresh water left in storage in the aquifer might be required for dilution of the total volume of brine which has entered the aquifer over the years.

The preceding case histories, only four out of the multitude that exist, provide some insight into how much damage improper brine disposal practices have done to aquifers. (Areas are measured in square miles, time required for natural rehabilitation ranges from tens to hundreds of years, and economic losses can amount to millions of dollars.)

In 1966, some 100 specialists of the Texas Railroad Commission investigated over 23,000 cases of surface-water and ground-water contamination caused by oil-field brines. [13] Other states that produce large amounts of oil (and brine) might similarly be expected to have a large number of problems. On a nationwide basis, only in those areas where ground water has been developed would ground-water contamination likely be noted. There are many other areas where surface-pit disposal of brine has taken place and aquifers have been contaminated, but lack of ground-water development precludes knowledge of the situation.

Looking into the future, one can expect a continuing decrease in the total amount of brine that reaches aquifers from unlined disposal pits as widespread regulations prohibit them. However, lined pits, secondary recovery and disposal wells, and abandoned wells and test holes not properly plugged, all of which can leak, will continue to contaminate fresh water. The existence of already contaminated aquifers will appear as a more and more critical problem as the demand for water supplies increases and wells are drilled in previously undeveloped areas.

TECHNOLOGICAL CONSIDERATIONS

Two very effective methods that exist for brine disposal, excluding use for chemical recovery which would presuppose that no residual wastes remain, are lined evaporation pits and injection wells. With proper siting, design, construction, and monitoring, these methods virtually assure that no ground-water contamination will take place.

Imperviously lined pits must be situated in areas where evaporation rates exceed precipitation rates, otherwise the pit will eventually fill up and overflow. In some areas, the pits must be deep enough to avoid overflow in the event of abnormally heavy rainfall. For efficient operation of the system, the brine or water surface must be free of oil, which retards evaporation. Under some conditions, one gal.

(3.8 litres) of oil can cover a 25-acre (10-ha) water body. 14) Spraying can be employed to increase evaporation rates. The residue in evaporation pits is a salt bed which, as long as the liner remains intact, has no potential to contaminate. If the bottom of the pit is always above the water table and the salt has dried out and been sealed over, the condition of the liner is of little concern. Water cannot reach the salt, dissolve it, and percolate downward to contaminate an aquifer.

A more universally applicable method is disposal into deep formations by means of injection wells. The receiving zone may be at a greater or lesser depth than the producing formation. If brines are reinjected into the producing zone, secondary recovery of oil may be possible, putting the brine to beneficial use.

The desirable characteristics for a waste injection formation are: an injection zone with adequate permeability and thickness; an areal extent sufficient to provide storage at safe injection pressures; and sufficiently impermeable overlying layers separating it from fresh-water zones. 14) Thus, geological factors impose a considerable influence on brine disposal wells. In some areas, formations are found that are perfectly suited as injection horizons.

In Kansas, the Arbuckle Formation, a siliceous limestone, takes immense volumes of water by gravity feed, and no injection pressure is needed. 15) On the other hand, some formations take little water even under extremely high injection pressures.

A major problem associated with this method of disposal is natural plugging of the well and formation commonly caused by solids, oil, muds, salt precipitates, sulfur and bacteria in the brine. 15) Corrosive products from the injection system may also contribute. In order to minimize plugging, the brines must usually be treated to make them compatible with the receiving formation, even where fluids are reinjected into the same formation, because chemical changes can occur between production and injection. Both open and closed disposal systems are in general use. In the open system, oxygen contact is prevented, but pressure and temperature variations which take place when the fluid is brought to the surface can alter its chemistry. However, with the closed system, the only treatment often needed is removal of oil and suspended solids and on occasion, the addition of biocides to prevent bacterial clogging of the formation. For the open system, treatment generally entails the removal of dissolved gases, suspended solids, some dissolved substances, and the addition of biocides.

The capacity of an injection well will likely still decrease with time despite treatment, and certain remedial measures can be taken to increase capacity. These measures include acidizing, hydraulic fracturing, backwashing, and the use of chlorine and other chemicals. [14]

Proper siting, design, and construction of injection wells are necessary for protection of aquifers. In addition, monitoring is needed to detect accidental discharge of brine to fresh-water zones through leaky casing. Injection pressure at the well head is a good indicator because sudden declines may indicate ruptured casing. When a significant pressure decrease is noted, the well can be examined and tested to determine the reason for change. Repairs, if needed, are made as soon as possible if the operator is conscientious, keeping the amount of brine escaping to a minimum.

The use of brine injection wells is widespread and will increase as regulations prohibiting other methods of disposal are promulgated and enforced. Currently, most of the wells inject brine for secondary recovery purposes rather than for disposal only. For example, in Texas, 27,749 wells were used for secondary recovery and 3,759 for disposal in 1974. [16] For California, the figures were 11,700 and 390, respectively. [3] There are probably 50,000 to 60,000 or more wells in the nation presently injecting brine.

The cost factors involved in conversion from other disposal methods to the use of injection wells are extremely variable. Abandoned oil and gas wells or exploratory holes that tap permeable formations and are located near the site of brine production can often be used. In this instance, expenses are very low, assuming that only minor treatment is needed, few well modifications are made, and only low injection pressures are required. At the other extreme, the capital expenditures and operating costs, related to newly drilled wells at great distances from the production wells with construction of lengthy pipelines and including extensive treatment and high pressure injection, can be prohibitive. Under moderate conditions, with some pre-injection treatment and amortization of the initial capital investment, disposal costs run from one to 2 cents per barrel. [15]

Because of the many inconsistencies associated with deep-well disposal, computer programs have been developed to predict relationships between physical conditions and injection costs, knowing input variables. One of the programs was formulated by Haynes and Grubbs, [17] and although keyed to costs dictated by the geology of Alabama, it is applicable to any area by modification of tables pertaining to drilling costs.

INSTITUTIONAL ARRANGEMENTS

In those states that have oil and gas production, measures to prevent ground- and surface-water contamination are an integral part of the law administered by the state oil and gas regulatory agency. Enactment of oil and gas laws has been primarily motivated by recognition of oil and gas producers of the need for orderly development of oil fields in order to prevent waste of the resource and to stop economic losses that result from unregulated competition. However, such laws reveal an awareness of the close relationship of petroleum activities to ground-water resources. They contain general and specific provisions prohibiting contamination of water, some of which are discussed herein. A summary of pertinent regulations can be found in a bulletin put out by the American Petroleum Institute. 18)

To generalize, in almost every state, disposal to streams, rivers, ditches and unlined pits is prohibited. Many states allow use of lined evaporation pits and most permit the use of brine injection wells. As mentioned previously, Arkansas encourages disposal wells through tax incentives.

It appears that Pennsylvania is the only state allowing discharge of brines to streams and rivers as an approved method of disposal. (It does require that all oils and residues are removed from the brine prior to discharge.) The logic behind condoning discharge in this manner is that many streams have sufficient dilutive capacity so that serious water-quality degradation does not occur.

One might argue that deep-well disposal would still be preferable to stream discharge in Pennsylvania. However, the situation here is somewhat unusual. In the northwestern part of the state, where oil and gas have been extensively developed, there are thousands of abandoned wells and test holes, few of which have been properly plugged. It is felt that injection would force brine up these conduits, to flow out on the land surface or to contaminate shallow aquifers. 19) The brine would be at full strength and possibly contain residual oil. Furthermore, its control might take years, during which damage would occur.

A survey made by the Interstate Oil Compact Commission (IOCC) found that all but a few states require installation of surface casing to protect all known fresh-water aquifers penetrated by oil- and gas-drilling operations, and this casing must be cemented to the surface. [1] (Cement is placed in the space between the outside of the casing and the sides of the bore hole to prevent fluids or gases from moving up or down outside the casing.) This type of regulation applies to wells used for exploration, production, disposal of waste, and secondary recovery.

According to the IOCC survey, most states require that tests be conducted to determine the adequacy of the cement job; a minority require that such tests be witnessed by state inspectors. [1]

Nearly all states insist that when the well is abandoned, a cement plug be emplaced, starting at the bottom of the surface pipe. The typical length of such a plug is 50 ft (15 m) but Florida requires a 200 ft (60 m) plug. [1] Almost all states also stipulate that a cement plug is to be set at the top of the surface pipe. Regulations usually require notification of intent to abandon a well, so that the administering agency may give instructions for plugging.

States generally allow use of earthen pits or lagoons for storage or disposal of brine produced in connection with oil or gas; however, some of these states severely restrict their use. Those that allow pits may require that they be impermeable, or that they be used only in limited situations, or that they not be used at all where they might cause contamination.

For typical examples of the various types of restrictions and controls, the following state regulation summaries are presented. Colorado has a detailed regulation on retention pits for the storage of produced water: [18]

> "Pits shall be kept free of surface accumulations of oil;
>
> Each operator shall file an Affidavit of Condition of Operator's Retaining Pits on the 10th day of each month;
>
> If the waters to be contained in any retaining pit are of such salt, brackish or other quality as to cause pollution if they were to reach other waters of the state, the pit shall be constructed, maintained and operated so as to prevent any surface discharge that directly or indirectly may reach the waters of the state and also lined so as to prevent seepage where the underlying soil condi-

tions are such as to permit such seepage reaching subsurface fresh waters."

The regulation then states that no statewide rule governing construction and lining of pits is adopted due to the varying conditions that may be encountered, but that early planning for non-contaminating disposal must be inaugurated.

Almost all states have a system of inspection or monitoring of injection wells or other disposal systems. [1] These regulations usually require only periodic reports by the operator, and only annual inspections by the administering agency.

On a nationwide basis, it appears that most states now have adequate regulations to protect ground water from contamination as a result of oil and gas exploration and production activities; some states do need better regulations.

The primary cause for concern is not dearth of regulations but lack of enforcement. An example is Oklahoma where unlined pits are prohibited, but several were noted during a recently completed investigation described earlier. [11] Conversely, with the exception of a few special cases, there has been no reported contamination of ground water from oil wells, gas wells and dry holes drilled in Michigan since 1925. [20] The degree of enforcement from state to state is difficult to assess.

In most instances, less than strict enforcement is a result of limited funds and manpower available to the regulatory agency. Also, because the oil and gas industry is usually treated as a separate entity which has its own regulatory group, more than one agency is faced with the related groundwater problems. This overlap makes coordination difficult, and every effort should be exercised at the state level to eliminate the ambiguity of authority among state agencies.

REFERENCES CITED

1. Interstate Oil Compact Commission. No date. Water problems associated with oil production in the United States. Oklahoma City, Oklahoma. 88 pp.

2. United States Water Resources Council. 1968. The nation's water resources. Washington, D. C. Page 3-2-8.

3. Division of Oil and Gas. 1975. Personal communication. Sacramento, California.

4. American Petroleum Institute. 1975. Annual statistical review. American Petroleum Institute. Washington, D. C. 79 pp.

5. Latta, B. F. 1963. Fresh water pollution hazards related to the petroleum industry in Kansas. Transactions of the Kansas Academy of Science, Vol. 6., No. 1, Pages 25-33.

6. State Water Resources Control Board. 1975. News and Views. VI(2):6. Sacramento, California.

7. Crain, L. J. 1969. Ground-water pollution from natural gas and oil production in New York. State of New York Conservation Department, Water Resources Division, Report of Investigation RI-5. Albany, New York. p. 10.

8. Fryberger, J. S. 1972. Rehabilitation of a brine-polluted aquifer. EPA-R2-72-014, Office of Research and Monitoring, U. S. Environmental Protection Agency, Washington, D. C. 61 pp.

9. Collins, A. G. 1970. Finding profits in oil-well waste waters. Chemical Engineering, September 21, 1970, Pages 165-168.

10. Oklahoma Water Resources Board. 1975. Salt water detection in the Cimarron Terrace, Oklahoma. EPA-660/3-74-033, National Environmental Research Center, Office of Research and Development, U. S. Environmental Protection Agency, Corvallis, Oregon. 166 pp.

11. Lehr, J. H. 1969. A study of ground water contamination due to saline water disposal in the Morrow County oil fields. Research Project Completion Report, Water Resources Center, Ohio State University, Columbus, Ohio. 81 pp.

12. Burnett, S. C., and R. L. Crouch. 1964. Investigation of ground-water contamination, P.H.D., Hackberry, and Storie oil fields, Garza County, Texas. Texas Water Commission, Austin, Texas. 77 pp.

13. Anon. 1967. Crack down on oil field pollution. Petroleum Engineer. July 1967. Pages 33-36.

14. Reid, G. W., and others. 1974. Brine disposal treatment practices relating to the oil production industry. EPA-660/2-74-037, Office of Research and Development, U. S. Environmental Protection Agency, Washington, D. C. 175 pp.

15. Reid, G. W., and L. E. Streebin. 1972. Evaluation of wastes from petroleum and coal processing. EPA-R2-72-001, Office of Research and Monitoring, U. S. Environmental Protection Agency, Washington, D. C. 205 pp.

16. Texas Railroad Commission. 1975. Personal communication. Austin, Texas.

17. Haynes, C. D., and D. M. Grubbs. 1969. Design and cost of liquid-waste disposal systems. Report 692, Natural Resources Center, University of Alabama, University, Alabama. 90 pp.

18. American Petroleum Institute. 1975. Environmental protection laws and regulations related to exploration, drilling, production, and gas processing plant operations. API Bulletin D18, first edition. American Petroleum Institute, Washington, D. C. 319 pp.

19. Water Quality Management Board. 1975. Personal communication. Harrisburg, Pennsylvania.

20. Eddy, G. E. 1965. The effectiveness of Michigan's oil and gas conservation law in preventing pollution of the state's ground water. Ground Water, Vol. 3, No. 2. Pages 35-36.

SECTION XII

DISPOSAL OF MINE WASTES

SUMMARY

All forms of mining can result in products and conditions that may contribute to ground-water contamination. The patterns of ground-water recharge and movement responsible for the distribution of contaminants are highly variable and almost entirely dependent upon the mining practice itself and such local conditions as geology, drainage, and hydrology. Although every mine is a potential contamination hazard, few studies of the effects on ground-water quality have been carried out.

With both surface and underground mining, refuse piles and slurry lagoons are probably the major potential sources of ground-water contamination. Where aquifers underlie these sources, water with a low pH (except in arid regions) and an elevated level of total dissolved solids can percolate to ground water.

Coal mining is a major industry in the United States. In 1973, 592 million tons (537 million tonnes) of bituminous coal product were produced. Another 108 million tons (98 million tonnes) were rejected from the preparation plants. Between 1930 and 1971, almost 200,000 acres (81,000 ha) were used for disposal of coal mining wastes, less than 27,000 acres (11,000 ha) of which have been reclaimed. Past surface mining has affected 1.3 million acres (0.5 million ha) of land, and about 4,900 active mines were disturbing 75,000 acres (30,000 ha) annually.

According to the U. S. Census Bureau figures, five states -- Pennsylvania, West Virginia, Alabama, Illinois, and Kentucky -- each have coal mining operations which discharged more than 5 billion gal. (19 million cu m) of waste water in 1972. Other states discharging high volumes of waste water are Ohio, Indiana, and Virginia.

Metal mining in the United States has also been substantial, and in 1972 the number of active mines producing crude metal ore was about 800. The quantity of tailings disposed of in ponds by the metal mining industry alone is estimated at 250 million tons (230 million tonnes) per year. Phosphate rock mines dominate the non-metal category and produced over 137 million tons (124 million tonnes) of crude ore and 426 million tons (387 million tonnes) of total material handled.

Procedures for the abatement of ground-water contamination from mining waste disposal practices can be divided into two broad categories. The first consists of methods for control of seepage and infiltration of surface water and ground water into the mine. The second is treatment to reduce levels of contaminants in the waste. All are very costly processes and have not been practiced to any significant degree.

Most states must rely on all-encompassing water pollution control statutes in order to regulate disposal of mine wastes. There are Federal regulations which pertain solely to the disposal of coal mine wastes. However, these focus primarily on worker safety and have little mention of water pollution, especially as related to ground water.

DESCRIPTION OF WASTE DISPOSAL PRACTICE

Two subcategories are discussed in this section: the coal mining industry and other mineral mining industries. A distinction is also made between surface and underground mining waste disposal practices.

Coal has recently received renewed interest as a major energy source; a development that promises to continue. Although most of the world's major coal fields have been discovered, the total reserve has not been fully determined. It is estimated to be approximately 6 trillion tons (5.5 trillion tonnes), approximately one-third to one-half of which is located in the United States. In 1973, over 592 million tons (537 million tonnes) of bituminous coal were mined in this country, about half by surface mining methods. 1,2,3)

Domestic mining for metallic and nonmetallic minerals (excluding organic fuels) was a $10 billion industry in 1974. All 50 states have such operations. In 1972, over 15,000 mines were producing crude ore. Of this total, clay mines numbered 1,398; sand and gravel operations, 7,110; crushed and broken stone operations, 4,716; dimension stone operations, 478; other nonmetal mines, 507; and metal mines, 792. 4)

While the problems generated by surface and underground mining techniques are somewhat different, waste disposal from both methods poses a potential threat to ground-water quality. Although there is no single waste-disposal practice common to all forms of mining, the various elements of the mining process produce products and conditions which may contribute to ground-water contamination. The patterns of ground-water recharge and movement which are responsible for the distribution of contaminants in the ground-water system are highly variable and almost entirely dependent upon the mining practice and such local conditions as geology, drainage, and hydrology.

Certain general waste disposal techniques are widely used by the mining industry. Liquid and slurry wastes are disposed of in a variety of ways, including tailing ponds, sumps and lagoons, injection wells, land application, and discharge to surface water. Solid wastes are commonly left in piles near the mine.

Tailing Ponds, Lagoons, and Sumps

Liquid waste disposal to surface depressions is very common in the mining industry. These surface depressions are called tailing ponds, sumps, or lagoons, depending upon whether they serve as collection points for process waste water, mine drainage, or waste water from support facilities. Nevertheless, the basic functions of these structures (hereafter collectively termed pond) are very similar. Waste fluids are pumped or drained to the pond via pipeline or drainage ditch. The suspended solids then settle to the floor of the pond, and the remaining portion (effluent) is either used industrially, discharged into local surface water, or spread on the land surface. As the solids settle out, the pond fills with sediment and is either abandoned or dredged out to create new storage space.

The ponds are located in natural depressions or excavated from native soil; perimeter dikes are constructed of mine waste rock, alluvial sand and gravel, clay, or other fill material. Seepage is the most prevalent source of ground-water contamination from ponded waste. Disposal ponds often contain fluid with a high concentration of contaminants such as nitrate, chloride, heavy metals, and radioactive substances. The ponds are constructed in the unsaturated zone and if unlined, as is common practice, seepage of the ponded fluid will occur. Some, but not all, of the contaminants are retained in the unsaturated zone; the remaining contaminants migrate into the ground-water system.

Other problems are pond overflow and dike leakage, both of which can recharge a local aquifer with contaminated water. Also, many of the procedures for completing a fully sedimented pond, prior to abandonment, do not decrease the potential for continuing contamination. These facilities are often insufficiently covered by waste rock or soil, or not covered at all, and rainfall passing through the highly concentrated contaminants in the pond, and then percolating to the water table, can contaminate ground water.

Injection Wells

The disposal of liquid wastes through injection wells is occasionally practiced by the mining industry. Injection-well disposal is employed when waste fluids are too toxic to meet quality standards for discharge into surface water and treatment is not selected as a viable alternative. Means through which such wells can contaminate are discussed in another section of this report.

Direct Application to Ground Surface

At many surface and underground mines, a large portion of the waste water and drainage is allowed to flow over the land surface and infiltrate into the ground-water system. The severity of contamination by this process is largely determined by geologic and hydrologic conditions and the nature of the waste.

Discharge into Surface Water

Direct discharge of liquid waste to surface water is a very common practice of the mining industry. Although the waste is disposed of directly into a surface-water body, significant ground-water contamination can result when the surface water is a source of recharge to an aquifer.

Spoil and Tailing Piles

Spoil and tailing piles result from disposal of solid wastes generated by mining activities. Spoil piles are composed of overburden from surface mining and waste rock from underground mining. Tailing piles are solid wastes from the on-site processing operations of cleaning and concentrating ore. With certain surface-mining techniques, spoil is used to reclaim the mine immediately following complete extraction of the ore. However, for the most part, spoil and tailing piles become permanent features of the landscape.

The source of ground-water contamination from these waste piles is the leachate produced when rainfall or runoff, percolating downward through the uncovered pile, dissolves various contaminants present in the waste. The contaminated water then percolates through the unsaturated layer beneath the pile and reaches the water table.

Dewatering Activities

Mine-dewatering activities, including both pumping and drainage, may cause artificial lowering of ground-water levels. Upon completion of mining, dewatering would be discontinued, allowing the mines to refill with water, and portions of the aquifer which were depleted might be replenished with water of poor quality.

In addition to these waste disposal practices, various mining techniques contribute to the ground-water contamination problem. These techniques are peculiar to either surface or underground mining methods.

Surface Mining

The practice in some regions of stripping ore in the vicinity of the outcrop and continuing the operation underground as the seam becomes progressively deeper, provides a path for contaminated water to reach aquifers. Although some restoration of surface mine districts can be accomplished by backfilling and grading, these actions do not insure protection of ground-water quality. Drainage can move selectively through the backfill material where it is more permeable than the mother rock.

Underground Mining

Contamination mechanisms characteristic of underground mining are as follows:

1. Mine tunnels and shafts can affect the rates and quantities of ground-water movement. Also, the fracture of rock by explosives increases its permeability. These changes can create pathways for the movement of contaminants into the ground-water system.

2. Unplugged wells and test borings which pass through the underground mine workings can interconnect high water-quality aquifers with those containing poor quality water. The same holds true when well casings fail due to corrosion by acid mine water.

3. Bodies of highly mineralized ground water occur naturally at depth beneath large portions of the Appalachian coal belt and the mid-continent region. The removal of substantial quantities of water during dewatering can lower the hydrostatic head in the shallow zones to an extent that upwelling of the deeper mineralized water takes place, contaminating shallow potable water aqui-

fers.

CHARACTERISTICS OF CONTAMINANTS

The principal ground-water contaminants from mine waste are acidity, dissolved solids, metals, radioactive materials, color, and turbidity. While many of the contaminants are not toxic, they can be present at levels in excess of U. S. Public Health Service and EPA drinking water standards. 5)

The most prevalent contamination problem associated with coal mining is the formation and discharge of large volumes of acid. Acid formation will occur when precipitation brings water into contact with pyrites (metallic sulfide). The exposure of pyritic minerals to air and water results in their oxidation to form sulfuric acid. Beyond the basic chemical relationship between pyrites, oxygen, and water, it is suspected that the acid formation may be influenced by complex biochemical reactions involving one or more types of bacteria. 6) Although not completely dependent upon acidity, the solubility of metals varies with pH, and acidic runoff and seepage may contain high metal ion concentrations.

The oxidation of pyrite in the presence of water results in generation of both sulfuric acid and ferrous sulfate. The sulfate levels in coal mine drainage and in the receiving aquifers may be as high as several thousand ppm and are quite commonly in excess of the recommended limit of 250 ppm. This limit has been set for reasons of taste; water containing higher concentrations can be safely consumed although certain sulfate salts function as laxatives. 5)

Following the initial oxidation of pyrite and the production of ferrous sulfate, subsequent oxidation will normally produce a ferric sulfate. The end result is the release, in associated aquifers, of substantial quantities of iron at concentrations in excess of the recommended limit of 0.3 ppm. 5) Iron at this level is not harmful to health, but it tends to impair taste and discolor water under certain conditions. Precipitates of iron and related iron bacterial colonies can clog plumbing, water transmission lines and water-supply wells. High levels of dissolved iron are commonly noted in ground water that has been contaminated from coal mining activities.

Acid dissolves many minerals, producing soluble salt solutions. In this way, the dissolved solids concentration can become quite high. Metals frequently associated with metallic ore deposits such as copper, zinc, cadmium, and manganese also are dissolved by acid mine drainage. The combination of acidity, high dissolved solids concentration, and metals makes the presence of the acid mine water very undesirable in either surface or ground water.

In regions where mining proceeds in alkaline rocks, such as limestone, dissolution of the rock may produce water with a high pH. Basic mine water has a relatively low complement of heavy metals because most form insoluble salts under conditions of high pH. However, the water can still be highly mineralized. Calcium and magnesium ions are frequently found in significant concentrations, making the water hard.

Although carbon is the most important element found in coal, as many as 72 other elements have been associated with some deposits. The ash formed by bituminous coals of West Virginia consists of about one percent each of sodium, potassium, calcium, aluminum, silica, iron, and titanium. In addition, 26 metals were present in trace amounts, including lithium, rubidium, chromium, cobalt, copper, gallium, germanium, lanthanum, nickel, tungsten, and zirconium. [1)]

In a recent coal survey, water-quality sampling was carried out at sites in Pennsylvania, West Virginia, Kentucky, and Indiana. The samples were collected from seeps and points of direct runoff from coal refuse piles and lagoons. Results show that the chemistry of the water is quite variable, but has generally low pH values and relatively high concentrations of both sulfate and certain metals. For several areas, the range of values for selected constituents are summarized below: [3)]

pH	2.7	to	7.5	
Conductivity	200	to	16,500	umhos/cm
Total Acidity	0	to	34,300	ppm
Sulfate	75	to	40,500	ppm
Sodium	6	to	780	ppm
Magnesium	3.4	to	664	ppm
Aluminum	1	to	1,014	ppm
Potassium	0.5	to	22	ppm
Calcium	13	to	450	ppm
Manganese	0.01	to	545	ppm
Iron	0.1	to	6,168	ppm
Nickel	0.3	to	1.7	ppm
Copper	0.04	to	0.14	ppm
Zinc	0.2	to	2.8	ppm

Methane and hydrogen sulfide are quite commonly found in the geologic formations associated with coal as well as in the various disposal areas; waters in the vicinity of coal mines may be highly charged with dissolved gases. [7)]

In many metal mining operations, the ore contains a large amount of worthless rock that must be separated from the mineral before the ore can be smelted or refined. This process of concentrating the ore, known as beneficiation, is usually performed at the nearest possible site to the mine to eliminate costly transportation of unwanted rock. The most common concentrating techniques are flotation and acid separation.

The waste from concentrating operations, referred to as tailings, can be in solid or liquid form. The solid is composed of minerals associated with the metallic compound being beneficiated. As an example, the waste from a typical copper beneficiating operation contains quantities of lead, zinc, gold, and silver; smaller amounts of arsenic, antimony, bismuth, selenium, tellurium, nickel, cobalt, and cadmium; and trace quantities of germanium, indium, tin, and thallium. [8]
Solid waste from the beneficiation of other metallic ores contains many of the same substances. Liquid waste from beneficiating operations contains water and acids (usually nitric or sulfuric), in addition to the minerals associated with the solid tailings. Thus, waste from the concentrating phase of ore processing contains large quantities of toxic substances and can be a serious source of ground-water contamination.

Underground and surface uranium mining practices have been found to greatly increase the concentration of dissolved radium-226 in ground water. This is believed to be the result of: (1) exposure and oxidation of the ore body, and (2) contact of mine drainage water with spilled ore and wastes within the mine. [9]

EXTENT OF THE PROBLEM

With the widespread and serious nature of contamination of rivers and streams caused by all types of mining operations, especially coal in the Appalachian region, little attention has been paid to potential and real problems of ground-water contamination. Also, in the worst areas, where mining has been active for 100 years or more, degradation of ground-water quality has become an accepted fact, and aquifers in the region have been written off as sources of water supply.

Although there has been no single effort to relate the impact of waste disposal practices of the mining industry on ground-water quality, the potential for ground-water contamination by mining wastes may be estimated indirectly, based on the quantity of these wastes generated. The total waste volumes may also be related to the size of the area devoted to waste disposal and the methods used.

Coal

In 1973, 592 million tons (537 million tonnes) of bituminous coal product were produced in the United States. Another 108 million tons (98 million tonnes) were rejected from the preparation plants. By 1985, an estimated 1.1 billion tons (1.0 billion tonnes) of bituminous coal will be produced, and 200 million tons (182 million tonnes) may be rejected as waste. Between 1930 and 1971, almost 200,000 acres (81,000 ha) were used for the disposal of coal mining wastes; less than 27,000 acres (11,000 ha) have been reclaimed. [2)]

With regard to the rate of acid production in these coal mining waste piles, it is reported that the average rate of acid formation at one site in southern Illinois is 198 lb/acre/day (222 kg/ha/day) of acidity (as $CaCO_3$). [10)] Acid production for the reclaimed acreage in the same area was measured at only 16 lb/acre/day (18 kg/ha/day). [11)] These rates are probably widely representative because climatic conditions are similar over a majority of the major coal mining regions. Thus, total acid production from coal mine waste piles could be more than 6 million tons (5.5 million tonnes) per year. It should be emphasized that this figure is an estimate only, because other factors such as sulfur content and form in the coal can control acid production.

There is a lack of published data regarding the impact of leakage from coal mining waste slurry lagoons on ground-water quality. From a theoretical analysis of other types of waste lagoons, presented in other sections of this report, it can be assumed that the slurry lagoons contribute significantly, if not substantially, to the total coal related ground-water contamination problem.

According to U. S. Census Bureau figures, five states -- Pennsylvania, West Virginia, Alabama, Illinois, and Kentucky -- have coal mining and processing operations which discharged more than 5 billion gal. (19 million cu m) of waste water in 1972. [12)] In three additional states -- Ohio, Indiana, and Virginia -- data regarding water disposal volumes were withheld. However, based on other U. S. Census Bureau figures that are available, the volumes of waste water discharged in these states is assumed to be large. Table 52 shows the states which reported water use by coal mining establishments, along with the volumes used, volumes discharged, and total number of establishments included in the U. S. Census Bureau survey. Figure 68 shows the states reporting significant water use by coal mining establishments.

Table 52. STATES IN WHICH SIGNIFICANT VOLUMES OF WASTE WATER ARE DISCHARGED FROM COAL MINING AND PROCESSING OPERATIONS 1972. after 12)

State	Total water used including recirculation (billion gal.)	Total water discharged (billion gal.)	Total number of establishments reporting water use
Pennsylvania	14.4 + D*	28.2	69
Ohio	D*	--*	10
Indiana	D*	--*	8
Illinois	24.8	7.1	30
Missouri	D	0.05	1
Kansas	D	--	1
Virginia	D*	--*	18
West Virginia	47.0	20.5	89
Kentucky	9.4	5.3	24
Tennessee	D	--	1
Alabama	6.3	7.4	16
Texas	D	--	1
Colorado	D	--	2
New Mexico	D	--	1
Arizona	D	--	1
Utah	1.1	D	4
Washington	D	--	1
Total (of above)	93.6	68.5	
Actual total	149.3	76.9	

D = Data compiled but not disclosed
-- = Data not available
Substantial volume is probable

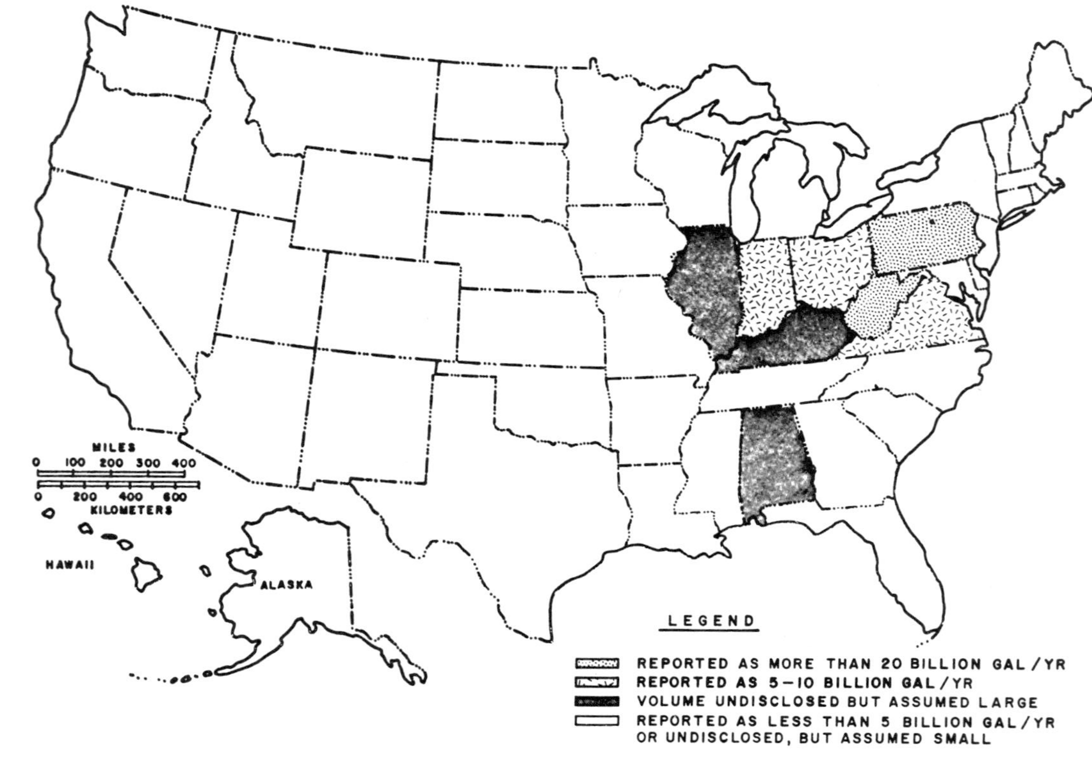

Figure 68. States in which significant volumes of waste water are discharged from coal mining and processing operations, 1972.

In 1974-75, the U. S. Department of the Interior and EPA estimated that past surface coal mining activities had affected 1.3 million acres (0.5 million ha) of land, and that about 4,900 active mines were disturbing 75,000 acres (30,000 ha) annually.[13,14,15] In order to evaluate the impact of these mining activities on the quality of ground water on a nationwide basis, and to identify those areas which are most affected, the United States is subdivided into four distinct regions where similar geologic and topographic conditions govern the type of mining practice. These regions are shown on Figure 69.

Estimates of total coal reserves in the United States vary. However, according to the survey represented on Figure 69, there are over 1,770 billion tons (1,600 billion tonnes) of total coal and lignite reserves in the United States with over 145 billion tons (132 billion tonnes) reported to be strippable, distributed as follows: 14)

Regions I and II	-	5.6 billion tons (5.1 billion tonnes)
Region III	-	29.2 billion tons (26.5 billion tonnes)
Region IV	-	110.9 billion tons (100.7 billion tonnes)

Region I -

Region I includes West Virginia, Tennessee, and eastern Kentucky. Total coal production in Region I in 1973 was 230,017,000 tons (208,809,433 tonnes) of which 52,674,113 tons (47,817,560 tonnes) were produced by surface mines. Mines in the region numbered 965 of which 323 were surface mines. The bulk of the coal mining in Region I is done in West Virginia with slightly more than half the 1973 production coming from that state.[14]

For the most part, the density of on-site supply wells throughout this region is light (0 to 10 wells per sq mi) to moderate (11 to 20 wells per sq mi). This density reflects the generally rural character of this region. The individual domestic water-supply well is most likely to be affected by mine-related contamination sources. Centralized public supply and industrial wells are more carefully located with respect to such contamination sources, and are therefore less likely to be affected.

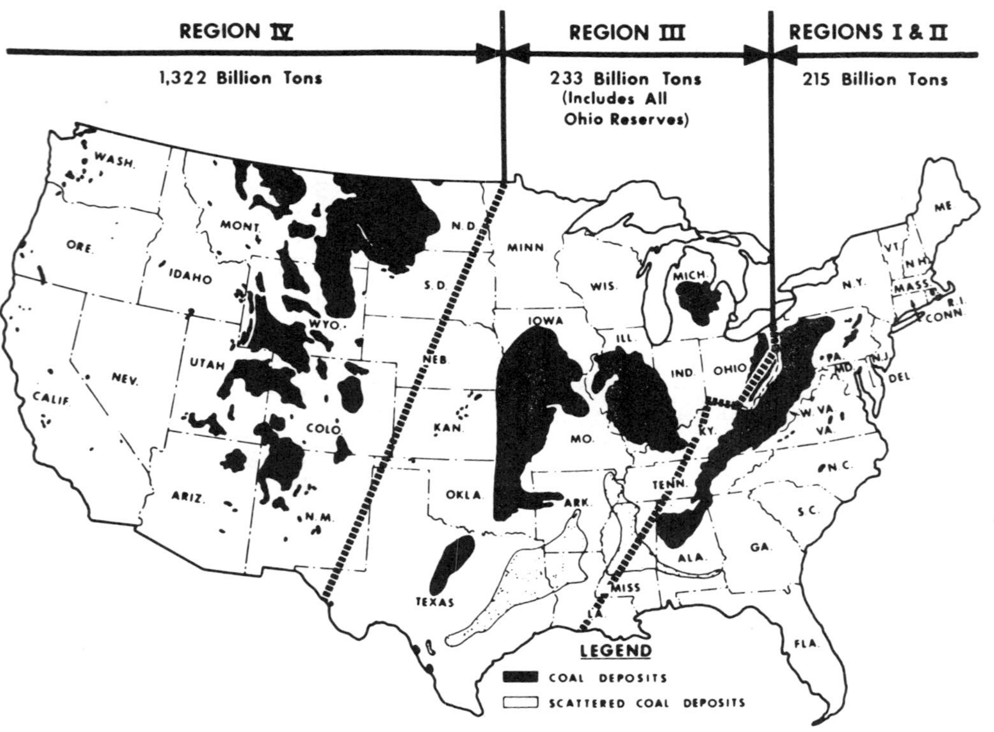

Figure 69. Total coal reserves of the conterminous United States. 14)

Note: Tons multiplied by 0.9078 equals tonnes.

Region II -

Region II includes Pennsylvania, Maryland, Alabama, and southeastern Ohio, and represents the northern and southern extremes of the Appalachian coal field. Total coal production in Region II in 1973 was reported to be 103,643,000 tons (94,087,115 tonnes) of which 39,735,251 tons (36,071,661 tonnes) were produced by surface mines. Of 406 mines, 281 were surface mines. Pennsylvania is by far the most productive state in this region. 14)

The region is mainly rural, with the exception of some major urban centers. Most households are served by individual wells. The density of such development is light (0 to 10 wells per sq mi) although a few areas of moderate (11 to 20 wells per sq mi) to heavy development (21 to 60 wells per sq mi) are noted in the plateau regions of western Pennsylvania and eastern Ohio. The portions of this region where ground-water contamination from coal mining sources would appear to be most likely are in southwestern Pennsylvania and to a lesser extent in eastern Ohio.

Region III -

This region includes Illinois, Oklahoma, Indiana, Missouri, Kansas, Iowa, and western Kentucky. Illinois has the largest known high volatile bituminous reserves of the nation. Total coal production in Region III in 1973 was reported to be 188,956,164 tons (171,534,406 tonnes), with 118,659,307 tons (107,718,919 tonnes) coming from surface mines. Of 354 coal mines, 268 were surface facilities. 14)

This region is almost exclusively rural, with the exception of a few urban centers, and is dependent upon local shallow aquifers and individual domestic-supply wells for water. The density of wells is light over all of Region III. There appear to be several areas in southern Illinois and western Kentucky where the level of activity is sufficiently high, in terms of coal production, to be a source of ground-water contamination. However, for the most part, these areas are associated with aquifers of low productivity.

Region IV -

Region IV, which includes Arizona, Colorado, Montana, New Mexico, North Dakota, and Wyoming, contains the potentially largest coal reserves in the nation, if not the world. 14)

The reserves that make Region IV so important range from lignite fields in North Dakota and Montana to high volatile bituminous coals in other fields. Mining in this region has only recently been initiated on a large scale to meet the growing demand for energy.

Total coal production was reported to be 50,805,378 tons (46,121,122 tonnes) in 1973, most of which was recovered from surface mines. The total number of mines in the region was 52, of which 34 were surface mines. Montana and Wyoming lead in production with approximately 11 and 15 million tons (10 and 14 million tonnes), respectively. 14)

Population densities in this region are among the lowest in the nation. Overall water requirements for domestic uses are therefore light, although other related demands exist for such uses as livestock watering and irrigation. The density of domestic supply wells per square mile is reported to be less than 10 over the entire region.

Precipitation is low, averaging less than 15 in. (38 cm) in Region IV, which limits the potential for generation and distribution of the various soluble contaminants. Although the water table varies in depth in accordance with local conditions, it probably occurs below the mineable coal formations in many of the strip-mining districts. In addition, the coal beds occur at a considerable interval above the major aquifers, which, for the most part, are overlain by relatively impermeable strata, further retarding significant vertical movement of water.

Because water is not readily available in this region, it is likely that measures will be taken to conserve whatever amounts are available in the vicinity of the mining operation for use in the processing and cleaning operations.

At this point, it would appear that the significant environmental problem associated with western coal mining which might influence the quality of ground water is the alkaline conditions associated with the mine wastes. The high soil alkalinity values which are normally associated with mine water and runoff are reported to inhibit the development of plant cover on the spoil banks. In the absence of vegetation, the weathering process is virtually uncontrolled and alkaline wastes continue to be generated and discharged into the surrounding environment.

Other Minerals

Excluding clay, sand and gravel, and stone operations, which are not considered significant ground-water contamination threats, there were 1,299 active mines in the United States in 1971, 61 percent of which were metal mines. [4] Table 53 shows the number of domestic ore-producing mines by commodity and volume of crude ore production. Table 54 shows the principal producing states for these commodities. Crude ore production ranged from a rate of less than one ton (0.9 tonne) per year to over 35 million tons (32 million tonnes) per year at a Utah copper mine. Seventeen mines produced over 10 million tons (9.1 million tonnes) of ore each: 7 were copper; 5, iron; 4, phosphate rock; and one molybdenum. 4)

The 25 leading nonmetal mines, with phosphate rock dominating the list, produced over 137 million tons (124 million tonnes) of crude ore and 426 million tons (387 million tonnes) of total material handled. This represented 17 percent of the total material handled at nonmetal mines. [15]

Mining Techniques

Surface mining accounted for 94 percent of crude ore production and 96 percent of total material handled in 1972. Underground mining accounted for substantial percentages of crude ore production in only six states in 1972: Colorado, 36; New Mexico, 33; Missouri, 26; Wyoming, 21; Louisiana, 17; and Tennessee, 16. Eighteen states reported no 1972 underground activity at all. [17]

Various factors contribute to the obvious predominance of surface mining. The two major advantages of surface mining that favor its use are the higher percentage of ore recovery (up to 100 percent in some cases), and the lower unit cost per ton of recovered ore (due primarily to the smaller amount of manpower required for extraction processes).

Materials Handled

Producers of metal and nonmetal minerals handled over 4 billion tons (3.6 billion tonnes) of ore and mine-waste rock in 1972; of this total, just under 1.5 billion tons (1.4 billion tonnes) were waste rock. Waste rock is by no means the total waste that must be disposed of by the mining industry. Mine operations must also dispose of mine waste water, mine drainage, and waste from support facilities. [4]

The handling of more than 100 million tons (91 million tonnes) of mined material was reported by ten states: Arizona, California, Florida, Illinois, Michigan, Minnesota, New Mexico, Texas, Utah, and Wyoming. Arizona and Florida led the nation in total quantity of material handled and in waste rock generated. [4]

Table 53. NUMBER OF DOMESTIC METAL AND NONMETAL MINES IN 1971, BY COMMODITY AND MAGNITUDE OF CRUDE ORE PRODUCTION.[4)]

Commodity Metals	Number of Mines	Thousand Tons [a)]					
		1	1 to 10	10 to 100	100 to 1,000	1,000 to 10,000	10,000
Bauxite	23	-	8	11	4	-	-
Copper	84	25	5	10	16	21	7
Gold: Lode	30	22	3	1	3	1	-
Placer	37	20	6	10	-	1	-
Iron Ore	75	-	7	11	23	29	5
Lead	45	19	7	3	11	5	-
Mercury	66	37	18	7	-	-	-
Silver	42	25	10	5	2	-	-
(Titanium) Ilmenite	6	-	-	-	1	5	-
Tungsten	66	59	5	1	1	-	-
Uranium	245	37	132	58	18	-	-
Zinc	52	10	5	14	23	-	-
Other [b)]	25	10	1	4	4	5	1
Total metals	792	264	207	135	106	67	13
Abrasives	14	3	7	3	1	-	-
Asbestos	9	3	-	1	5	-	-
Barite	35	1	4	18	12	-	-
Boron Minerals	2	-	-	1	-	1	-
Diatomite	11	2	1	7	1	-	-
Feldspar	40	2	22	12	4	-	-
Fluorspar	21	4	10	5	2	-	-
Gypsum	65	2	6	18	39	-	-
Mica	20	8	5	5	2	-	-
Perlite	14	2	5	5	2	-	-

Table 53 (continued). NUMBER OF DOMESTIC METAL AND NONMETAL MINES IN 1971, BY COMMODITY AND MAGNITUDE OF CRUDE ORE PRODUCTION. 4)

Commodity Metals	Number of Mines	Thousand Tons a)					
		1	1 to 10	10 to 100	100 to 1,000	1,000 to 10,000	10,000
Phosphate Rock	55	1	9	8	17	16	4
Potassium Salts	8	-	-	-	1	7	-
Pumice	107	16	36	45	8	-	-
Salt	20	-	3	2	9	6	-
Sodium Carbonate (Natural)	3	-	-	-	1	-	-
Talc, Soapstone, Pyrophyllite	55	7	22	25	1	-	-
Vermiculite	3	-	1	-	2	-	-
Other c)	25	10	3	5	7	-	-
Total Nonmetals	507 d)	63	134	160	114	32	4
Grand Total	1299	327	341	295	220	99	17

a) Tons multiplied by 0.908 equals tonnes.

b) Includes antimony, beryllium, magnesium, manganiferrous ore, molybdenum, nickel, platinum-group metals, rare earth metals, tin and vanadium.

c) Aplite, graphite, green sand marl, iron oxide pigments (crude), kyanite, lithium minerals, magnesite, millstones, olivine, wollastonite and zeolite.

d) In addition, there were 1395 clay mines, 7110 sand and gravel operations, 4715 crushed and broken stone operations and 478 dimension stone operations, but specific data on these operations are not available.

Table 54. MINERALS PRODUCED IN THE UNITED STATES BY PRINCIPAL PRODUCING STATES. 4)

Mineral	Principal Producing States in Order of Quantity
Antimony	Idaho, Montana, Nevada
Aplite	Virginia
Asbestos	California, Vermont, Arizona, North Carolina
Asphalt (native)	Texas, Utah, Alabama, Missouri
Barite	Missouri, Nevada, Arkansas, Georgia
Bauxite	Arkansas, Alabama, Georgia
Beryllium conc.	Utah, South Dakota, Colorado
Boron Minerals	California
Bromine	Arkansas, Michigan, California
Brusite	Nevada
Calcium-Magnesium Chloride	Michigan, California
Cement	California, Pennsylvania, Texas, Michigan
Clays	Georgia, Texas, Ohio, North Carolina
Cobalt	Pennsylvania
Copper	Arizona, Utah, New Mexico, Nevada
Diatomite	California, Nevada, Washington, Arizona
Emery	New York, Oregon
Feldspar	North Carolina, California, Connecticut, South Carolina
Fluorspar	Illinois, Colorado, Kentucky, Montana
Garnet (abrasive)	New York, Idaho
Gold (mine)	South Dakota, Nevada, Utah, Arioona
Graphite	Texas
Gypsum	Michigan, California, Texas, Iowa
Iron Ore	Minnesota, Michigan, California, Missouri
Kyanite	Virginia, Georgia, Florida
Lead	Missouri, Idaho, Utah, Colorado
Lime	Ohio, Pennsylvania, Missouri, Texas
Lithium Minerals	North Carolina, Nevada, California
Magnesite	Nevada
Magnesium Chloride	Texas
Magnesium Compounds	Michigan, California, Texas, New Jersey
Manganese Ore	Montana
Marl greensand	New Jersey, Maryland
Mercury	California, Nevada, Texas, Idaho
Mica, scrap	North Carolina, Alabama, Georgia, South Carolina
Mica, sheet	North Carolina, Colorado

Table 54 (continued). MINERALS PRODUCED IN THE UNITED STATES BY PRINCIPAL PRODUCING STATES. [4]

Mineral	Principal Producing States in Order of Quantity
Molybdenum	Colorado, Arizona, Utah, New Mexico
Nickel	Oregon
Olivine	Washington, North Carolina
Perlite	New Mexico, Arizona, California, Nevada
Phosphate Rock	Florida, Idaho, Tennessee, North Carolina
Platinum-Group Metals	Alaska
Potassium Salts	New Mexico, California, Utah
Pumice	Arizona, Oregon, California, Hawaii
Pyrites Ore and Concentrate	Tennessee, Pennsylvania, Colorado, Nevada
Rare Earth Metal Concentrates	California, Georgia
Salt	Louisiana, Texas, Ohio, New York
Sand and Gravel	California, Michigan, Illinois, Minnesota
Silver	Idaho, Arizona, Utah, Colorado
Sodium Carbonate (natural)	Wyoming, California
Sodium Sulfate (natural)	California, Texas
Staurolite	Florida
Stone	Pennsylvania, Illinois, Ohio, California
Sulfur (Frasch Process)	Louisiana, Texas
Talc, Soapstone, Pyrophyllite	New York, Texas, Vermont, California
Tin	Colorado, Alaska
Titanium Concentrate	New York, Florida, New Jersey, Georgia
Tripoly	Illinois, Oklahoma, Arkansas, Pennsylvania
Tungsten Concentrate	California, Colorado, North Carolina, Nevada
Uranium	New Mexico, Wyoming, Colorado, Texas
Vanadium	Colorado, Arkansas, Idaho, Utah
Vermiculite	Montana, South Carolina
Wollastonite	New York
Zinc (mine)	Tennessee, New York, Colorado, Maine
Zircon Concentrate	Florida, Georgia

The mining industry is widespread throughout the United States. Virtually every mine contaminates ground water through its waste disposal practices, although the nature and extent of the contamination differs as a result of the type of mineral mined, the characteristics of the environment from which it is extracted, and the methods of waste disposal.

The most potentially serious ground-water contamination problems are those related to waste disposal practices at metal mines. Uranium and some copper mines have unique waste products, due to the nature of the ore or the mining procedure. An extremely serious problem in all aspects of waste disposal from uranium mining is the presence of high concentrations of dissolved toxic materials such as selenium, molybdenum and arsenic, and toxic particulate matter, as well as radioactive materials.

Several copper deposits are being mined by in-place leaching. Using this technique, the deposits are first cut by a system of tunnels, then fractured by explosive or hydraulic methods, and leached of copper by sulfuric acid introduced through pipes at the upper part of the deposit. The copper-rich fluids are withdrawn from the lowest levels of the mine and pumped to the surface for processing. This method produces large quantities of highly toxic sulfuric acid, rich in heavy metals and other contaminants.

Solution mining of uranium ore is becoming increasingly important in New Mexico and Texas. As with the copper leaching procedures, this technique presents a serious threat to ground-water quality. Stringent controls to protect ground-water quality are now in effect in Texas; however, the efficacy of such controls is presently unknown.

In the mining industry there are numerous processes which generate wastes disposed of in ponds. In the metals mining industry, the concentrator tailings are generated by flotation, vat leaching, and cyclone processes; the slurries by flotation processes; mine water and waste rock by ore excavation; sludges by waste-water treatment; and dusts by crushing and grinding operations. All of these wastes go to tailing ponds.

The quantity of concentrator tailings deposited in ponds by the metals mining industry alone is estimated at 250 million tons (230 million tonnes) per year. 18) While information regarding the volume of waste water discharged to ponds by the mining industry is not available, the U. S. Census Bureau has surveyed the total volume of waste water discharged.

Table 55 lists the total volumes of water used and discharged by mining establishments (excluding coal and petroleum).[12] States with significant waste-water discharges are shown on Figure 70. For some, the volume of discharge was withheld to protect the rights of individual companies. However, from other data included in the U. S. Census Bureau survey, it is possible to determine which of these states probably has substantial waste-water discharge from mining operations. These are also shown on Figure 70.

Case Histories

Midwest Region -

Studies began at the New Kathleen Coal Mine in southern Illinois in 1968 to determine the rates of acid production from a mine refuse pile.[10] The investigation was limited to a site which occupies an area of approximately 40 acres (16 ha), together with a neighboring slurry lagoon of approximately 50 acres (20 ha). The site forms part of an abandoned coal mining operation, active from 1945 to 1955, which included the strip mining of a coal seam approximately 110 ft (34 m) below land surface, together with a coal cleaning and processing operation.

This type of coal mining is common throughout the midwest and produces two distinct types of solid-waste products. The first is the coarse refuse, composed of overburden and soil, which is generally redeposited in the strip pit at the completion of mining by backfilling. The second and major waste product, termed "gob," is generated by the coal-cleaning operations that are performed to remove impurities from the coal. The fine reject material from the mining operation is transported in slurry form by pipeline to a lagoon. The remaining portion, which is composed largely of coal, intermixed with pyrites, sandstone, clays, and shales, is normally stored in refuse piles. These studies indicated that the refuse piles are the major source of contaminated water.

At the New Kathleen Mine, the refuse pile occupied an area of 40 acres (16 ha), with a maximum height of 65 ft (20 m) and a volume of approximately 2 million cu yd (1.5 million cu m). The investigators examined the weathering profile of the refuse pile and found that the process of waste-water formation is largely a near-surface process. Three distinct zones were noted. The outermost zone, approximately 4 to 10 in. (10 to 50 cm) thick, contained virtually no fine-grained materials and was relatively permeable and open to the free circulation of both air and water. The second zone is a

Table 55. STATES IN WHICH SIGNIFICANT VOLUMES OF WASTE WATER ARE DISCHARGED FROM MINING AND ORE PROCESSING OPERATIONS (EXCLUDING COAL AND PETROLEUM) 1972. after 12)

State	Total water used including recirculation (billion gal.)	Total water discharged (billion gal.)	Total number of establishments reporting water use
New York	. D	D	44
Pennsylvania	16.2 + D	D*	43
Ohio	12.3	24.4	76
Indiana	3.3	6.5	22
Illinois	13.1	5.7	22
Michigan	161.4 + D	D*	24
Minnesota	681.0	D*	31
Iowa	16.3	D*	29
Missouri	18.6 + D	D*	15
Maryland	4.9	D	7
Virginia	9.8 + D	4.2	21
North Carolina	D*	D*	14
South Carolina	D	3.1	34
Georgia	30.1	15.1	15
Florida	394.7 + D	76.3 + D	28
Tennessee	D*	D*	21
Alabama	3.6 + D	0.8	8
Mississippi	12.6	7.1	10
Arkansas	8.9	D	14
Louisiana	23.2	12.6	10
Texas	35.2 + D	26.6	41
Idaho	5.6	5.0 + D	9
Wyoming	24.5	14.2	9
Colorado	12.6 + D	6.9 + D	18
New Mexico	35.3 + D	8.4 + D	14
Arizona	240.8 + D	25.3 + D	25
Utah	81.5	D*	8
Nevada	38.8	D*	10
Washington	2.9 + D	D	9
California	37.0	D*	60
Total	1924.2	242.2	
Actual total	2307.9	861.5	

D = Not Disclosed
* = Substantial volume is probable
Note: Billion gallons multiplied by 3.785 equals million cubic meters.

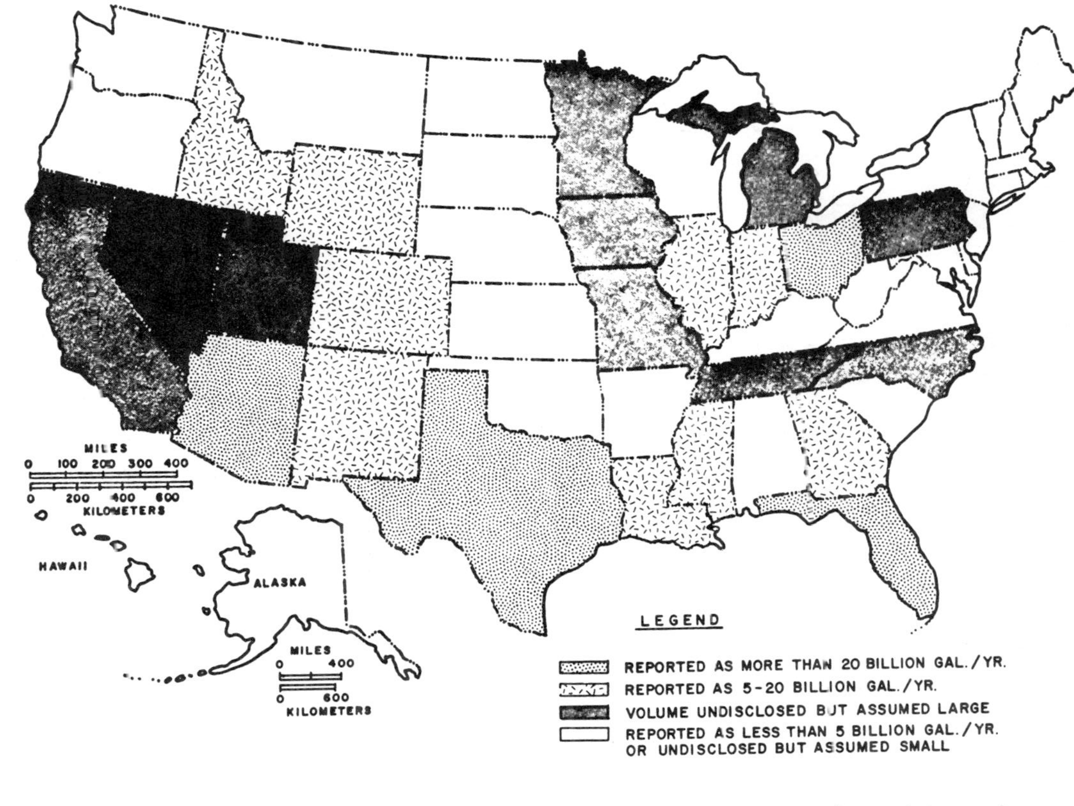

Figure 70. States in which significant volumes of waste water are discharged from mining and ore processing operations (excluding coal and petroleum), 1972.

clean layer of clay, approximately one in. (2.5 cm) thick, thought to have been compacted by rain action. This may be a zone of accumulation for portions of the clay fraction, as well as certain precipitates. It appears to serve as a partial barrier to both the circulation of air and water, although a detailed examination revealed numerous discontinuities which would permit the entry of water into the main body of the refuse pile. The third zone is essentially unweathered material extending into the interior of the pile.

It was concluded that the bulk of the acid production from this refuse pile occurs in the outermost layer, which is subject to rapid erosion. As a result, fresh pyrite continues to be exposed, providing a source for ongoing production of sulfuric acid. The near-surface deposits create relatively inhospitable soil conditions which hamper the establishment of vegetation, without which weathering and erosion processes are virtually uncontrolled.

Between rains, pyrite oxidation proceeds at a relatively modest but constant rate, with the acid products accumulating in the outer reactive mantle. When precipitation occurs, approximately 54 percent of the rainfall appears immediately as acidic runoff, while the remainder either infiltrates into the interior of the pile, later reappearing as seepage, or eventually returns to the atmosphere by evaporation. The average rate of acid formation was found to be 198 lb/acre/day (222 kg/ha/day) of acidity (as CaCO3).

In the Phase II report on the control of mine drainage for the New Kathleen Mine in southern Illinois, the authors discuss the results of various field experiments in connection with the control of mine drainage. 11) Studies were conducted where various abatement techniques were employed under different field conditions. The results were carefully evaluated. The study provides excellent background data on the various aspects of mine drainage contamination, including the results and costs of abatement techniques. These experiences could probably be applied to other mining districts in the midwest.

The major finding of the study was that the acid runoff from the refuse piles could be controlled by covering the wastes with soil, establishing a vegetative cover, and providing adequate drainage to minimize erosion. No significant differences were noted in the rates of acid formation from individual test plots covered with 1, 2, or 3 ft (0.3, 0.6, or 0.9 m) of top soil. The average rate of acid formation for the entire restored refuse pile was estimated at 16 lb/acre/day (18 kg/ha/day) of acid (as $CaCO_3$). This compares with the 198 lb/acre/day (222 kg/ha/day) of acid which was the reported discharge rate from the unrestored refuse pile. The cost per acre for refuse pile restoration necessary to achieve these results would be approximately $6,000/acre ($14,800/ha). It is reasonable to conclude that the restoration of the refuse piles, and associated reduction in acid production, would substantially reduce the possibility of ground-water contamination by direct percolation.

Appalachian Region -

In the Redbank Creek watershed of northwestern Pennsylvania, the strip mining of bituminous coal has contaminated both surface and ground water. During a study, analyses of water were made from 6 springs, 15 wells (including 13 flowing abandoned oil and gas wells), and 31 surface-water sampling stations. Iron and manganese levels exceeded drinking water standards in all samples. The highest trace metal levels were observed in the ground-water samples, with zinc, chromium, and cadmium present in several. The principal points of contaminated ground-water discharge included springs and improperly plugged, abandoned, flowing oil and gas wells. [19]

Another study in this region discussed the effects of coal mining on ground water in the Toms River drainage basin, also in northwestern Pennsylvania. [20] The area is supplied with ground water from a deep multi-aquifer system. Numerous abandoned oil and gas wells penetrate the aquifers and serve as the major conduits for the introduction of contaminated mine water into the producing aquifers. The report provides data on water quality for the principal aquifers and other sources. These results are summarized in Table 56.

The report concludes that drainage in the vicinity of strip mines moves down to the deeper aquifer through natural rock openings and improperly plugged, abandoned oil and gas wells. The contaminated water moves laterally through the aquifer system and is discharged back to the surface through other abandoned oil and gas wells and natural seeps and springs. The mine drainage affects ground-water quality largely by increasing the iron and sulfate content.

There are many well documented cases of ground-water contamination from the waste-disposal practices of mines other than coal mines. Three of these cases are briefly described below, with special emphasis placed on the quality of the affected ground water.

Table 56. GROUND WATER QUALITY IN THE TOMS RIVER DRAINAGE BASIN. [20)]
(in ppm, except pH)

Aquifers	pH	Sulfate (SO_4)	Iron (Fe)	Chloride (Cl)
Burgoon Sandstone and Lower Sandstone Unit of the Connoquenessing Formation	5.2 - 6.8	1 - 10	0.0 - 16.0	4 - 10.0
Upper Sandstone Unit of the Connoquenessing				
Nonmining areas	6.3 - 6.8	4 - 13	3 - 16	1 - 7
Areas near mining	2.9 - 5.4	30 - 620	25 - 160	1 - 7
Homewood Formation				
Nonmining areas	6.5 - 6.7	10 - 15	10 - 15	0 - 2
Areas near mining	3.0 - 5.5	39 - 80	20 - 70	0 - 2
Discharges from coal mines	2.6 - 3.3	2,450 - 4,400	200 - 500	0 - 2
Discharges from abandoned oil and gas wells	2.9 - 6.3	21 - 678	10 - 140	5 - 8

Grants Mineral Belt, New Mexico -

In 1975, the U. S. Environmental Protection Agency, Office of Radiation Programs conducted a study of the Grants Mineral Belt, a major uranium producing area. [9] One of the objectives of the study was to assess the impacts of waste discharges from uranium mining and milling on surface waters and ground waters.

Ground water is the principal source of water supply in the study area. Both the shallow alluvial and the underlying limestone aquifers have been developed to supply the needs of agriculture, the uranium mills, and public water supply systems. The nearby city of Gallup receives its water supply primarily from deep wells in the Gallup sandstone.

The study found that shallow ground-water contamination in the area is the result of the infiltration of waste-water effluents, mine drainage, and discharge of tailings from the mines and mills. In addition, deep well injection of wastes into the potable limestone aquifer is practiced.

Company data show that seepage from one tailing pond in the area averaged 48.3 million gal./yr (183 million l/yr) for 1973 and 1974. It is estimated that, over the period 1960 to 1974, 0.41 curies of radium have entered the shallow, potable aquifer from seepage alone. The average volume injected for the same period was 91.9 million gal./yr (348 million l/yr).

At another mill in the same area, seepage from the tailing ponds occurs at the rate of 130 million gal./yr (149 million l/yr). At this second site, 0.7 curies have entered the ground-water system over the same 14-year period.

While the sorptive capacity of the soils in the area is effective in removing most of the radium, abnormally high concentrations were found in many wells. Background radium-226 levels in the area average 0.16 pCi/l, and levels as high as 6.6 pCi/l were found in shallow ground water near contamination sources. By way of comparison, the EPA proposed drinking water regulations would limit the radium in finished drinking water to 5 pCi/l. Mine and mill effluents contain as much as 178 pCi/l, and mine drainage contains as much as 75 pCi/l. Elevated levels of total dissolved solids, ammonia, nitrate, and chloride in the ground water have also resulted from mining in the area.

Although widespread ground-water contamination from the mining and milling operations was not observed during this study, the scarcity of adequate monitoring wells and the lack of historical data precludes positive conclusions with regard to ultimate spread of the contaminants. In addition, trends of increasing water quality degradation in some wells indicate that present contamination levels are not representative of the ultimate severity of the problem.

Galena, Illinois - 21)

A serious ground-water contamination case occurred in the lead-zinc mining area of northwestern Illinois. The operation consisted of extensive underground workings that required dewatering by pumping large quantities of water from the aquifer below the ore vein. During the dewatering operation, it was necessary to deepen a number of farm wells located within approximately a one-mile (1.6-km) radius, some to a depth of 300 ft (90 m), in order to insure a continuing supply of water for farm and home use.

In January 1966, the Illinois Department of Public Health prohibited discharge of this waste to the Galena River; subsequently, the liquid waste was disposed of through direct application to land surface. Almost the entire volume of discharge water from the mill-sedimentation ponds flowed across the land surface and entered old mine workings. This toxic waste then infiltrated the aquifer tapped by wells of the nearby farms. The discharge continued for almost two years until the affected farmers began to voice complaints of water-quality deterioration.

Unfortunately, no chemical analyses were performed during the interim period on either the processing waste or the contaminated ground water. A series of samples of the mill waste water was taken in mid-1968 and revealed the following ranges in the concentration of chemical constituents:

total dissolved solids	-	2,940	to	4,120	ppm
hardness	-	2,680	to	2,960	ppm
iron	-	2.2	to	20	ppm
copper	-	0.01	to	0.09	ppm
lead	-	0.0	to	2.4	ppm
zinc	-	0.9	to	19	ppm
cyanide	-	0.07	to	0.2	ppm
sulfate	-	308	to	850	ppm
pH		7.2	to	7.5	

Table 57 shows the composition of the water in each of the affected wells.

The farmers filed suit against the company that operated the

Table 57. CONCENTRATIONS OF CHEMICAL CONSTITUENTS IN WELLS INSIDE (FARMS 1-4) AND OUTSIDE (FARM 5) THE AFFECTED AREA IN PPM.[22]

Farm No.	TDS	Hardness ($CaCO_3$)	Iron (Fe)	Copper (Cu)	Lead (Pb)	Zinc (Zn)	Cyanide (CN)	Sulfate (SO_4)
1	6,020	2,460	33	0.02	0.2	5.6	0.0	1,250
2	4,340	2,850	30	0.06	0.2	5.0	0.0	1,050
3	2,940	2,400	33	0	0.2	6.8	0.1	925
4	1,960	1,640	16	0.08	0.2	2.4	0.0	400
5	480	380	0.6	0.0	0.0	0.1	0.0	23

Note: The pH of these water samples ranged from 6.5 to 7.5

mill and were awarded a total of $69,250 in damages.

Coeur d'Alene District, Idaho -

More studies of contamination from mining activities have been conducted in the silver, lead, zinc, and antimony mining region of the Coeur d'Alene district of northern Idaho than in any other area of the United States. The data discussed below concern the effect on ground-water quality of acid mine drainage from a large silver, lead, and zinc mine. 21) The case history in the following paragraph illustrates only one of many contamination problems associated with waste disposal in this heavily worked mining area.

Since the opening of the first tunnels of this mine in 1885, ore has been extracted from a rich vein by various techniques. Drainage from workings has been collected in sumps or allowed to flow freely from the mine. The quality of this water is extremely poor, with high concentrations of heavy metals, and an average pH of 2.3. The drainage contains iron, zinc, manganese, and magnesium, in average concentrations of 10,000, 8,000, 2,000, and 1,500 ppm, respectively, as well as significant concentrations of cadmium, copper, lead, and aluminum. This drainage infiltrates directly to the local aquifer, and an analysis of a ground-water sample taken near the mine showed only slightly diluted concentrations of the materials found in the mine drainage.

TECHNOLOGICAL CONSIDERATIONS

Today's level of mining and environmental technology appears to be capable of dealing with most sources of ground-water contamination caused by coal mining activities. Procedures for the abatement of ground-water contamination from mining waste disposal practices can be divided into two basic categories. The first consists of methods that control seepage, infiltration, and other hydrologic sources of contamination. The second includes treatment to reduce levels of contaminants in the waste. In addition, careful consideration should be given to the method of mining relative to the actual site conditions such as topography, drainage, and ground-water hydrology. Many specialists in this field recommend detailed pre-mining studies which would include an assessment of the potential for ground-water contamination by the mining operation and collection of detailed information on the hydrogeology of the site.[13)]

Control of Liquid Waste

Reducing the volume of water entering the mine is the key to controlling mine drainage. To reduce seepage of ground water into surface mines, impermeable barriers are sometimes constructed along the walls and floor. These barriers may be composed of clay, concrete, or concrete block. Barrier construction is not common; it is costly and therefore limited to specific problems.[23)]

Diversion of surface water and storm runoff away from mine sites by ditches can also reduce the amount of mine-drainage water that has to be handled. Although most effective around surface mines, the diversion of surface water from the vicinity of underground mines can reduce flow into the mine through cracks, fissures, and sinkholes. It can locally lower the water table, which in turn, will cause a reduction of inflow. Diversion ditches are very common, both around the mine itself and around waste piles. Underdrains are most effective when installed prior to the creation of a waste pile, although they may be constructed under existing piles at additional cost.

An effective contamination control is lining lagoons, sumps, and tailing ponds with impermeable barriers. The materials most commonly used include concrete, asphalt, and clay. In addition, rubber and plastic liners have been used.

Depressions, caused by subsidence or other means, often collect large quantities of surface water and allow it to percolate slowly into underground mine workings. One control method designed to reduce the volume of water entering underground mines involves eliminating depressions and grading the surface, in order to increase runoff from the site. The effectiveness depends on individual site conditions, as do the costs, which are extremely variable. In most cases, this method has been found to significantly reduce infiltration.

The sealing of exploratory bore holes from earlier mineral exploration and fracture zones encountered during the mining operation helps prevent infiltration into underground mine workings, reducing the quantity of mine drainage. Bore holes and fracture zones act as water conduits where they penetrate underground mines. Bore holes can be located and plugged to prevent passage of water. This procedure is more effective if performed from within the underground mine itself. The permeability of fracture zones can be reduced by drilling holes into the zone and pressure grouting. Various types of grout are available (concrete is most commonly used). The effectiveness of this technique has not been documented.

Control of ground water in the vicinity of a mine may be accomplished by dewatering wells and drainage of mine water into sumps. During the course of mining, attention should be given to the prevention of subsidence and rock fracturing under streams and rivers. Sufficient material can be left in place to prevent caving in those sections of the mine. The rapid removal of accumulated water from a mine is important in preventing the formation of acid water, and may be accomplished by underground pumping plants, wells, or drainage tunnels.

Various methods have been suggested for sealing underground mines to exclude air and thus limit formation of contaminants. Steps commonly taken are controlled flushing or sluicing of mine refuse into selected portions of the mine through a network of specially designed pipes and waterways, including the pumped-slurry method where wastes are pumped into the workings through bore holes in a manner similar to grouting. Concrete plugs, bulkheads, and packers have been used for this purpose. Underground mines may also be sealed by removing the ceiling supports at the completion of mining, allowing the ceiling to collapse.

Inert gases such as nitrogen, injected into underground mines to displace oxygen, limit the generation of acid. It has been demonstrated experimentally that by reducing oxygen levels to 0.4 percent or lower, the amount of acidity generated could be reduced by as much as 97 percent.[24)]

Common methods of treating liquid waste generated by mining activities are neutralization and removal of solids, both suspended and dissolved. Several alkaline materials are available for neutralizing acid mine drainage and waste water. These include, among others, lime, hydrated lime, limestone, caustic soda, soda ash, and ammonium hydroxide.[23)] The treatment agent employed depends on the quantity of water treated, and the quality of effluent desired, as well as on local cost factors.

Neutralization processes can also be used to reduce concentrations of heavy metals. As pH increases, precipitation of the metal ions from solution in the form of insoluble hydroxides occurs, generally reducing concentrations to one ppm or less, depending on the metal.

Removal of the principal radiochemical contaminant, radium, from waste waters associated with uranium mining and milling is through the use of double-liming or barium chloride to cause precipitation. Industry is also experimenting with bio-uptake methods of contaminant removal, using algae growths in tailing ponds. 9)

There are various techniques used by the mining industry to remove solids from the liquid waste. These include flash evaporation, freezing (ice crystals are formed with a lower mineral content than the original water), and settling ponds (by far the most common method). With ponds, the suspended solids are allowed to settle and "clean" water is skimmed off the surface of the pond. These processes do result in a solid-waste residual with a contamination potential.

Control of Solid Waste

The most prevalent method for the handling of solid waste generated by mining activities is by reclamation. The reclamation of strip-mined land involves a variety of tasks, with an overall objective to restore the landscape to either the original configuration or something useful or attractive. The average cost of reclaiming land disturbed by active coal strip mining in 1964 averaged $230/acre ($568/ha) for complete reclamation and $149/acre ($368/ha) for partial reclamation. The cost for restoring abandoned strip-mined land (orphaned land) is considerably higher, with costs for complete reclamation, including grading, water control, and revegetation, ranging from $1,800 to $4,000/acre ($4,448 to $9,884/ha). 23)

Both soil preparation and revegetation are part of the overall reclamation process. Prior to soil preparation and planting, the land surface must be properly graded and compacted as soon after the completion of mining operations as possible so that the surface is not subject to rapid erosion. Addition of top soil is widely recommended to supply a viable medium for final covering of the restored surface. Limestone is commonly added to spoil materials to neutralize acid conditions.

In arid regions, where the spoil materials are commonly alkaline, soil scraping, mulching and water are used to improve the spoil in preparation for seeding. Successful restoration includes the selection of the correct plant type(s) for the particular site, and proper timing for planting.

INSTITUTIONAL ARRANGEMENTS

A few states have specific provisions regulating handling of mine wastes, but even in states with substantial mining activity, all-encompassing, water-pollution control statutes are frequently relied upon for this type of regulation. State regulations vary widely, but provisions for protection of water resources are generally directed only at controlling surface-water quality. Rarely is the ground-water contamination potential of mining operations recognized. Where ground-water provisions exist, they cover the handling of ore, soil, and wastes to prevent acid production.

Illinois has detailed regulations applied by the state Environmental Protection Agency. The applicant must show that the activity for which the permit is sought will not cause, threaten, or allow pollution of the air or waters of the state during or after active mineral production. One of the requirements for obtaining a permit is that the applicant describe the proposed method of mining and mine refuse disposal. The procedures that will be integrated into such methods and the procedures that will be taken upon abandonment to prevent air and water pollution must also be described. An operator under such a permit is specifically required to notify the state of any emergency situation at the mine which causes or threatens to cause a sudden discharge of contaminants into waters, and to undertake necessary corrective measures. The regulations contain sections on mine operations, including requirements for plugging all holes; mine refuse disposal, including requirements for subsoil so that leachate will not pollute water; abandoned areas; and monitoring and reporting. 25)

The Pennsylvania statute prohibits operation of a mine or allowing a discharge from a mine into waters of the Commonwealth unless authorized or under permit. It authorizes the state to require the operator to post a bond insuring compliance with the law and regulations and conditions of the permit, including provisions insuring that there will be no polluting discharge after mining operations have ceased. 26) Although "waters of the Commonwealth" include ground water, most effort is directed toward protecting rivers and streams.

Surface-mining reclamation laws commonly contain provisions directed at preventing water pollution as one of several objectives of carrying on strip-mining operations with minimal damage to the environment. Provisions include the requirement that the operator obtain a license for the area to be mined, and that he first obtain approval of a plan for reclaiming the area. For the administering agency to approve the plan, it must find (among other things) that the plan does not pose a threat of water pollution. The statute will also require the operator to prevent pollution while mining and reclaiming, and condition approval of reclamation and release of his bond upon proper reclamation including prevention of water pollution.

In 1974, Ohio enacted a law applicable to surface mining of minerals such as sand and gravel, clay, and limestone. That statute requires that the plan filed by the operator include a statement of the measures the operator will perform during mining and reclamation to insure that contamination of underground water supplies is prevented. 27)

A proposed Wisconsin regulation governing metallic mineral mining and reclamation would require that the mining and reclamation plan be approved by the Department of Natural Resources. In addition to details of tailings production and handling and ground- and surface-water management techniques, the plan must contain procedures for long-term maintenance of the project sites, including monitoring of wastes and water quality. 28)

A rule of the Florida Department of Environmental Regulation specifies requirements for constructing, operating, and inspecting earthen dams used in phosphate mining and processing. This rule is directed at the problem of dam failure which releases large volumes of pollutants. 29)

Colorado's guidelines for mill tailing ponds have a similar objective. Although the guidelines by their terms apply also to prevention of ground-water contamination, the details are directed at construction and maintenance to prevent washout into surface waters. 30)

Michigan is one of the states that relies upon its water-pollution control law without specific regulations applicable to mining. Mine-waste disposal activities are monitored, and a permit is required if it appears that pollution may occur. 31)

Minnesota's regulation prohibits deposit of industrial waste in such a place, manner, or quantity that effluent or residue may actually or potentially limit the use of underground waters as a potable water supply, or contaminate underground waters. Persons responsible for industrial waste or deposits or operations from which residues may reach underground waters are required to submit monthly reports on operation of the system, waste flow, and characteristics of the influent, effluent, and underground waters of the vicinity. 32)

A Wyoming statute requires to be submitted with the application for a license to mine: a plan for insuring that all acid forming, or toxic materials, or materials constituting a fire, health or safety hazard uncovered during or created by the mining process are promptly treated or disposed of during the mining process in a manner designed to prevent pollution of surface or subsurface water or threats to human or animal health and safety. Such methods may include, but not be limited to covering, burying, impounding or otherwise containing or disposing of the acid, toxic, radioactive or otherwise dangerous material. [33])

At the present time, Pennsylvania has the most stringent regulations to control the surface disposal of coal waste. Among other things, the regulations adopted in May of 1973 require: [34])

1. Operators of existing coal waste piles to apply for a permit within 6 months after adoption of the regulations.

2. Operators to obtain permits before creating new refuse piles.

3. The applicant to provide proof of his financial responsibility with such responsibility continuing for up to 10 years following the completion of the disposal operation to insure proper closure of sites.

4. The application to include maps and other technical information pertaining to soils, geologic and ground-water characteristics of the area and the possibility of subsidence from past and future mining beneath the disposal area. The plan must provide for the prevention of water pollution, the stability of the disposal area and the prevention of air pollution. Drainage from the coal waste areas must meet the water quality standards established for the particular receiving streams.

5. Approval for the construction of impoundments or silt lagoons on coal waste areas under a disparate permit subject to the state's existing regulations for design and construction of impoundments.

6. The side slopes of a waste pile not to exceed 33 percent (18°). A vegetation plan for the entire area is mandatory. All earth moving and revegetation activities must comply with the state's existing regulations for erosion and sediment control.

7. Concurrent compaction of the coal refuse pile to minimize permeability and air entraining capabilities.

REFERENCES CITED

1. Longwell, C. R., R. F. Flint, and J. E. Sanders. 1969. Physical geology. John Wiley & Sons, New York.

2. Martin, J. F. 1974. Quality of effluents from coal refuse piles. Pages 26-37 _in_ National Coal Association. First symposium on mine and preparation plant refuse disposal. Washington, D. C.

3. Falkie, T. V., J. E. Gilley, and A. S. Allen. 1974. Overview of underground refuse disposal. Pages 128-144 _in_ National Coal Association. First symposium on mine and preparation plant refuse disposal. Washington, D. C.

4. U. S. Bureau of Mines. 1973. Minerals yearbook 1972. U. S. Government Printing Office, Washington, D. C. Pages 50-65.

5. U. S. Department of Health, Education and Welfare. 1962. Public Health Service drinking water standards. Revised 1962. 61 pp.

6. Barnes, I., and F. E. Clark. 1964. Geochemistry of ground water in mine drainage problems. U. S. Geological Survey, Professional Paper 473-A. 6 pp.

7. Collier, C. R., and others. 1964. Influences of strip mining on the hydrologic environment of parts of Beaver Creek, Kentucky, 1955-59. U. S. Geological Survey, Professional Paper 427-B. 85 pp.

8. Katari, U., G. Isaacs, and T. W. Devitt. 1974. Trace pollutant emissions from the processing of metallic ores. U. S. Environmental Protection Agency Report EPA 650/2-74-115. Pages 4-11.

9. Kaufmann, R. F., G. G. Eadie, and C. R. Russell. 1975. Summary of ground-water quality impacts of uranium mining and milling in the Grants Mineral Belt, New Mexico. U. S. Environmental Protection Agency, Technical Note ORP/LV-75-4.

10. Consolidated Coal Company, Truax-Traer Coal Company Division. 1971. Control of mine drainage mineral wastes, phase I, hydrology related experiments. U. S. Environmental Protection Agency, Project 14010 DDH. 361 pp.

11. Kosowski, Z. V. 1973. Control of mine drainage from coal mine mineral wastes, phase II, pollution abatement and monitoring. U. S. Environmental Protection Agency Project 14010 DDH. 81 pp.

12. Bureau of the Census. 1975. 1972 census of mineral industries. U. S. Department of Commerce. MIC72(1)-2. 58 pp.

13. Grimm, E. C., and R. D. Hill. 1974. Environmental protection in surface mining of coal. U. S. Environmental Protection Agency Project 670/2-74-093. 273 pp.

14. Skelly and Loy, Engineers/Consultants. 1975. Economical engineering analysis of the U. S. surface coal mines and effective land reclamation. U. S. Bureau of Mines Contract, Report SO241049. 559 pp.

15. Averitt, P. 1975. Coal resources of the United States, January 1974. Geological Survey Bulletin 1412. U. S. Geological Survey. 131 pp.

16. U. S. Department of the Interior, Water Resources Research Office. 1975. Acid mine water, a bibliography. 564 pp.

17. U. S. Environmental Protection Agency. 1973. Methods for identifying and evaluating the nature and extent of nonpoint sources of pollutants. Report 430/4-73-014. Page 182.

18. Bendersky, D. 1975. A study of waste generation, treatment and disposal in the metals mining industry. Midwest Research Institute, Kansas City, Missouri. Unpublished draft. Contract No. 68-01-2665.

19. Gang, M. G., and D. Langmuir. 1974. Controls on heavy metals in surface and ground waters affected by coal mine drainage; Clarion River-Redbank Creek watershed, Pennsylvania. Pages 39-69 in National Coal Association. Fifth symposium on coal mine drainage research. Washington, D. C.

20. Emrich, G. H., and G. L. Merrit. 1969. Effects of mine drainage on ground water. Ground Water (3):27-32.

21. Walker, W. H. 1973. Where have all the toxic chemicals gone? Ground Water 11(2):51-62.

22. Ralston, D. R., and others. 1973. Solutions to problems of pollution associated with mining in northern Idaho. U. S. Bureau of Mines, U. S. Government Printing Office. Pages 16-71.

23. Skelly and Loy, Engineers/Consultants. 1973. Processes, procedures, and methods to control pollution from mining activities. U. S. Environmental Protection Agency Report EPA 430/9-73-011. Pages 61-268.

24. West Virginia University, Coal Research Bureau. 1970. Underground coal mining methods to abate water pollution, a state of the art literature review. U. S. Environmental Protection Agency Project 14010FKK. 50 pp.

25. Illinois Pollution Control Board Regulations, Chapter 4. Mine related pollution.

26. Pennsylvania Act of June 22, 1937, PL 1987, as amended. Section 315.

27. Ohio Revised Code, Chapter 1514.

28. Wisconsin Department of Natural Resources. June 13, 1975. Proposed Code Chapter NR 131. Metallic mineral mining and reclamation.

29. Florida Department of Pollution Control. Rules, Chapter 17-9. Minimum requirements for earthen dams, phosphate mining and processing operations.

30. Colorado Department of Health, Water Pollution Control Commission. 1968. Guidelines for the design, operation and maintenance of mill tailing ponds to prevent water pollution.

31. Michigan Water Resources Commission, Bureau of Water Management. 1975. Telephone communication.

32. Minnesota Pollution Control Agency. Rules and Regulations, Chapter WPC 22. Classification of underground waters of the state and standards for waste disposal.

33. Wyoming, 1973 Cumulative Supplement, Section 35-502.24 (b) (ix).

34. Pennsylvania Regulations. 1973. Chapter 125: coal refuse disposal areas. Pennsylvania Bulletin 3(23): 1045.

SECTION XIII

WASTE DISPOSAL THROUGH WELLS

SUMMARY

Industrial waste, sewage effluent, spent cooling water, and storm water are discharged through wells into fresh- and saline-water aquifers in many parts of the United States. The greatest attention in existing literature has been given to deep disposal of industrial and municipal wastes through wells normally drilled a thousand feet or more into saline aquifers. A total of 322 such wells have been constructed in 25 states, and 209 are operating. They pose a comparatively small potential for contamination when compared to the tens of thousands of shallow wells injecting contaminants directly into fresh-water aquifers.

Irrigation and storm-water drainage wells and septic tank effluent disposal wells total about 15,000 in Florida, Oregon, and Idaho alone. On Long Island, New York, approximately 1,000 diffusion wells inject about 80 million gpd (300,000 cu m/day) from air conditioning or cooling systems into two of the principal aquifers tapped for public water supply. Thousands more are used for disposal of storm-water runoff. In a few areas, principally in limestone and basalt regions where openings in the rock are large enough to transmit high volumes of liquid, the practice of discharging raw sewage and sometimes industrial waste in shallow fresh-water aquifers has not been uncommon.

Of wells used for disposal of industrial and municipal wastes in saline aquifers, few failures have been reported. This is due to the strict regulation and permit system generally enforced by public agencies in those states which allow construction of this type of well. On the other hand, shallow wells completed in potable-water aquifers and used for waste disposal have received little attention. This has resulted in a number of documented cases of severe ground-water contamination, frequently from the illegal use of wells for the disposal of various types of hazardous wastes.

Under general water pollution control laws, most states automatically rule out the use of wells for injection of either sewage or industrial wastes into fresh-water aquifers. In a few states, where drainage wells have been a popular means for disposal of domestic waste water, storm runoff and irrigation runoff, programs are underway to eventually eliminate this practice. Federal regulations only cover industrial injection wells, municipal sewage disposal wells, irrigation drainage wells, and storm-water disposal wells.

DESCRIPTION OF THE PRACTICE

A waste disposal practice which has been in use for many years and has become increasingly popular is the injection of waste (liquid or gaseous) through wells. In some instances, subsurface disposal can be the most economic waste disposal alternative and from the view of the regulatory agency in charge, can be the most environmentally acceptable.

Any more or less vertical shaft (where depth is significantly greater than diameter) used for the introduction of waste fluids into the subsurface may be termed an injection well. The hole may or may not penetrate to the zone of saturation. Where it does not, it is called a "dry well." Injection wells can only accept fluids, meaning materials that flow.

Two general types of fluids are being handled by injection wells -- low volume, high toxicity fluids which are hazardous even in small quantities; and high volume, low toxicity fluids.

Considerable controversy has been generated over the use of injection wells. For example, in a 1969 report, it was suggested that underground disposal should be reserved for particularly toxic wastes that can not otherwise be treated and readily disposed of. 1) Conversely, in a 1974 study, it was stated that "underground migrations of injected waste can not be determined accurately; therefore highly toxic compounds, such as cyanide, should not be injected." 2)

Some potential consequences associated with subsurface injection through wells are: 3)

1. ground-water contamination
2. surface-water pollution
3. alteration of formation permeability
4. land subsidence and/or earthquakes
5. contamination of mineral resources

Herein, the major concern is the potential contamination of usable ground water. Of all the purposes to which injection wells have been put, the one which has been most frequently inventoried is that of industrial and municipal waste disposal. As of July 1975, at least 322 industrial and municipal waste-water injection wells had been constructed in 25 states as shown in Table 58. Of the 322 wells constructed, 209 were operating, and of these, nearly 60 percent were

Table 58. DISTRIBUTION OF 322 INDUSTRIAL AND MUNICIPAL WASTE-WATER INJECTION WELLS AMONG 25 STATES HAVING SUCH WELLS IN 1973. [4)]

State	Number of wells	State	Number of wells
Alabama	5	Mississippi	1
Arkansas	1	Nevada	1
California	5	New Mexico	1
Colorado	2	New York	4
Florida	7	North Carolina	4
Hawaii	3	Ohio	9
Illinois	6	Oklahoma	14
Indiana	13	Pennsylvania	9
Iowa	1	Texas	98
Kansas	30	Tennessee	3
Kentucky	3	West Virginia	7
Louisiana	65	Wyoming	1
Michigan	28		

used by the chemical, petrochemical and pharmaceutical industries. [4] These firms commonly produce toxic wastes that are difficult to treat.

Injection rates are relatively low. Eighty-six percent of all such wells inject less than 400 gpm (25 litre/sec) and 59 percent inject less than 100 gpm (6 litre/sec). [5,6] The receiving reservoirs are nearly equally distributed between sand, sandstone, and carbonate rocks, which are also the three most common aquifer types.

When compared with the total number of disposal wells nationally (tens of thousands), and the volumes of fluids involved (billions of gallons), one might conclude that the potential danger of contamination from industrial waste injection is relatively insignificant. It is, however, the extremely hazardous nature of the injectants which emphasizes the need for closer scrutiny.

Because of the toxic chemical concentrations often present in industrial wastes, injection zones are usually deep salaquifers. Of 262 wells for which a depth of an injection zone is reported in 1973, only 6 percent were at less than 1,000 ft (305 m). Seventy-eight percent were between 2,000 and 6,000 ft (600 and 1,800 m). [6] Salaquifers are generally found in the thick sediments of coastal plains and in deep geologic basins. These aquifers are zones of slow circulation; this feature makes them attractive for disposal because injected wastes tend to migrate slowly toward discharge points.

The disposal of oil-field brines and the related practice of repressuring of petroleum reservoirs through wells, detailed in another section, are injection well uses. Another source of brine which requires disposal is the presently small volume of hot brine produced from geothermal energy production. Volumes of hot brine are expected to increase with the development of the geothermal resource. A rapidly developing source of brine for disposal is from the impending construction of desalination plants in water-short areas. However, in terms of volumes injected and the number of wells involved, oil-field brine injection is far and away the most common use for injection wells.

Injection wells are also used to dispose of radioactive wastes, domestic and municipal sewage effluent, storm-water runoff, excess surface water, and irrigation return flows. They may be used for artificial recharge to increase the volume of water in storage in depleted fresh-water aquifers or to provide barriers against intrusion of sea water. The intent of artificial recharge is to provide beneficial use but can be coupled with the disposal of excess fluids. Storage, in saline or brackish water aquifers, of excess volumes of treated sewage effluent, surface water or runoff which can not be effectively stored in surface reservoirs, is also included here as a use of artificial recharge. Although an attempt to recover the injectant may never be made, potential recovery is often used to justify operation of the facility.

Injection wells can cause ground-water contamination through the following mechanisms:

1. Direct emplacement into potable water zones

2. Escape into potable aquifer by well-bore failure

3. Upward migration from receiving zone along outside of casing

4. Leakage through inadequate confining beds

5. Leakage through confining beds due to unplanned hydraulic fracturing

6. Leakage through deep abandoned wells

7. Displacement of saline water into potable aquifer

8. Injection into salaquifer eventually classified as a potable water source

9. Migration to potable water zone of same aquifer

The first seven of these contamination problems have been known to occur; the eighth is expected to occur with the reclassification of potential water sources, and the ninth is a likely occurrence in the future.

However, properly constructed and monitored industrial waste injection wells can be operated with reasonably little danger of ground-water contamination. Pre-construction feasibility studies, test drilling, and water-quality testing can eliminate the problem of direct emplacement into potable water zones, as well as determine whether a production well will initially operate efficiently and economically. Nonreactive casing, properly joined, with an acceptable thick and complete cement grout between the casing and bore hole will prevent leakage of injectant into potable aquifers by well failure or upward migration through the bore hole. Exploratory drilling will help determine whether confining beds are capable of resisting upward leakage. After the commencement of injection operations, water-level and water-quality data from monitoring wells assist in early detection of system failure.

CHARACTERISTICS OF CONTAMINANTS

Fluids injected into wells can range from nearly pure rain water, through treated sewage effluent, to highly toxic chemical wastes and radioactive substances. Depending upon the source of the injectants, they can chemically, physically, or biologically degrade ground water. A partial list of injected wastes is shown in Table 59.

Chemical degradation is a major concern. Little is known about the chemical reactions of injectants underground, both with the formation and the formation fluids. Many reactions occur under the heat and pressure in the subsurface which do not at room temperature and atmospheric pressure.

Similarly, liquid radioactive wastes are disposed of by means of wells. Also radioactive solids have been incorporated into cement or asphalt slurries and injected into fractured shale. These practices have been carried out at several Federal sites.

Physical contamination of the subsurface also can occur. The most common physical degradation results from temperature change. For example, air-conditioning systems circulate ground water through the system, transferring heat from the air to the water. This water, at an elevated temperature but not significantly changed chemically, is often returned to the ground through wells. The degradation of ground water by increasing its temperature may decrease its value, either for cooling and air conditioning, for drinking, or for industrial use. Raised temperatures of ground water discharging to surface-water bodies could promote algal growth.

Injected water has been reported to degrade the physical properties of wells and aquifers. Occasionally, productive water wells have been used to artificially recharge aquifers. Lake and river water, and often storm-water runoff, in surplus at certain times of the year, are recharged through water wells and extracted during water-deficient periods. The recharge water, if unfiltered, is usually high in suspended solids, and these solids clog the pore spaces in the vicinity of the well bore. When the well is later pumped, a significant loss in efficiency may have occurred, which may only be partially recoverable.

Table 59. INJECTED WASTES.[5,6]

acetaldehyde
acetate ammonia
acroline
activated sludge
alcohols
aldehydes
aluminum hydroxide
ammonia liquor
ammonium chloride
ammonium sulphate
acids
  acetic
  adipic
  chromic
  formic
  hydrochloric
  sulphuric
benzene
bicarbonates
boiler water
BOD waste
butadiene waste
butanol
brines
bromides
calcium chloride
calcium carbonate particles
calcium sulphate
chloromycetin
chlorinated hydrocarbons
chlorinated organics
chromates
chromium
clay particles
COD waste
coke quench water
colloidal compounds
contaminated storm drainage
cooling tower water
cresols
cyanides
caustic
detergents
diatomaceous earth
drilling muds
ethynol
ferric chloride
ferrous chloride
ferrous sulphate
hexamethylediamine chlorates
heavy metal salts
hydrocarbons
ketones
lime sludge
laundromat waste
magnesium sulphate
mineral acids
methyl cellulose
mercaptans
magnesium chloride
methyldichlorophosphine
nitriles
naphthalene
natural plasticizer wastes
nitroles
oils
oil refinery waste
organic phosphorus
organic solvents
organic nitrogen
photo process waste
phosphorus trichloride
pharmaceutical process waste
phenol
polyethylene waste
pulping liquor
paint removers
propylene oxide
silica
steam drain
steroids
sodium hydroxide
sodium sulphate
sodium chloride
uranium mill and radioactive laboratory wastes

Even fresh water, low in suspended solids, has been known to degrade aquifer properties. For example, in Norfolk, Virginia, treated potable water was injected into a shallow brackish-water aquifer (total dissolved solids of 3,000 ppm). As a result of the change in the electrolytic concentration, clay dispersion (plugging) in the sediments resulted, which caused a loss in hydraulic conductivity that was not completely recoverable. 7)

Injectants often contain biological contaminants which are health hazards, sometimes greater than those of the chemical constituents in the fluid. Industrial waste water is often high in biological oxygen demand which is a direct cause of noxious tastes and odors in the effluent. Raw, primary treated, and secondary treated sewage effluent, all of which may be put into injection wells, are very active biologically. Storm-water runoff picks up biological and chemical contaminants from streets and roofs. Surface-water bodies are biologically active too. Bacteria and viruses from human and animal wastes are assumed to be removed naturally after injection into a porous medium with an anaerobic environment. However, bacteria in the subsurface have been known to travel hundreds of feet, and little is known about the movement of viruses.

Such a wide variety of fluids are injected that it would be impossible to give a complete characterization of each one. However, a number of them are described in some detail in other sections of this report.

EXTENT OF THE PROBLEM

No comprehensive data are available on the number of instances of ground-water contamination from injection wells, the degree to which aquifers are being affected, or the number of water wells lost or threatened. Interestingly enough, in House of Representatives committee hearings on the Safe Drinking Water Act of 1973, documented instances of contamination were read into the record for 22 states. 8) The cases were collected in a very short time span. To this list can be added known instances in nearly every state, totaling into the hundreds, plus probably thousands of unknown cases where an injection system has caused contamination as yet undetected.

Some estimate of the number of contaminating injection wells might be made if the assumption that all wells conducting untreated fluids to the zone of aeration or shallow fresh-water aquifers contaminate. These include all drainage and septic tank effluent disposal wells. The total number in

Florida, Oregon, and Idaho alone is about 14,500.[9,10] Undoubtedly, many thousands more could be added to this amount. In fact, thousands of wells injecting into the zone of aeration have been estimated to be in use in the northeastern states.[11] Thus, nationally, there are probably many tens of thousands.

Disposal through wells into fresh-water zones is most prevalent in those regions where the receiving aquifer consists of limestone or basalt or any other highly porous rock which will take relatively large volumes of waste water without the need for injection under artificial pressure. As indicated on Figures 27 and 28, many states in addition to Florida, Oregon, and Idaho are underlain by such highly porous formations. As described above, deep injection wells for municipal and industrial wastes also have the potential to contaminate shallow fresh-water aquifers. However, how many are doing so has not been surveyed in detail.

Case Histories

A great number of well-documented case histories of contamination have not been developed, as the disposal practice has only recently become generally popular. However, a few cases have been outlined here to illustrate the range and nature of the contamination. Because the rate of ground-water movement is usually very slow in the deeper subsurface and monitoring facilities are limited, contamination may be occurring undetected. Disposal in shallow zones is often so prevalent that no effort is made to explore individual cases.

Long Island, New York -

Fear of the threat of sea-water intrusion and a concern for conservation of ground-water resources led New York State in 1933 to require the return to the ground of water pumped for industrial air conditioning purposes. All new industrial wells with capacities of 100,000 gpd (378 cu m/day) or more used to supply air conditioners were to return water "in an uncontaminated condition through diffusion wells or other approved structures." Over 1,000 recharge wells on Long Island inject about 80 mgd (302,800 cu m/day) of heated air conditioning return water. As far back as 1937, reports of thermal pollution in western Long Island (Kings County, also known as Brooklyn) were noted. The natural ground-water temperature is 52° to 56°F, but water in the glacial aquifer near one diffusion well showed a 14°F rise. The temperature of the recharge water was 83°F. Water in a few wells in other parts of the county has a temperature over 80°F.[11]

Wilmington, North Carolina -

In the spring of 1968, a chemical company received permission to inject acid waste waters derived from the manufacture of dimethyl terephthalate (DMT) into a zone of interbedded sand, silt, clay and limestone. The only fresh-water aquifer in the area is the surficial sand to a depth of about 75 ft (23 m). The several aquifers below, the "300-ft," "500-ft," "700-ft," and "900-ft" sands are progressively more saline.

In the injection zone from 850 to 1,025 ft (259 to 312 m), the chloride content ranges from 8,500 ppm to 12,500 ppm. The artesian pressure is very high in the injection zone (about 90 ft or 27 m above sea level). Heads in the upper zones are progressively lower as land surface is approached. Aquifers are confined by units in which clay predominates.

Injection of the effluent, containing acetic acid, formic acid, and methanol, began in May 1968 at a rate of 200 gpm (757 litre/min).

With the initiation of injection, pressure in the receiving zone rose rapidly. By September 1968, head pressure in the injection well had reached the equivalent of 300 ft (91 m) above sea level, and in the observation wells, about 165 ft (50 m). Pressure continued to build until it exceeded 450 ft (137 m) in the injection well (Well 6) in June 1969. Because it was no longer possible to continue waste injection at the 200-gpm rate, at the permissible injection pressure limit of 150 psi (10 kg/sq cm), and as the observation wells were no longer of use because the waste front had passed them, permission was granted by the North Carolina Department of Natural Resources to operate Wells 4 and 5 as emergency injection wells. The pressure surface established during this period is shown in Figure 71, indicating upward leakage at Well 6 into shallower aquifers. This condition was known for a long time, as indications of the waste had been first noted at Observation Well 3 in February 1969. Leakage also was occurring upward at Well 1. Although a new injection well was constructed and placed in operation in January 1971, it was concluded that the system was not feasible. A conventional waste treatment facility was designed, and the injection of waste ceased in November 1972. [12)]

Orange County, Florida -

Drainage wells are extensively used to convey storm-water runoff and excess surface water to the Floridan aquifer, the principal aquifer for peninsular Florida. There are an esti-

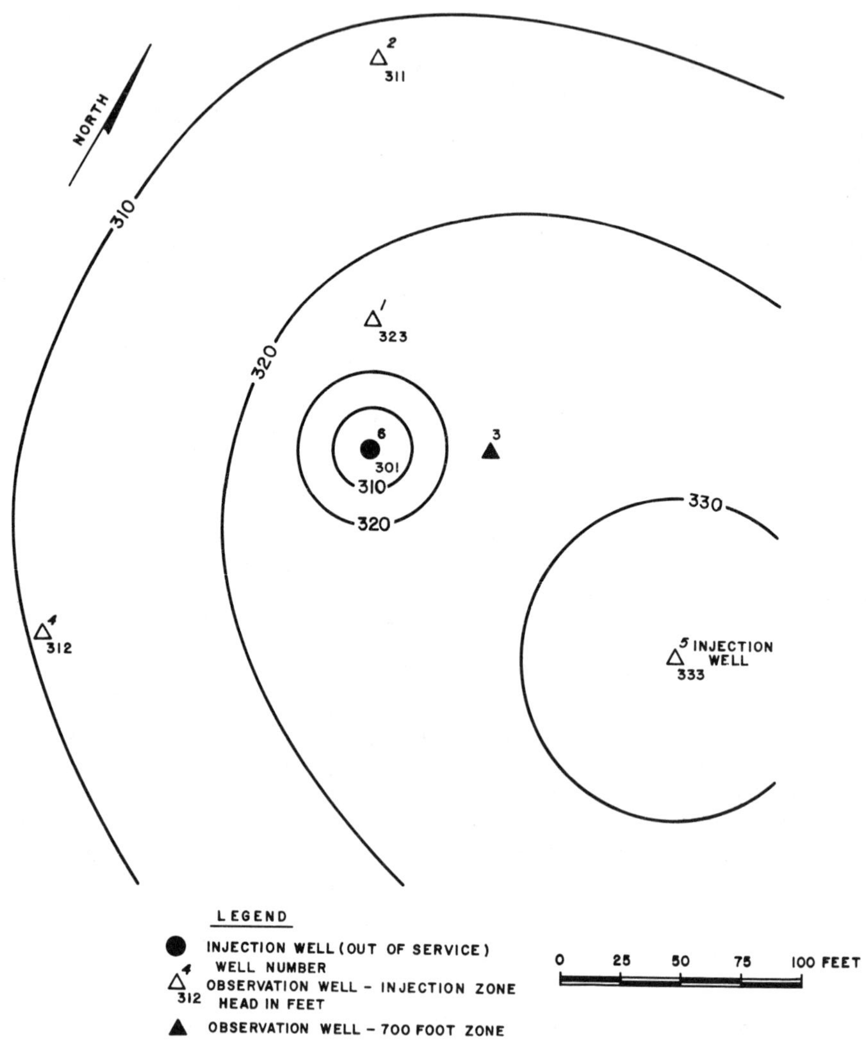

Figure 71. Map of pressure surface in feet above sea level, September 1, 1970, with Well 6 out of service and Well 5 in use as injection well. 12)

mated 6,500 of these wells in Florida. [9] In Orange County alone, 400 such wells have been drilled since 1904, most of them in the Orlando-Winter Park area. [13]

The potential capacity of these wells in the Orlando area ranges from 100 to 9,500 gpm (380 to 36,000 litre/min). Although the total volume of water recharged through drainage wells is unknown, it is undoubtedly high and the reason that no appreciable cone of depression has developed in the Orlando-Winter Park area despite combined pumpage exceeding 50 mgd (190,000 cu m/day) at times. [14]

Widespread contamination, primarily bacterial, has been reported in the upper part of the aquifer. In the lower zone, from which most public supplies draw their water, only local, isolated problems have been reported. However, the potentiometric head of the lower zone is always below that in the upper zone in the area. Therefore, it may only be a matter of time until downward movement of contaminated water is detected.

The danger of contamination to the aquifer could be considerably reduced by the elimination of the present drainage wells. However, expensive drainage canals and pumping stations would have to be built to replace the drainage wells and conduct runoff to the ocean. The resulting decrease in recharge could encourage the upward movement of salty water from deeper sections of the aquifer at centers of heavy pumping. The upper limit of this salty water is now reportedly only 500 to 1,000 ft (150 to 300 m) below the bottoms of municipal supply wells in the area. [14]

Two alternatives to maintain recharge without restricting development are transporting excess water to natural recharge areas, allowing surficial sediments to treat the water, and treating runoff and then injecting it into recharge wells. The latter would require considerable study to determine optimum treatment methods, plant design, and geochemical effects.

Pacific Northwest -

An activity which is widely known to occur but is only regionally recognized as a waste disposal practice is the use of wells to dispose of domestic and sanitary waste effluent. This procedure is used in isolated instances nationally and is a recognized waste disposal practice in the northwestern states, where extensive areas of nearly flat-lying beds of lava and sedimentary rock are found.

There are an estimated 3,000 disposal wells in Oregon alone, the majority of which are used for domestic waste disposal.[10] In 1968, a comprehensive investigation was made of disposal wells in the lava terrane of central Oregon. The disposal wells in this area are tied into septic tank systems and are generally drilled to the underlying lava, to a depth where cracks and crevices can receive and disperse the effluents. Most disposal wells in the area are 100 to 300 ft (30 to 90 m) deep and water levels are 400 to 600 ft (122 to 183 m) deep.[15] A schematic diagram of a typical sewage disposal well in lava terrane is shown on Figure 72. No major ground-water quality deterioration was noted during the survey, but little or no filtration is provided by the lava rock and the threat of contamination exists. Deep, uncased water wells were found to be particularly vulnerable. Figure 73 illustrates the mechanism of potential contamination of an uncased water well by septic effluent discharged into a disposal well. Recent laws passed in Oregon require permits for waste disposal wells as of January 1, 1975, and prohibit construction or use of such wells after January 1, 1980.[16]

In Idaho, there are some 5,000 disposal or drain wells in the Snake River Plain disposing of sewage effluent, street runoff, irrigation excess, and industrial wastes into lava and interbedded sediments. Approximately 3,000 drain wells are concentrated in Lincoln, Jerome, and Gooding Counties. Results of an investigation by the Idaho Bureau of Mines indicate that drain wells are used where geologic conditions at or near the ground surface render conventional disposal systems impractical. In many areas soil cover is thin and often impermeable, and septic tanks cannot function properly. In addition, it was found that many people believe that wastes will be rendered harmless by natural purification in the subsurface.[16]

The numerous openings, fractures, channels, and lava tubes in the basalt make it possible to discharge large volumes of waste fluid without any apparent effects. For this reason, disposal wells are seldom cased beyond the first basalt layer and are almost never grouted. Most are used to dispose of septic effluent. Others are used to remove excess irrigation water or to dispose of street runoff where storm sewers are absent.

Contamination of a public water-supply well by a drain well was detected in 1960 in the city of Idaho Falls. Fluorescein dye was injected to trace the contaminant and appeared in the supply well located 190 ft (57 m) from the drain well in 90 minutes.[16] Bacterial contamination of domestic

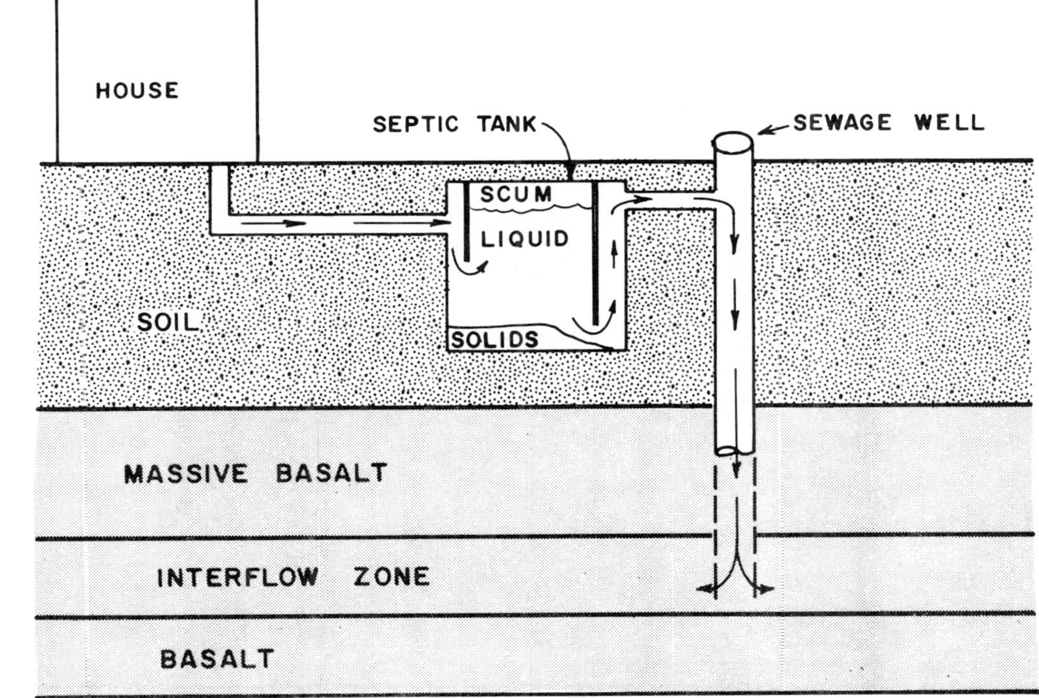

Figure 72. Schematic diagram of a typical sewage disposal well in lava terrane. 16)

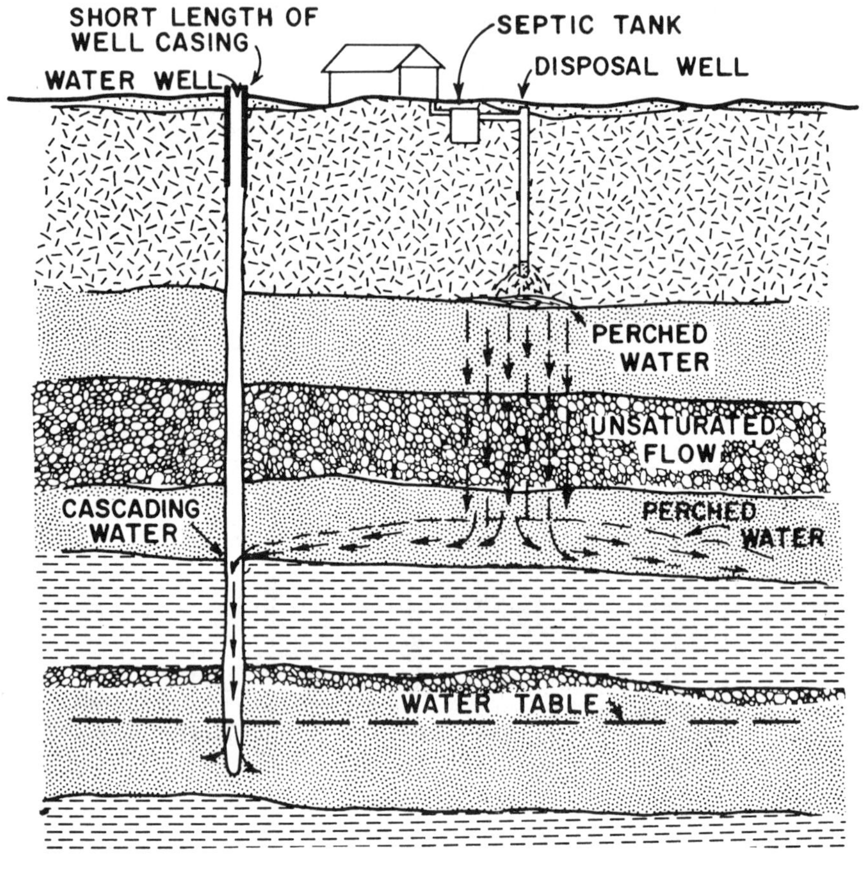

Figure 73. Potential contamination of ground water from perched water entering uncased water well. 15)

wells by waste water discharged into irrigation drain wells has been documented south of Idaho Falls.

In spite of the large volume of waste being discharged to the Snake River Plain aquifer, reported incidents of serious contamination are few. This is attributed to the excellent permeability of the aquifer which allows rapid movement of ground water to carry away the fluid wastes.

Also in the Snake River Plain, disposal well operations at the Idaho National Engineering Laboratory have resulted in the discharge to the subsurface of 1.6×10^{10} gal. (6.1×10^{7} cu m) of liquid waste containing 7×10^{4} curies of radioactivity and 1×10^{8} lb (4.5 million kg) of chemicals. This has contributed to plumes of chloride and tritium covering an area of 15 square miles.

TECHNOLOGICAL CONSIDERATIONS

Technology is presently available to construct an injection well which will not contaminate usable ground water because of a flaw in well construction. Historically, construction has been far from adequate, primarily due to lack of regulation. As regulation to protect ground water increases, the number of poorly constructed injection facilities will decrease. The threat of contamination from high volumes of low toxicity wastes will decline as these wastes are treated before injection or are disposed of in other ways. At the same time, with technological advances, the number of monitored waste facilities will increase as technicians become more confident in their abilities to control wastes underground.

The two areas of major concern in siting, design and construction of injection wells are acceptable operation and adequate ground-water protection. Although primary interest in this report is the protection of ground water, the successful physical operation of any waste disposal practice decreases the urgency, at least from the waste producer's viewpoint, of searching for alternatives.

The ideal injection formation is sufficiently porous and permeable to receive the volume and rate of flow involved, is itself chemically compatible, and contains fluid which is chemically compatible with the injectant so as to avoid problems of precipitation and other reactions. In addition, formations which receive undesirable fluids should be overlain by sufficiently impermeable units to prevent the upward migration of fluids into shallow potable water zones and zones with recoverable mineral resources.

In some cases, environmentally acceptable disposal can be attained with a minimum of treatment. Therefore, wastes are often injected with little pretreatment. Commonly, pretreatment has been employed solely to maintain the operation of the well. Plugging of the bore hole and formation wall is a major cause of early operational failures. To maintain flow rates without excessive injection pressures, some of the pretreatment practices are as follows:

1. Filtration to remove fine suspended particles.

2. Aeration to prevent the formation of ferrous hydroxide precipitate and the growth of iron bacteria on and in the vicinity of the well bore.

3. Removal of oils to prevent permeability reduction.

4. For wastes containing high concentrations of biochemical oxygen demand, treatment, filtration and sterilization with a suitable bactericide to avoid the growth of bacteria.

Two other treatment methods are occasionally employed to insure continued operation. Where waste waters are incompatible with fluid or rock in the injection zone and reactions might cause reduced permeability, a non-reactive buffer solution may be injected ahead of the waste. In some locations, where highly acidic waste is injected into limestone, the waste is neutralized to make it compatible with the formation, which is alkaline. However, this practice is no longer followed where it has been noted that permeability and, therefore, operational efficiency can be increased by introducing acids directly to cause dissolution of the limestone.

Ground water can become contaminated by injection well operations in many ways. The greatest potential for contamination occurs in the immediate vicinity of the well bore. Here the natural geologic structure which is relied upon to contain the waste water has been breached by the well construction.

To reinforce the geologic structure, cement grout is forced around the casing. In the case of the most noxious and corrosive fluids, multiple well casings are telescoped and cemented down to the receiving formation. Still, this cement seal is not always as impenetrable as the original natural structure. Thus, the well bore becomes the weak point of the system. Acidic wastes can corrode the casing. Excessively high injection pressures can rupture the casing joints and crack the cement envelope. These events can cause the waste to move out into shallower and possibly potable water-bearing zones. Excessive injection pressure can cause leaks at the well head itself, allowing effluent to flow across the land surface, perhaps contaminating ground water and surface water on the way.

Even beyond the well bore itself, the subsurface structure may be influenced by injection. For example, dynamite used to speed up drilling in one section of the hole might produce a fracture zone which could conduct fluids upwards toward potable water. Fluids can migrate along natural faults and fractures under the influence of injection pressures. The lubrication of faults and fractures by fluid pressure at the Rocky Mountain Arsenal produced an increased frequency of earthquakes near Denver, Colorado.

Other wells in the vicinity of an injection well drilled to or through the receiving formation, like unplugged abandoned oil and gas wells, can provide a conduit for the migration of fluids. Injection can reverse the hydraulic relationship such that wells with corroded casings, which may have been transferring shallow fresh water downward into the deeper saline formations, will now transfer saline water upward into potable zones. For example, the development and improper abandonment of the Lima-Indiana petroleum field has eliminated the Trenton limestone for injection. In the late 1800's and early 1900's, nearly 75,000 wells were drilled in that field. Many of the locations are still unknown. [17)]

The best security is to not install injection wells in areas of extensive petroleum exploration. Nonetheless, these areas have been most popular for injection wells because of the availability of subsurface information. As can be seen in Table 58, industrial injection wells proliferate in major petroleum producing states such as Kansas, Louisiana, Michigan, Oklahoma, and Texas.

Monitoring of injection operations has been grossly inadequate. For many types of injection, like drainage wells and sewage effluent wells, no monitoring has occurred. Even for industrial waste and radioactive waste injection, monitoring is meager. Operations are monitored to make sure that the proper amount of effluent is going into the well safely, but that is usually the limit of monitoring. Only infrequently is monitoring of the injection zone or shallower zones undertaken. Such information is critical, as can be seen in the case history for Wilmington, North Carolina. Obviously, monitoring facilities cost money but they are a necessity to maintain surveillance over the disposal operation and to provide data on potential or occurring contamination.

As has been noted, technology exists to overcome problems inherent with the construction and operation of injection wells, assuming that they are located in geologically suited areas. Thus, injection wells can be an effective and environmentally sound disposal method that presents no more than a minimal threat to usable ground water.

Cost Factors

As was stated previously, one factor which makes injection wells so attractive as a waste disposal method is the relatively low cost. Generally, construction and operation of injection wells is less than for other waste disposal facilities. The acceptability of a lesser degree of treatment for the effluent will also lower overall costs.

Obviously the cost of construction and operation is highly variable, depending upon the type of injection well. For the domestic or small industrial waste producer with an abandoned well available, the cost may be practically nil. For the individual homeowner in Oregon trying to dispose of domestic sewage effluent in an area where conventional disposal systems can not easily be constructed, drilling an effluent drainage well may be the cheaper alternative.

A breakdown for industrial waste injection wells is provided in Tables 60, 61, and 62. Table 60 compares costs of constructing facilities to pump equal volumes of filtered, noncorrosive waste into a 5,000-ft (1,524-m) deep well. Rock is hardest in the Great Lakes area, hence the higher drilling cost. Table 61 indicates average installation costs for existing facilities in the three areas, and Table 62 indicates average operating costs. Table 63 compares capital and annual operating costs for four actual plants operating both well and surface treatment facilities. Capital investments are usually less for wells, and operational costs of well facilities represent a clear saving. [18)]

For comparison with the above figures, Reichhold Chemicals, Inc., allocated $675,000 for drilling and testing of a 5,500-ft (1,676-m) test well in Alabama. Supervised by the Alabama Geological Survey, this well was used to develop data on porosity, confinement potential, fluid and formation compatibility and other conditions relative to waste injection with complete environmental safety. [18)] Standard Oil of Ohio had been incinerating an industrial waste stream at a cost of $600,000/year. The estimated operating cost of a 3,000-ft (914-m) injection well was $100,000/year. [19)]

In a feasibility study for the deep-well injection of waste

Table 60. COMPARATIVE COSTS. 18)

	Great Lakes	Mid-Continent	Gulf Coast
Drilling	$122,000	$ 94,000	$ 62,000
Equipment	45,000	45,000	63,000
Surface facilities	142,000	98,000	98,000
Total	$309,000	$237,000	$223,000

Table 61. AVERAGE INSTALLATION COST. 18)

Area	Number of wells	Depth, ft.	Cost per well	Cost per ft.
Great Lakes	12	3,290	$294,400	$ 90.00
Mid-Continent	6	3,580	175,300	50.00
Gulf Coast	17	3,200	362,400	110.00

Table 62. AVERAGE OPERATING COST. 18)

Area	Number of wells	Injection per well per year million gal.	Annual cost, million gal.
Great Lakes	10	62.3	0.42
Mid-Continent	5	68.2	0.16
Gulf Coast	14	69.0	1.17

Table 63. ECONOMIC COMPARISON OF WELL VS. SURFACE SYSTEM. 18)

	Capital		Annual Operating Cost	
Plant	Well	Surface	Well	Surface
A	$225,000	$ 500,000	$ 20,000	$100,000
B	300,000	140,000	52,000	178,000
C	468,000	1,250,000	62,000	195,000
D	270,000	*	100,000	600,000

Plant A - Mid-Continent
Plant B - Gulf Coast
Plant C - Great Lakes
Plant D - Great Lakes

* Not available

brine from inland desalting plants, it was stated that oil-field brine injection cost was $0.25 to $0.70/1,000 gal. (3,785 liters). By comparison, industrial waste injection cost was $1.00 to $2.00/1,000 gal. with satisfactory pre-treatment. [20]

The cost for construction and operation of other injection facilities lies between that of industrial waste and brine injection wells. In West Palm Beach, Florida, a 3,500-ft (1,667-m) test well for municipal sewage effluent injection cost nearly $500,000. A cost breakdown per acre-foot of water in fiscal year 1966-67 for the successfully operated West Coast Basin Barrier Project, Los Angeles County, California, was as follows: [21]

operation and maintenance	$13.00/acre-ft ($10.50/1,000 cu m)
cost of filtered Colorado River water	$24.00/acre-ft ($19.00/1,000 cu m)
capital outlay for facilities	$10.00/acre-ft (8.00/1,000 cu m)

Other, less obvious costs sometimes need to be considered. In areas where petroleum exploratory and abandoned production wells penetrate the injection zone, an expensive plugging program may be necessary. In the Hubbard Creek Watershed in Appalachia, 60 abandoned oil and gas wells were plugged for $1,500 each in 1963-65. In a 1972 study, four abandoned wells in the Appalachian area were plugged at costs of $8,600 to $14,000 each. [22]

INSTITUTIONAL ARRANGEMENTS

Cost may no longer be a consideration in alternative waste disposal practices. According to EPA guidelines, injection shall be considered only if no more environmentally acceptable solution is available. Presently, in the short term, injection wells are often the least costly method of disposal. However, it is expected that more stringent pretreatment and monitoring requirements may equalize costs with other practices.

In keeping with the continuing doubts expressed about the practice of waste-water injection, not only about the acceptable types of wastes which may be injected, but whether wastes should be injected at all, no approach has been taken to protect ground-water quality on a national scale. Individual states have completely different views on the subject of injection well regulation and control. The views may differ in the administrative organization of the regulatory agencies, the form of controls, and the substance of controls. [23]

Four major concerns have been expressed over injection facilities which affect their regulation. The first concern is social, that injection may proceed more rapidly than the assessment of public policy and the adequacy of regulatory procedures. Secondly, the technical concern is for the limited extent of knowledge about subsurface conditions and high-pressure hydraulics. The managerial consideration of marshaling sufficiently experienced personnel is another problem. The fourth concern is a legal one -- the question of the definition of underground trespass and of subsurface "public" waters. [19]

Injection wells have traditionally been regulated by statutes and administrative organizations. Only about one-half of the states have regulations controlling the construction and operation of injection well systems. Few distinguish between reinjection of natural brines and the injection of wastes. [24] In fact, only three of the 50 states have regulated the use of injection for industrial waste. No state prohibits such injection, although nine states reject applications for industrial waste injection wells and otherwise discourage them. [19]

Even among those states with defined procedures, programs have been generally weak. Some of the deficiencies are: [19]

1. Inadequate legislative guidelines and technological criteria

2. Poorly defined jurisdictional linkage between various agencies

3. Staff unfamiliarity with many aspects of injection well practice

The confusion over the general acceptability of waste injection method and the permissible types of wastes has led to striking incongruities. For example, disposal of wastes into salaquifers more than 1,000 ft (305 m) deep has occurred in only 2 of the 11 states in the northeast region, New York and Pennsylvania. The attitudes of representatives of environmental agencies in all 11 northeastern states toward injection wells were very negative; the principal reason given was that geologic conditions are unfavorable. In most of these states, injection wells are not even considered. In

others, rigid constraints would nearly rule out consideration of this alternative for waste disposal. [11]

On the other hand, shallow wells, less than 1,000 ft (305 m) deep and completed in fresh-water aquifers, are used to dispose of a variety of liquid wastes including storm water, sewage, cooling water, and industrial effluent in the same 11 states. Shallow wells recharging inadequately treated sewage effluent or industrial waste water are known to exist but are considered to be illegal. In some areas, public agencies have encouraged experimentation or the use of shallow wells for the disposal or recharge of storm-water runoff, cooling water from air conditioners and tertiary treated sewage effluent. In the mid-states, injection of industrial effluent to the deepest aquifers is strictly controlled yet disposal of brines to the shallowest salaquifers continues.

A basic public policy issue underlies the whole discussion of injection wells: Under what circumstances should society find it reasonable to trade off the potential environmental risk for the benefits of injection? [19]

The first mention in national legislation of disposal through wells occurred in the Water Pollution Control Bill S.2770 (1970) where brief references were made to "disposal in wells and subsurface excavations." [25] The first statement of national policy on deep well disposal was made by the Federal Water Quality Administration in 1970 before it became part of EPA. It was stated that the Federal government was opposed to deep well disposal "without a clear demonstration that such disposal will not harm present or potential subsurface water supplies....or otherwise damage the environment." The conclusion of the policy statement recognized subsurface injection as a technique limited in space and time to be used carefully and only until better methods of disposal are developed. [25]

A restatement of EPA policy is that "subsurface injection will be used only after all alternative measures have been explored and found to be less satisfactory in terms of environmental protection." [26] To meet its responsibilities to provide environmental protection, EPA is developing a program under Part C of the Safe Drinking Water Act of 1974 (PL 93-523), which is a Federal/state cooperative effort based on Federally set minimum standards and regulations administered by the states.

REFERENCES CITED

1. Haynes, C. D., and D. M. Grubb. 1969. Design and cost of liquid-waste disposal systems. Natural Resources Center, University of Alabama, Report 692, Tuscaloosa, Alabama.

2. Donaldson, E. C., R. D. Thomas, and K. H. Johnston. 1974. Subsurface waste injection in the United States: 15 case histories. U. S. Bureau of Mines, Information Circular 8636.

3. Stallman, R. W. 1972. Subsurface waste storage - the earth scientist's dilemma. Pages 6-10 in American Association of Petroleum Geologists. Memoir 18 - underground waste management and environmental implications. Tulsa, Oklahoma.

4. Reeder, L. R., and others. 1975. Review and assessment of deep well injection of hazardous waste. Solid and Hazardous Waste Research Laboratory, National Environmental Research Center, Cincinnati, Ohio. Open file.

5. Paul, J. R., and T. S. Talbot. 1969. Subsurface disposal of solids. Presented at 98th Annual Meeting of American Institute of Mining, Metallurgical and Petroleum Engineers, New York.

6. U. S. Environmental Protection Agency, Office of Water Program Operations, Oil and Special Materials Control Division, Special Sources Control Branch. 1974. Compilation of industrial and municipal injection wells in the United States. U. S. Environmental Protection Agency Report EPA-520/9-74-020.

7. Brown, D. L., and W. D. Silvey. 1973. Underground storage and retrieval of fresh water from a brackish-water aquifer. Pages 379-419 in Jules Braunstein, ed. Underground waste management and artificial recharge, Vol. 1. American Association of Petroleum Geologists, Tulsa, Oklahoma.

8. 1973. Known incidents of ground-water pollution from subsurface injection. Presented at Hearings before the Subcommittee on Public Health and Environment of the Committee on Interstate and Foreign Commerce, House of Representatives, Ninety-third Congress, first session on HR 5368, H.R. 1059, H.R. 5348 and H.R. 5995, March 8 and 9, 1973.

9. Vernon, R. O. 1970. The beneficial uses of zones of high transmissivities in the Florida subsurface for water storage and waste disposal. Bureau of Geology, Division of Interior Resources, Florida Department of Natural Resources, Information Circular 70.

10. van der Leeden, F., L. A. Cerrillo, and D. W. Miller. 1975. Ground-water pollution problems in the northwestern United States. U. S. Environmental Protection Agency Report EPA-660/3-75-018.

11. Miller, D. W., F. A. DeLuca, and T. L. Tessier. 1974. Ground water contamination in the northeast states. U.S. Environmental Protection Agency Report 660/2-74-056.

12. Peek, H. M., and R. C. Heath. 1973. Feasibility study of liquid-waste injection into aquifers containing salt water, Wilmington, North Carolina. Pages 851-875 in Jules Braunstein, ed. Underground waste management and artificial recharge, Vol. 2. American Association of Petroleum Geologists, Tulsa, Oklahoma.

13. Lichtler, W. F., W. Anderson, and B. F. Joyner. 1968. Water resources of Orange County, Florida. Division of Geology, State Board of Conservation, State of Florida, Report of Investigations 50.

14. Lichtler, W. F. 1972. Appraisal of water resources in the east central Florida region. Bureau of Geology, Division of Interior Resources, Florida Department of Natural Resources, Report of Investigations 61.

15. Sceva, J. E. 1968. Liquid waste disposal in the lava terrane of central Oregon. Technical Projects Branch, Northwest Region, Federal Water Pollution Control Administration, Report FR-4.

16. Abegglen, D. E. 1970. The effect of drain wells on the ground-water quality of the Snake River Plain. Idaho Bureau of Mines and Geology Pamphlet 148.

17. ORSANCO Advisory Committee on Underground Injection of Wastewaters. 1973. Recommendations for underground injection of wastewaters in the Ohio Valley Region. Ohio River Valley Water Sanitation Commission (ORSANCO), Cincinnati, Ohio.

18. Wright, J. L. 1969. Underground waste disposal. Industrial Waste Engineering, May 1969.

19. Cleary, E. J., and D. L. Warner. 1969. Perspective on the regulation of underground injection of wastewaters. Ohio River Valley Water Sanitation Commission (ORSANCO), Cincinnati, Ohio.

20. Beogly, W. J., and others. 1969. The feasibility of deep-well injection of waste brine from inland desalting plants. Division of Technical Information, U. S. Atomic Energy Commission. Report TID-25081.

21. Bruington, A. E. 1969. Control of sea-water intrusion in a ground-water aquifer. Ground Water 7(3):9-14.

22. U. S. Environmental Protection Agency, Office of Air and Water Programs. 1973. Ground-water pollution from subsurface excavations. U. S. Environmental Protection Agency Report EPA-430/9-73-012.

23. Walker, W. R., and W. E. Cox. 1973. Legal and institutional considerations of deep well disposal. Pages 3-19 in Jules Braunstein, ed. Underground waste management and artificial recharge, Vol. 1. American Association of Petroleum Geologists, Tulsa, Oklahoma.

24. Ballantine, R. K., S. R. Reznek, and C. W. Hall. 1972. Subsurface pollution problems in the United States. U.S. Environmental Protection Agency Report TS-00-72-02.

25. Greenfield, S. M. 1972. U. S. Environmental Protection Agency - the environmental watchman. Pages 14-18 in American Association of Petroleum Geologists Memoir 18 - underground waste management and environmental implications. Tulsa, Oklahoma.

26. Keeley, J. W. 1971. Need for ground-water protection in subsurface disposal and surface impoundments of petrochemical wastes. Presented as testimony before the Senate Subcommittee on Air and Water Pollution, April 5, 1971.

SECTION XIV

DISPOSAL OF ANIMAL FEEDLOT WASTE

SUMMARY

The generation and disposal of large quantities of animal waste at locations of concentrated feeding operations is a relatively new environmental problem. Case histories of actual contamination of ground water caused by animal feeding operations are almost non-existent. However, because such practices are relatively new, assessment of potential problems is still underway.

There are three primary mechanisms of ground-water contamination from animal feedlots and their associated treatment and disposal facilities: (1) runoff and infiltration from the feedlots themselves, (2) runoff and infiltration from waste products collected from the feedlots and disposed of on land, and (3) seepage or infiltration through the bottom of a waste lagoon. The principal contaminants are phosphate, chloride, nitrate, and in some cases, heavy metals.

Cattle are the most serious potential problem in terms of the volume of waste produced but sheep, poultry, and hog feeding operations also represent potential sources of ground-water contamination. During its 120- to 150-day stay in the feedlot, each beef animal will produce over one-half ton (0.45 tonne) of manure on a dry weight basis. In January 1975, there were almost 10 million cattle in feedlots of more than 1,000-head capacity.

The two leading cattle feedlot regions, the Corn Belt and the Northern Plains, form a grain-farming and livestock-growing belt that extends easterly from the south-central part of the Northern Plains, traverses the Missouri and Mississippi Rivers and terminates in western Ohio. Other significant feedlot areas are found in California, Arizona, New Mexico, Texas, and Washington. Principal states for poultry raising are located in the south, for hogs in the midwest, and for sheep in the southwest and in the far west.

Application of manure to land for its fertilizer and soil conditioner value is the classic system through which manure has been utilized. Several methods have been proposed for converting manure to energy products, the principal one involving thermochemical processes for conversion to methane, oil, and/or synthesis gas.

"Concentration animal feeding operations" are regulated under the Federal Water Pollution Control Act Amendments of 1972, and thus may be required to have a permit as a "point source" under the NPDES. State animal-feedlot regulations typically apply to the situation where the ratio of the number of animals to land area is high.

DESCRIPTION OF WASTE DISPOSAL PRACTICE

The generation and disposal of large quantities of animal waste is a relatively new environmental problem. With the increasing demand for more and better quality meat, livestock producers have responded with thousands of large concentrated feeding operations. 1) Cattle feedlots are the most serious problem in terms of volume of waste produced, and are given major emphasis in this report. However, sheep, poultry and hog feeding operations also represent potential sources of ground-water contamination.

Until 10 or 15 years ago, most beef animals were raised on pasture land where wastes were easily assimilated into the soil without significant surface-water or ground-water contamination. In recent years centralized feeding operations have increased sharply and by January 1975, there were almost 10 million cattle in feedlots of 1,000- to 50,000-head capacity. 2)

During its 120- to 150-day stay in the feedlot, each beef animal will produce over one-half ton (0.45 tonne) of manure on a dry weight basis. The heavy concentrations of animal wastes can overtax the natural assimilative capacity of the soil, and runoff and infiltrating rainfall can carry high concentrations of contaminants to both ground- and surface-water bodies. Special collection, treatment and other control facilities are normally employed in an attempt to reduce the environmental impact of the wastes.

The ground-water contamination problems associated with animal feeding operations are diverse. Considerable research and numerous demonstrations of waste management techniques have been conducted in recent years. Some of the procedures which were developed for the collection and handling of animal wastes from feeding areas are shown in Figure 74. Essentially, solids and liquids are separated by a mechanical device or settling pond, or removed from the site as a slurry. If separated, the solid and liquid portions are removed to their respective treatment, disposal, or reuse facilities.

Generally, only the larger installations (more than 1,000-head capacity) have waste treatment facilities. The various

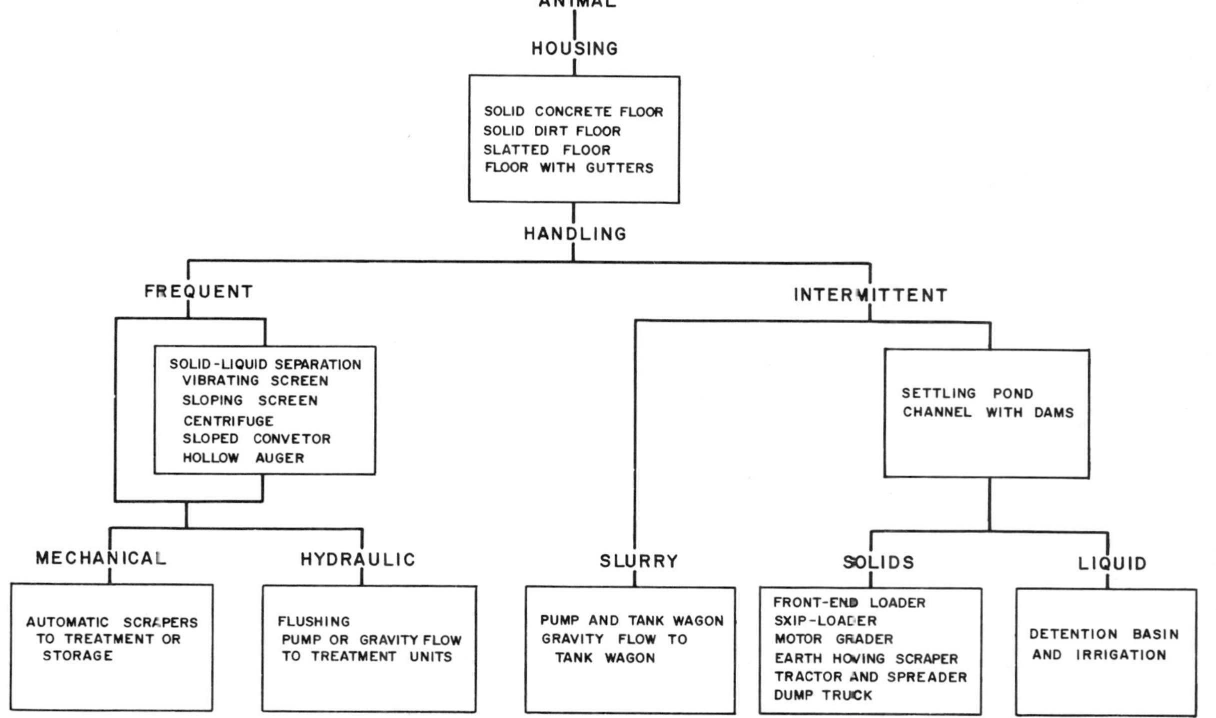

Figure 74. Procedures for the collection and handling of arimal wastes.[3]

alternatives for treatment are shown in Figure 75, and alternatives for the utilization or disposal of solid wastes and treated effluent, are shown in Figure 76.

There are three primary mechanisms of ground-water contamination from animal feedlots and their associated treatment and disposal facilities:

1. Runoff and infiltration from the feedlots themselves.

2. Runoff and infiltration from waste products disposed on land.

3. Seepage or infiltration under a feedlot or through the bottom of a waste lagoon.

CHARACTERISTICS OF CONTAMINANTS

The characteristics of the contaminants generated by animal feeding operations are quite variable. A comparison of these contaminants is shown in Table 64. Data from a number of sources have been compared and certain constituents listed. The concentrations of contaminants in the runoff from a feedlot vary by a factor of as much as two in relation to the surface slope of the lot, and whether it is paved or unpaved. They are also influenced by the feeding method.

Table 65 shows analyses of runoff from unpaved (dirt) and paved surface lots, where different rations are fed to the cattle. The concentrations of contaminants in all cases shown on both Tables 64 and 65 are significant and pose a potential ground-water contamination problem. Major contaminants that have been indentified and are monitored quite closely are: the nitrogen compounds, which influence the concentration of nitrate in ground water; phosphate which leads to eutrophication in surface water; fecal coliform bacteria, which are not permissible in drinking water; chloride; and in some cases, heavy metals.

EXTENT OF THE PROBLEM

The exact extent of the ground-water contamination problem or potential problem from animal feeding operations is a much discussed issue in recent literature. To date, general inventories directed toward evaluating the effects of feedlots have not been conducted. Furthermore, the practice of mass feeding of livestock is a relatively new phenomenon and if problems are developing, they have not yet been recognized on any significant scale. Case histories of actual contamination instances are almost non-existent.

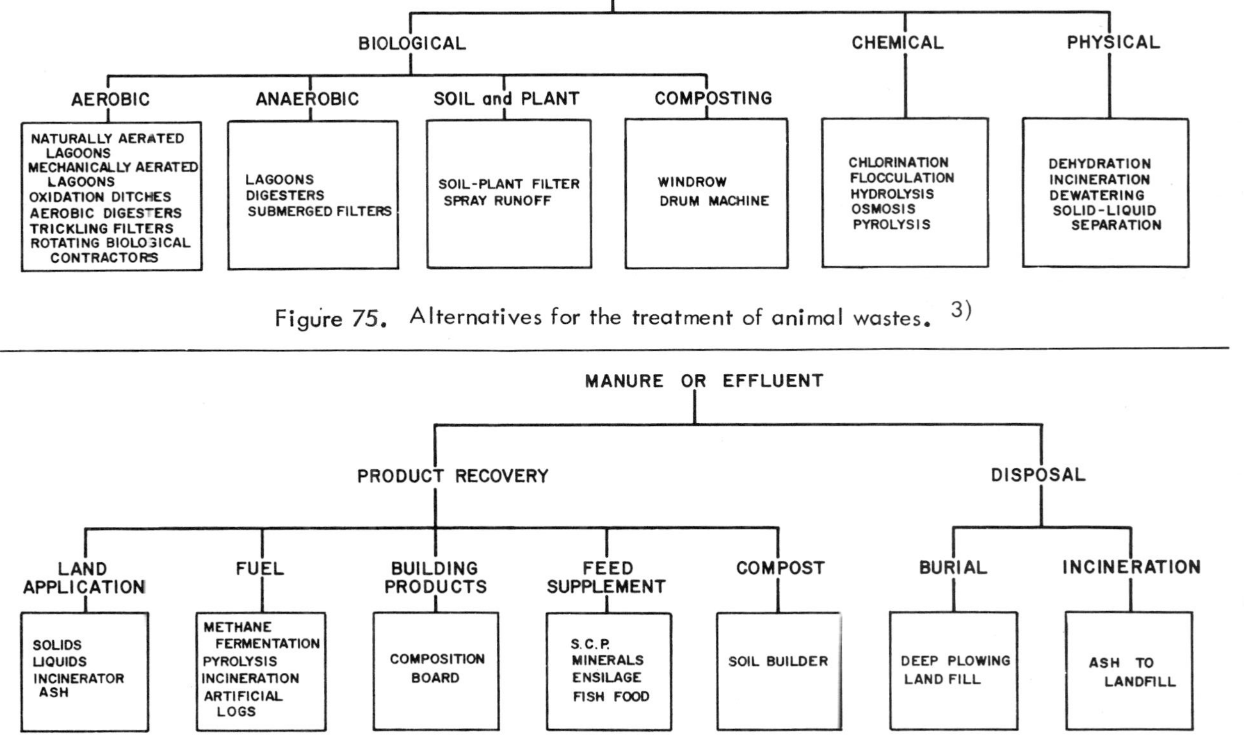

Figure 75. Alternatives for the treatment of animal wastes. 3)

Figure 76. Alternatives for the disposal and utilization of animal wastes. 3)

Table 64. COMPARISON OF CONTAMINANT CHARACTERISTICS REPORTED FROM FEEDING OPERATIONS. [4,5,6,7,8]
(all concentrations in ppm, unless noted)

Constituent	Average runoff (paved surface)	Average runoff (unpaved surface)	Average runoff (surface undefined)		
Sodium	235	1,000	2,280	1,675	2,490
Chloride	2,455	1,040	-	-	-
Nitrate	113	68	-	-	-
Total solids	11,260	4,900	-	-	-
Volatile solids	4,014	1,460	-	-	-
Suspended solids	4,250	1,096	-	-	-
Volatile suspended solids	2,720	619	-	-	-
Nitrogen	-	-	1.04 [a)]	3.12 [a)]	0.89 [a)]
Ammonia nitrogen	-	-	-	-	-
Potassium	-	-	1.09 [a)]	0.39 [a)]	0.97 [a)]
Calcium	-	-	7,757	5,580	9,800
Magnesium	-	-	3,934	4,450	4,200
Phosphorus	-	-	4,166	7,100	5,700
Zinc	-	-	66	-	56
Iron	-	-	8,825	-	4,810

a) As percent of total solids

Table 65. COMPARISON OF FEEDLOT RUNOFF WATER QUALITY. [9)]

Date	pH	BOD (ppm)	COD (ppm)	NO_3 (ppm)	NH_3-N (ppm)	ORG-N (ppm)	Alkalinity (ppm)
Concentrations of contaminants in runoff from concrete-surfaced lot on which cattle were fed roughage-concentrate ration							
8-24-69	6.70	10,067	28,450	875	519	300	2,104
8-26-69	6.15	8,500	32,800	320	515	384	2,056
9- 9-69	6.62	12,750	32,172	97	774	301	2,402
9-22-69	6.75	5,270	11,514	22	100	132	1,632
10-21-69	6.67	10,250	20,868	140	406	114	170
10-26-69	6.85	3,300	8,400	70	33	35	86
11- 1-69	7.00	5,566	16,252	0	115	115	336
Concentrations of contaminants in runoff from concrete-surfaced lot on which cattle were fed all-concentrate ration							
8-24-69	6.30	7,355	28,929	1,270	340	610	1,030
8-26-69	5.90	8,900	38,400	280	395	302	1,584
9- 9-69	5.60	10,400	30,230	228	518	797	1,174
9-22-69	6.90	4,424	10,080	36	460	650	1,380
10-21-69	6.35	8,300	20,742	386	304	293	116
10-26-69	6.70	12,000	23,000	350	140	235	236
11- 1-69	7.30	2,395	4,971	0	120	250	116
Average concentrations of contaminants in runoff from dirt lot on which cattle were fed roughage-concentrate ration							
8-24-69	7.68	1,758	8,280	103	79	200	1,238
8-26-69	7.40	1,010	2,964	24	77	25	928
9- 9-69	7.70	1,340	6,316	3	75	50	856
9-22-69	7.63	1,400	28,000	3	71	40	1,400
10-21-69	7.50	1,620	4,400	28	85	67	99
10-26-69	7.45	1,100	8,000	0	2	6	183
11- 1-69	7.70	2,200	8,795	0	3	117	436
Average concentrations of contaminants in runoff from dirt lot on which cattle were fed all-concentrate ration							
8-24-69	7.95	1,400	5,160	163	48	118	955
8-26-69	7.15	1,350	7,212	96	48	33	1,602
9- 9-69	7.62	1,145	6,220	6	83	25	864
9-22-69	7.30	1,580	4,817	62	75	20	746
10-24-69	7.35	1,390	4,042	24	50	22	70
11- 1-69	7.10	3,210	9,942	0	30	17	360

Some idea of the potential extent of the problem can be obtained from the number, capacity, and distribution of cattle feedlots. Table 66 lists the number and capacity (under or over 1,000 head) of feedlots in 23 states for 1974. The distribution of feeding operations by county and number of cattle is shown in Figure 77.

The two leading feedlot regions, the Corn Belt and the Northern Plains, form a grain farming and livestock growing belt that extends easterly from the south-central part of the Northern Plains, traverses the Missouri and Mississippi Rivers, and terminates in western Ohio. The rainfall in the two regions ranges from moderate in the west to abundant in the east. It is estimated that from 1962 to 1983, more than 0.8 billion tons (0.7 billion tonnes) of cattle feedlot wastes will have been generated in these two regions. (The Federal government has developed a variety of estimates on the volume of manure generated by feedlot cattle; these range from 4.5 to 11.7 tons/yr/1,000 lb, or 0.009 to 0.02 tonnes/yr/kg, live steer weight.) This represents about half of the United States total cattle feedlot waste during the projected period in regions that represent less than 15 percent of the nation's total area; i.e., a concentration of about six times that of the rest of the country. Table 67 shows the cattle production, feedlot acreages, and waste deposits in the three leading regions with projections, from 1962 through 1983.

The U. S. Department of Agriculture estimated in 1969 that 1.7 billion tons (1.5 billion tonnes) of cattle wastes are generated annually.[10)] This total probably exceeds waste production from any other segment of the national agricultural, commercial, and domestic complex. Of this total, only about 5 percent is deposited in feedlots, but the environmental threat of waste concentrated on feedlots is disproportionately large relative to the total cattle wastes.

The potential for ground-water contamination from all types of animal wastes is substantial even when compared to potential problems from human wastes. Table 68 shows the 1975 animal population in the United States and its corresponding man-equivalents of waste. Table 69 lists the principal states producing poultry, sheep and hogs. The volume of animal wastes is equal to about ten times that generated by the human population.

There are several potential contaminants in manure, but only one is frequently encountered in ground water -- nitrate. Nitrate is the oxidation product of organically bound nitrogen, ammonium, and nitrite. Ground water is vulnerable to

Table 66. NUMBER OF CATTLE FEEDLOTS AND CAPACITY, BY STATES - 1974. 2)

State	Under 1,000 head feedlot capacity		Over 1,000 head feedlot capacity		Total all feedlots	
	Number of lots	Cattle marketed (1,000 head)	Number of lots	Cattle marketed (1,000 head)	Number of lots	Cattle marketed (1,000 head)
Arizona	6	1	41	894	47	895
California	28	13	139	1,989	167	2,002
Colorado	425	131	188	1,761	613	1,892
Idaho	502	11	72	333	574	344
Illinois	14,445	755	55	95	14,500	850
Indiana	10,477	336	23	25	10,500	361
Iowa	31,835	2,710	165	387	32,000	3,097
Kansas	5,660	400	140	1,840	5,800	2,240
Michigan	1,667	177	33	65	1,700	242
Minnesota	10,970	795	50	69	11,020	864
Missouri	11,979	348	21	52	13,000	400
Montana	211	26	65	161	276	187
Nebraska	14,510	1,330	460	2,025	14,970	3,355
New Mexico	7	1	41	354	48	355
North Dakota	880	53	20	31	900	84
Ohio	8,175	328	25	58	8,200	386
Oklahoma	358	36	42	530	400	566
Oregon	305	22	26	104	331	126
Pennsylvania	5,997	114	3	9	6,000	123
South Dakota	9,123	407	77	178	9,200	585
Texas	1,001	85	199	3,814	1,200	3,899
Washington	165	33	21	268	186	301
Wisconsin	7,084	149	16	31	7,100	180
Total:	135,810	8,261	1,922	15,073	137,732	23,334

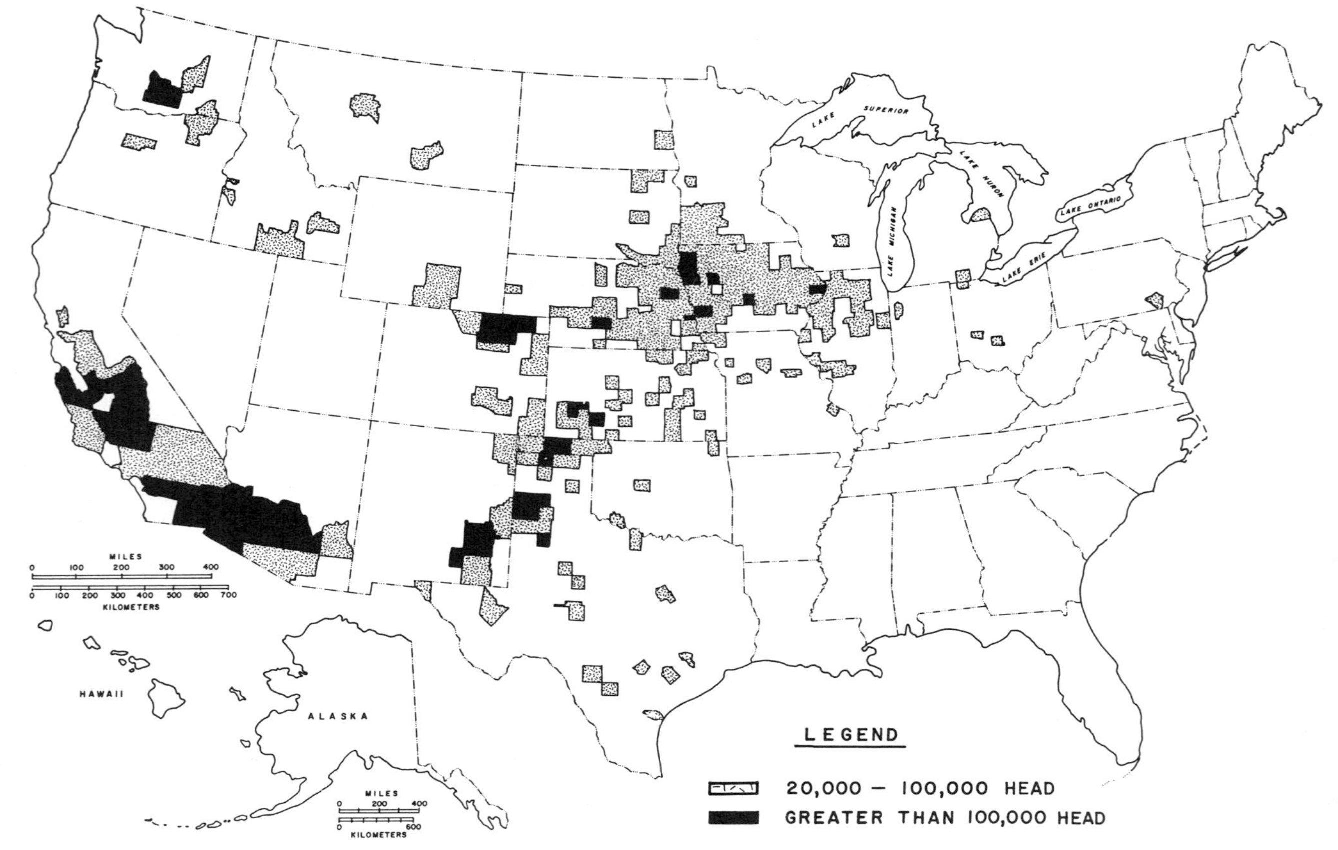

LAKE SUPERIOR
LAKE MICHIGAN
LAKE HURON
LAKE ERIE
LAKE ONTARIO
MILES
0 100 200 300 400
0 100 200 300 400 500 600 700
KILOMETERS
HAWAII
ALASKA
MILES
0 200 400
0 600
KILOMETERS
LEGEND
20,000 – 100,000 HEAD
GREATER THAN 100,000 HEAD

Table 67. GRAIN-FED BEEF CATTLE PRODUCTION, FEEDLOT ACREAGE, AND WASTE DEPOSITS OF THE THREE LEADING FEEDLOT REGIONS, 1962-1983. 6)

Region	1962	1968	1971	1975	1979	1983
Corn Belt						
Millions of cattle marketed and fraction of national total (percent)	5.23 (35)	7.28 (32)	6.64 (26)	7.42 (26)	8.23 (26)	9.04 (26)
Millions of tons of waste deposits	15.05	20.96	19.13	21.38	23.69	26.02
Thousands of feedlot acres	9.99	11.54	12.69	14.20	15.79	17.27
Northern Plains						
Millions of cattle marketed and fraction of national total (percent)	3.18 (21)	5.56 (24)	6.39 (25)	7.14 (25)	7.91 (25)	8.69 (25)
Millions of tons of waste deposits	9.17	16.02	18.39	20.55	22.77	25.01
Thousands of feedlot acres	6.08	10.63	12.20	13.63	15.11	16.60
High Plains						
Millions of cattle marketed and fraction of national total (percent)	1.07 (7)	2.71 (12)	4.58 (18)	5.12 (18)	5.67 (18)	6.23 (18)
Millions of tons of waste deposits	3.08	7.79	13.19	14.74	16.33	17.94
Thousands of feedlot acres	2.05	5.17	8.75	9.78	10.84	11.90
United States						
Millions of cattle marketed	14.96	23.04	25.70	28.72	31.82	34.96
Millions of tons of waste deposits	43.08	66.36	74.01	82.70	91.64	100.68
Thousands of feedlot acres	27.39	42.35	47.36	52.94	58.68	64.44

Table 68. UNITED STATES LIVESTOCK NUMBERS AND MAN WASTE EQUIVALENTS, 1975.[11,12]

Livestock	Thousands	Man equivalent per head	Thousands Total man equivalents
Cattle	131,826	16.4	2,161,946
Sheep	14,538	2.45	35,618
Hogs	55,062	1.90	104,618
Chickens	382,793	0.14	53,591
		Total man equivalents:	2,355,693

Table 69. PRINCIPAL STATES PRODUCING POULTRY, SHEEP AND HOGS IN 1973-1974. [13)]

Poultry (more than one billion broilers produced in 1973)	Hogs (more than one million on farms in 1973)	Sheep (more than one million on farms in 1974)
Alabama	Georgia	California
Arkansas	Illinois	Colorado
Delaware	Indiana	Texas
Georgia	Iowa	Wyoming
Maryland	Kansas	
Mississippi	Kentucky	
North Carolina	Minnesota	
Texas	Nebraska	
	North Carolina	
	North Dakota	
	Ohio	
	South Dakota	
	Texas	
	Wisconsin	

nitrate contamination because nitrate is soluble in water, and its concentration is essentially unchanged by contact with the soil matrix. Bacteria and phosphate, other manure-borne contaminants, are generally highly attenuated by soils and thus do not constitute a serious threat to ground water.

Factors which influence the susceptibility of ground water to contamination from all types of feedlots are stocking rate, manure removal management, depth to the water table, soil texture and structure, and volume of water going to ground-water recharge. [14] Low stocking rates and frequent manure removal allow better aeration of deposited wastes resulting in a high proportion of nitrate production. This is accompanied by greater water infiltration which can leach nitrate to ground water. [15] Such feedlots located where ground water is relatively close to the surface tend to contribute more nitrate to ground water than those located over a deep water table. [14]

Soil texture is actually less important as a factor in nitrate transport to ground water than is feedlot management which controls conditions at the soil surface. Heavy manure accumulations tend to produce a somewhat impermeable mat. The mat of manure creates its own physical characteristics which overcome those of the underlying soil. Anaerobic conditions, which allow denitrification, are likely within and below the mat. Nitrate may thus be volatilized, and little infiltration of precipitation will occur, thus limiting nitrate leaching. The potential for surface-water contamination from runoff is enhanced, however. Should runoff contaminate farm ponds, ground water could be contaminated by pond infiltration. [16]

Case Histories

An example of the principles described above is discussed in a report from a dairy in the Spokane Valley in Washington. [17] Soils were cored beneath a corral area and analyzed for chemical and bacterial contamination. Total coliform, fecal streptococci, and fecal coliform bacteria disappeared a few feet from the soil surface. Nitrate was evident throughout the length of the soil cores. However, the soils were well below field capacity in moisture content. These observations led to the conclusion that nitrate migration was a phenomenon which occurred early in the farm operation and was arrested as organic mass formed at the soil surface.

A survey of Holt County, Nebraska, reported a wide range of nitrate concentrations collected from 71 wells. [18] The concentrations ranged from 0.1 to 409 ppm (as NO_3). Ground water pumped close to feedlots was generally more enriched in nitrate than it was when pumped from more distant wells.

A study of several feedlots in the High Plains of west Texas showed that storage of wastes in unlined ponds can be a hazard to local ground-water quality.[16] Contamination of shallow aquifers was indicated in areas where direct runoff from feedlots was introduced onto agricultural lands and into unlined surface storage areas.

Poultry waste can cause disposal difficulties which are in some ways more acute than those from cattle feedlots. The wastes are generally more concentrated because of animal physiology. Poultry can be housed in areas of relatively dense human population which restricts disposal options while simultaneously increasing pressure for odor abatement. An example of a location stressed by poultry waste production is the Delmarva Peninsula, Delaware. There, 140 million broilers produce more solid waste annually than all of the citizens of New York City.[19] Most of the waste is recycled into the soil, but the available area is inadequate for proper utilization by crops. These conditions may lead to excessive soil salinity, poor crop yields, and leaching of nitrate.

TECHNOLOGICAL CONSIDERATIONS

Land Application

Application of manure to land for its fertilizer and soil conditioner value is the classic system through which manure has been utilized. A 1975 survey by EPA of ten major cattle feeding states has revealed that high costs and shortages of other types of fertilizers have increased the use of manure and even depleted some stockpiles.[20] Table 70 summarizes the results of the survey.

Thermochemical Conversion

Several schemes which have been proposed for converting manure to useful products by thermochemical processes are conversion to methane, conversion to oil, and conversion to synthesis gas.[21,22,23] While these experimental processes are not currently economically feasible for large-scale use, on-going research and increasing volumes of animal wastes concentrated at feedlots may change the situation in the future.

Table 70. SUMMARY OF SURVEY INFORMATION ON MANURE UTILIZATION.

State	Assessment of Amount of Manure Presently Being Utilized	How Utilized
Texas	Demand is greater than production. Stockpiles from previous years have been depleted.	As a fertilizer on grain sorghum, corn silage and wheat.
Iowa	100%	As a fertilizer on corn and soybeans.
Nebraska	Disposal of feedlot wastes is no longer a problem.	As a fertilizer on corn.
Kansas	Utilized at rate of generation.	As an organic fertilizer on corn for its nutrient and organic content.
California	Majority of wastes are being applied on cropland. This trend is increasing.	As a fertilizer on truck crops, vineyards and orchards.
Colorado	Utilization has increased two-fold in recent months.	As an organic fertilizer on corn for its organic as well as nutrient content.
Arizona	Stockpiles of manure from previous years are gone.	As a fertilizer on truck farms and citrus groves.
Illinois	A significant amount is presently utilized and its demand is increasing.	As a fertilizer on cropland.
Minnesota	The bulk of the wastes are utilized.	As a fertilizer on corn, sorghum and soybeans.
South Dakota	The majority of manure is being utilized. This rate has increased in recent months.	As a fertilizer on cropland.

Refeeding

Refeeding of manure to cattle is not sanctioned by the Food and Drug Administration in the United States, although it is permitted in Canada and the United Kingdom. Cattle defecate directly in feed troughs, and they more or less routinely "graze" on the manure contained in the feedlot.

Refeeding is considered by some investigators as a viable waste management technique. [24] Organic waste from ruminants has a chemical constitution similar to the feed ingested, and in addition, is enriched by the presence of an abundance of rumen microbial matter. Organic waste as it is voided by ruminants is a fermentation product which is biologically safe for animals, and it has none of the characteristics of organic waste products generally classified under the heading "filth."

Some attempts are being made to utilize manure for the production of flies, fly pupae, or fly larvae, which are all very rich in protein, and can provide a valuable protein supplement for animal feeds. [25] In the larval stage, flies will remove up to 80 percent of the organic matter of manure; converting what remains to a dry, reasonably stable, practically odorless product which retains value as a fertilizer or soil conditioner. [26] Other workers are utilizing microorganisms, rather than insects, as a means of converting manure into a usable protein. The advantage of microorganisms is that they reproduce much more rapidly than do insects. Thus, the degradation that can be accomplished in several days by insects can generally be accomplished in several hours by bacteria.

Costs -

Considerable amounts of research have been performed on costs required to provide acceptable waste management systems for disposal of animal wastes. Based on hearings before a subcommittee of the Committee of Government Operations, on the control of pollution from feedlots, [27] estimates of the total investment required, the investment per head, annual cost per head, and cost per hundredweight of beef marketed by size class have been tabulated in Table 71. The runoff control costs for fed-beef operations showing the range in cost increase per head marketed are given in Table 72. Figure 78 shows the cost per animal per day and investment cost for evaporation lagoons, which may be an acceptable method of waste management in the high evaporation climates of the southwest.

Table 71. TOTAL INVESTMENT, INVESTMENT/HEAD, ANNUAL COST/HEAD AND PER CWT OF BEEF MARKETED BY SIZE CLASS REQUIRED TO CONTAIN SURFACE-WATER RUNOFF FOR A 10-YEAR, 24-HOUR STORM EVENT ON FARMS JUDGED TO HAVE SURFACE-WATER POLLUTION PROBLEMS.[28)]

Size class (head)	Total investment (million dollars)	Total investment/ head (dollars)	Annual cost/ head (dollars)	Annual cost/ cwt of beef [a)] (dollars)
Eastern States [b)]				
100	91.8	145.20	21.20	4.24
100- 199	12.4	21.00	3.20	0.64
200- 499	10.1	11.65	1.85	0.37
500- 999	3.7	8.20	1.30	0.26
1,000	5.2	3.15	0.70	0.14
Western States [c)]				
1,000	7.4	21.70	5.80	1.16
1,000- 7,999	0.8	2.90	0.55	0.11
5,999-15,000	0.4	1.60	0.40	0.08
16,000+	0.9	1.40	0.35	0.07

a) Assuming 500 pounds of gain/animal marketed
b) Pennsylvania, Ohio, Indiana, Illinois, Iowa, Missouri, Michigan, Wisconsin, North Dakota, South Dakota, Nebraska, and Kansas
c) Oklahoma, Texas, Colorado, and California

Table 72. RUNOFF CONTROL COSTS FOR FED-BEEF OPERATIONS IN DOLLARS.[28]

Capacity (head)	Capital outlay per head (weighted average)	Range in capital outlay per head [a]	Cost increase per head marketed (weighted average)	Range in cost increase per head marketed [a]
< 100	145.20	-	21.17	-
100-199	21.00	13.42 - 46.84	3.19	2.57 - 6.55
200-499	11.60	8.66 - 37.61	1.84	0.64 - 5.04
500-999	8.18	4.44 - 20.73	1.28	0.68 - 2.49
1,000 and over	3.13	2.62 - 30.71	0.69	0.40 - 4.03

a) The range in these per head costs reflects differences in housing type, location, and per-unit charges for excavation and construction. In addition, as feedlot size was determined by marketings, the actual sizes of feedlots within a particular capacity category may vary among states, especially for smaller size classes; therefore, the per head cost increase range has been omitted for operations of less than 100 head.

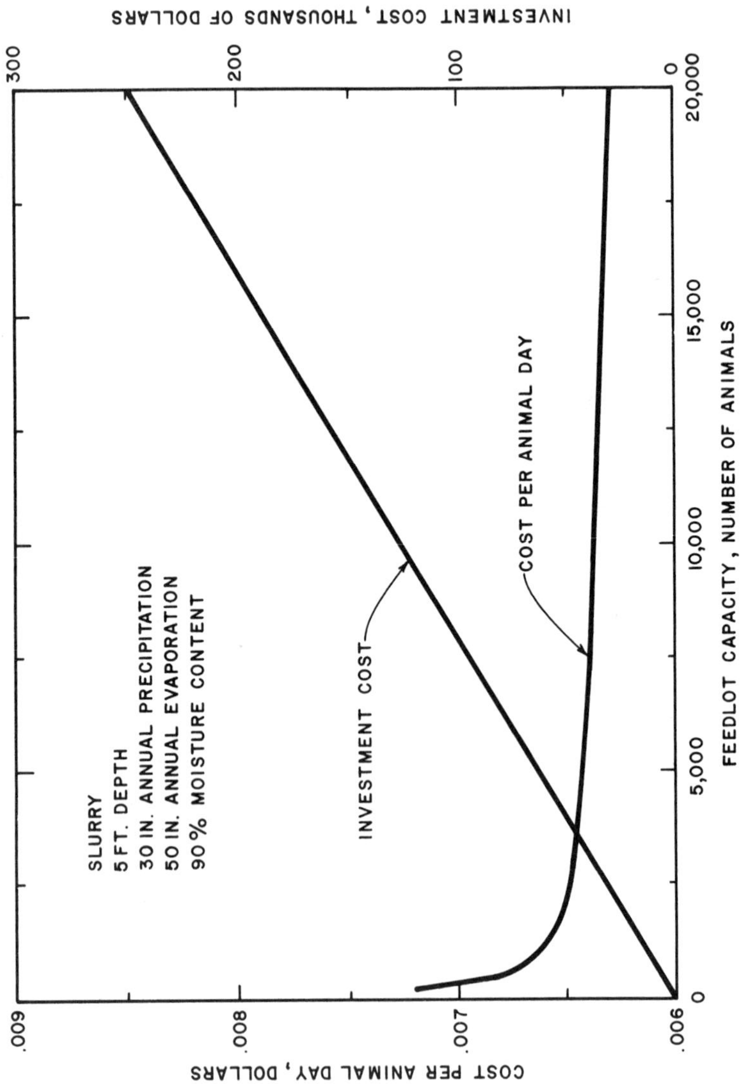

SLURRY
5 FT. DEPTH
30 IN. ANNUAL PRECIPITATION
50 IN. ANNUAL EVAPORATION
90% MOISTURE CONTENT

INVESTMENT COST
COST PER ANIMAL DAY

COST PER ANIMAL DAY, DOLLARS
.009
.008
.007
.006

INVESTMENT COST, THOUSANDS OF DOLLARS
300
200
100
0

0
5,000
10,000
15,000
20,000
FEEDLOT CAPACITY, NUMBER OF ANIMALS

Many waste management techniques have been and are being evaluated. Present price trends of fuel and feed are making these techniques more appealing than a few years ago. As more wastes are reused and recycled, the contamination potential will tend to decrease.

INSTITUTIONAL ARRANGEMENTS

"Concentration animal feeding operations" are regulated under Sec. 502(14) of the Federal Water Pollution Control Act Amendments of 1972, and thus may be required to have a permit as a "point source" under the National Pollutant Discharge Elimination System. 29) According to current EPA estimates, as affected by a recent court decision invalidating previous exclusion of small feedlots from permit requirements, as many as 95,000 livestock feeding operations may be covered by the Act.

State animal feedlot regulations typically apply only to situations where the ratio of the number of animals to land area is so high that concentrations of waste threaten to cause water pollution, in which case a permit is required. Basically, the regulations call for surface runoff to be diverted away from the lot. A settling pond and lagoon must be provided, with additional treatment of the effluent, if necessary. Some also specify the manner in which waste from the lot may be stored or disposed of on land. The regulations vary in detail. For example, Montana's regulation is brief, and general in statement, 30) whereas the Iowa and Oregon regulations are relatively detailed. 31,32)

Operations Subject to Regulation

The Iowa confined feeding regulation includes an open feedlot only where the animal population exceeds a specified number and the square feet of lot area per animal is less than a specified number. For instance, the regulation applies to beef cattle where animal population exceeds 100 and lot area per animal is less than 600 sq ft (56 sq m). The Iowa regulation separately treats a "confinement feeding operation" -- one having a roofed or partially-roofed enclosure where wastes are removed as liquid or semi-liquid. In such a case, area is not involved and the regulation applies by number of animals, e.g., for beef cattle, 50; for sheep, 600.

Registration of an open feedlot is required if one or more of the following conditions exist:

1. The number of animals exceeds 1,000 (beef cattle).

2. The feedlot contributes to a watercourse draining more than 3,200 acres (1,295 ha) of land above the lot and the distance from the feedlot to the nearest point on the watercourse is less than 200 ft (61 m) per 100 animals (beef cattle).

3. The runoff water from the feedlot (or collection facility) flows directly into a tile line or other buried conduit, well, hole, pit, lake, or pond.

Registration of a "confinement feeding operation" is required if:

1. The number of animals exceeds 100 (beef cattle).

2. Overflow contributes to a watercourse.

3. The runoff water from the feedlot (or collection facility) flows directly into a tile line or other buried conduit, well, hole, pit, lake, or pond. 31)

Examples of where regulations apply in other states are as follows: Kansas - 300 or more head of cattle, swine, sheep, or horses, any operation using a lagoon, or any operation having a water-pollution potential; 33) Minnesota, 34) Montana, 30) and Nebraska 35) - feeding any livestock in a confined area not normally used for raising crops or as pasture; Oregon - feeding or holding areas in buildings, pens, or lots where the surface has been prepared with concrete, rock, or fibrous material to support animals in wet weather or where the concentration of animals has destroyed the vegetative cover and the natural infiltrative capacity of the soil. 32)

The Iowa regulation contains factors to be considered in determining whether a facility will constitute a pollution problem, such as location relative to water sources; type of surface, soil and slope; hydrological and geological conditions; permeability of retention structure to control excessive seepage; control of discharge in proportion to stream flow; animal density; anticipated waste load; distance to structures occupied by humans; direction of prevailing winds; applicable water-quality standards; and proposed methods for waste disposal. Despite other criteria in the law or regulations, the law will be applied or waived in a particular situation depending upon these factors. 31)

<u>Information Required for a Permit</u>

Iowa requires the location to be sketched on an aerial photograph. Information must include details on: [31]

1. Building and lot areas.

2. Lagoons or waste-holding pits.

3. Direction of surface drainage from site.

4. Location of wells and dwellings within 1,000 ft (305 m) of site

5. Adjacent properties.

6. Land area set aside for waste disposal.

Minnesota and Oregon have similar requirements, but Minnesota also requires a description of geologic conditions, soil types, and ground-water elevations, a plan of operational procedures, location of treatment works, and quantity and type of effluent to be discharged. [34] Oregon requires climatological data and details of feed preparation and handling; a location map showing ownership, zoning, and use of adjacent lands; and location of the proposed operation in relation to residences and domestic water-supply sources. [32]

Prohibited Locations

The Nebraska regulation prohibits location of a livestock waste-control facility (e.g., a detention pond) within 100 ft (30 m) of any well used for domestic purposes, or within 1,000 ft (305 m) of a municipal water supply well unless the operator can show that it will not result in ground-water contamination. [35]

Minnesota prohibits location of new livestock feedlots within shoreland or a floodway (protected under other statutes), within 1,000 ft (305 m) of a public park, in sinkholes or areas draining into sinkholes, or within one-half mi (0.8 km) of the nearest point to a concentration of ten or more private residences. [34]

Facility Requirements

Subject to waiver when not necessary, or additional requirements when necessary, the Iowa regulation specifies that the minimum water-pollution control facilities for an open feedlot shall be terraces or retention basins capable of containing 4 in. (10 cm) of runoff. Diversion of surface drainage above the feedlot is required and a settling basin is to be provided where necessary. For a "confinement feeding operation," the minimum facility is a tank or basin capable of holding 120 days' waste. [31]

The Kansas regulation works in a similar manner, requiring facilities if a potential water-pollution problem exists. For cattle, the retention pond must hold 3 in. (8 cm) of runoff. [33]

The Nebraska regulation requires a detention structure capable of holding runoff from a 10-year, 24-hour storm. For "housed" operations, the requirement, as in Iowa, is the capacity to hold 120 days' accumulation. [35]

The Oregon regulation is considerably more detailed, consisting not only of capacity requirements, but method of diking, requirement of overflow relief structures, prevention of erosion, and other regulations including solids handling systems and conveyance and disposal facilities. [32]

Operation of Facilities

The regulations also contain certain operating requirements, which in essence require that contamination be prevented. Montana requires the operator to provide personnel who have adequate skill and time to maintain and operate the facility consistent with the approved application. [30] The Nebraska regulation states: "Caution should be exercised to insure that a thin layer of manure remains on the lots during scraping and that the soil-manure interface not be disturbed." That regulation also instructs the operator in keeping the feedlot surface aerobic so that production of odors is curtailed. [35] The California Water Resources Board has issued guidelines specifically addressed to the protection of ground water; among these is the recommendation that salt in animal rations be limited to that required to maintain animal health and optimum production. [36]

Storage, Transportation, and Disposal

Individual state regulations also may apply to storage, transportation, and disposal of collected wastes. In essence, the regulations state that these activities shall be conducted so that contamination does not occur, and pollution-control laws are complied with.

The Oregon regulation contains requirements for liquid-manure spreading, including: a plan of uniform coverage, plan of rotation for liquid-manure irrigation systems, selection of equipment, provision of adequate land for year-round disposal, type of land to use, harvesting of vegetative cover, livestock grazing, and prohibition of seepage basins except where the operator can demonstrate that ground-water contamination will not result. [32)]

The Iowa Water Quality Commission has adopted a policy on land disposal of animal wastes concerning a maximum average nitrogen application rate of 250 lb/acre (280 kg/ha). The policy also addresses such items as phosphorus limits, waste disposal on snow-covered land, land subject to flooding, land near watercourses, and odor control. [31)]

REFERENCES CITED

1. Scalf, M. R., J. W. Keeley, and C. J. LaFevers. 1973. Ground water pollution in the south central states. U. S. Environmental Protection Agency Report EPA-R2-73-268. 181 pp.

2. U. S. Department of Agriculture, Statistical Reporting Service, Crop Reporting Board. 1975. Cattle on feed. Mt An 2-1 (1-75). 13 pp.

3. Smith, R. J., and J. R. Miner. 1975. Livestock waste management with pollution control. Pages 4, 5, and 8, Table 3 in North Central Regional Research Publication 222.

4. Coleman, E. A., and others. 1971. Cattle feedlot pollution study. Page 6 in Texas Water Quality Board Report 2.

5. Manges, H. L., L. S. Murphy, and E. H. Goering. 1972. Disposal of beef feedlot wastes onto cropland. Pages 2 and 3 in American Society of Agricultural Engineers. Paper No. 72-961.

6. Karubian, J. E. 1974. Polluted groundwater: estimating the effects of man's activities. Pages 84, 85, and 95 in U. S. Environmental Protection Agency Report 680/4-74-002.

7. Clark, R. N., C. B. Gilbertson, and H. R. Duke. 1975. Quantity and quality of beef feedyard runoff on the Great Plains. U. S. Department of Agriculture.

8. State Hygienic Laboratory, Limnology Division. 1973. Storm water quality as affected by cattle feedlot runoff. Page 5, Table 2 in Iowa Department of Environmental Quality Preliminary Report.

9. Wells, D. M., and others. 1972. Characteristics of wastes from southwest beef cattle feedlots. Pages 13, 15, 16, 22, and 24 in Cornell Agricultural Waste Management Conference.

10. Secretary of Agriculture. 1969. Animal wastes. Pages 25-34 in U. S. Department of Agriculture, Control of agriculture - related pollution.

11. U. S. Department of Agriculture. April 1975. Livestock and meat situation.

12. Wadleigh, C. H. 1968. Wastes in relation to agriculture and forestry. Agricultural Research Service, U. S. Department of Agriculture. Miscellaneous publication No. 1065.

13. U. S. Department of Commerce, Bureau of the Census. 1974. Statistical abstracts of the United States. U. S. Government Printing Office, Washington, D. C. 1,028 pp.

14. McCalla, T. M., and F. G. Viets, Jr. 1969. Chemical and microbial studies of wastes from beef cattle feedlots. Agricultural Research Service, Nebraska Soil and Water Conservation Research Division. Page 24.

15. Walter, M. E., G. D. Bubenzen, and J. C. Cenvase. 1974. Movement of manurial nitrogen in cool, humid climates. American Society of Agricultural Engineers Paper 74-2018.

16. Wells, D. M., and others. Control of water pollution from southwestern cattle feedlots. Texas Water Quality Board. Page 21.

17. Crosby, J. W., III, D. L. Johnstone, and R. L. Fenton. 1971. Migration of pollutants in a glacial outwash environment. Water Resources Research 7(1):204-208.

18. Engberg, R. A. 1967. The nitrate hazard in well water with special reference to Holt County, Nebraska. Page 18 in University of Nebraska Conservation and Survey Division. Nebraska Water Survey Paper 21.

19. Gillies, N. P., editor. 1975. Ground Water Newsletter. Water Information Center, Port Washington, New York. 4:17.

20. Anderson, R. K. 1975. Feedlot wastes. Systems Management Division, Office of Solid Waste Management Programs, U. S. Environmental Protection Agency. Unpublished report.

21. Wells, D. M., G. A. Whetstone, and R. M. Sweazy. 1973. Manure, how it works. Page 18 in American National Cattlemens Association. Conference proceedings.

22. Feldman, H. F. 1971. Pipeline gas from solid wastes. American Institute of Chemical Engineers 69th National Meeting. New York.

23. Halligan, J. E., and R. M. Sweazy. 1972. Thermochemical evaluation of animal waste conversion processes. American Institute of Chemical Engineers 72nd National Meeting. New York.

24. Anthony, W. B. 1969. Cattle manure: the re-use through wastelage feeding in Cornell University Animal waste management conference proceedings. New York.

25. Culvert, C. C., R. D. Martin, and N. O. Morgan. 1969. Dual roles for house flies in poultry manure disposal. Poultry Science 48(5).

26. Anon. Feed recycling showing promise. California News.

27. House of Representatives, 93rd Congress, 1st Session. Control of pollution from animal feedlots. Hearings before a Sub-committee of the Committee on Government Operations.

28. Butchbaker, A. F., and others. 1971. Evaluation of beef cattle feedlot waste management alternatives. Oklahoma Agricultural Experiment Station, Oklahoma State University for U. S. Environmental Protection Agency.

29. Federal Water Pollution Control Act Amendments of 1972. PL 92-500.

30. Montana Board of Health and Environmental Sciences. 1972. Regulation for the control of water and air pollution from confined livestock feeding.

31. Iowa Department of Environmental Quality. Water quality commission regulation 1.3(455B).

32. Oregon Department of Environmental Quality. Regulations relating to water quality control in Oregon. Oregon Administrative Rules, Chapter 340, Division 5.

33. Kansas Board of Health Regulations 28-18-1 et seq.

34. Minnesota Pollution Control Agency. 1971. Regulations for the control of wastes from livestock feedlots, poultry lots and other animal lots.

35. Nebraska Department of Environmental Control. 1972. Rules and regulations pertaining to livestock waste control.

36. California State Water Resources Control Board. 1973. Minimum guidelines for protection of water quality from animal wastes. Sacramento, California.

SECTION XV

PRINCIPAL SOURCES OF GROUND-WATER CONTAMINATION NOT RELATED TO WASTE-DISPOSAL PRACTICES

SUMMARY

Aside from the possibility of contamination of ground water from present-day, waste-disposal practices, there are numerous other sources that can cause degradation of water quality. Few regional and national assessments of ground-water contamination problems have been undertaken. However, without exception, the number of documented cases reported is evenly divided between incidents related to waste-disposal practices and those related to non-waste disposal problems. Spills rank highest in reported incidents, with abandoned oil and gas wells, water wells, and highway deicing salts also of prime importance. Only salt-water encroachment in coastal regions has received major attention from regulatory agencies and because of this is adequately controlled in most critical areas.

INTRODUCTION

Aside from the possibility of contamination of ground water from present day waste disposal practices there are numerous other sources of contaminants that can cause degradation of ground-water quality. The principal ones are: (1) spills and leaks, (2) mine drainage, (3) salt-water intrusion, (4) water wells, (5) oil and gas wells, (6) surface water infiltration, (7) agricultural activities, (8) highway deicing salts, and (9) atmospheric contaminants.

The severity of ground-water contamination and the actual volume of ground water degraded by these sources is difficult to assess. Nevertheless, their total impact on ground-water quality might be as great as that caused by the waste disposal practices discussed previously.

Few inventories of ground-water contamination cases have been conducted to date. However, in 1972 and 1973, personnel from the U. S. Geological Survey made a general reconnaissance of ground-water contamination problems in the United States.[1,2,3,4,5] The investigation relied principally on published information, backed up by a limited number of personal interviews with representatives of Federal, state, and local agencies. Cases of ground-water contamination were compiled on a state by state basis for all 50 states. Of a total of more than 800 cases inventoried, about half were not related to waste-disposal practices.

A breakdown of these cases by source of contamination is shown on Table 73 (Column 1). Most problems reported were from salt-water intrusion, followed by leaks from pipelines and storage tanks. It is interesting to note that salt-water intrusion was the earliest form of ground-water contamination recognized by workers in the field and therefore, presently dominates the literature.

Results of an intensive investigation of ground-water contamination carried out in the central and lower Susquehanna River basin in Pennsylvania are an interesting comparison to the U. S. Geological Survey inventory (Table 73, Column 2). A total of 236 cases of ground-water contamination were inventoried in this relatively small region, 111 of which were not related to waste disposal practices. Spills and leaks were found to be by far the most prevalent sources of contamination, followed by highway deicing salts and agricultural practices. The region is not coastal, and salt-water intrusion is not a factor.

In 1975, an informal survey of 20 states was conducted by Geraghty & Miller, Inc. to determine the status of abandoned water wells and the principal reasons for abandonment.[7] It was found that, in the states contacted, detailed records of this nature are not maintained. However, based on the information obtained, most domestic, public supply, industrial, and irrigation wells are abandoned because of decline in yield. Many others are abandoned because of a rise in concentration in the well water of such constituents as iron and chloride, with the reasons for the change in water quality rarely fully investigated. The principal contaminant reported to public agencies appeared to be gasoline, which had leaked from buried storage tanks.

Other inventories of ground-water contamination problems are represented by a series of regional studies being sponsored by the EPA. To date, four investigations covering 26 states have been completed.[8,9,10,11] Of the principal sources of ground-water contamination not related to waste disposal practices evaluated in these studies, the most important from the standpoint of degrading ground-water quality in at least three of the four regions, were abandoned oil and gas wells, irrigation return flows, spills, and leaks from buried tanks and pipelines. The principal contaminants reported were chloride, nitrate, hydrocarbons, and heavy metals.

To provide a better understanding of these instances of

Table 73. PREVALENCE OF GROUND-WATER CONTAMINATION NOT RELATED TO WASTE DISPOSAL

Source of contamination	U.S.G.S. National Inventory 1,2,3,4,5)	Central and lower Susquehanna River basin, PA. 6)
Total number of cases	809	236
Related to waste disposal (non differentiated)	373	125
Not related to waste disposal		
Spills	34	31
Leaks (pipelines and storage tanks)	65	45
Mine drainage	12	5
Salt-water intrusion	114	-
Water wells	40	-
Oil and gas wells	38	-
Surface-water infiltration	39	3
Agricultural activities		
General	11	5
Fertilizers	17	1
Pesticides	5	4
Irrigation return flows	16	-
Highway deicing salts		
Storage	26	14
Application	19	3
Atmospheric contamination	0	0
Total	436	111
Percent of total cases	54	47

ground-water contamination, they are briefly discussed in the following sections. Non-referenced case histories that are quoted appeared in the four EPA regional studies referred to above.

SPILLS AND LEAKS

Accidental spills of liquid wastes, toxic fluids, gasoline, and oil occur in every region, accompanied by the risk that the contaminant can migrate down to the saturated sediments, and degrade ground-water quality. By far the most prevalent contaminants reported as affecting ground-water quality from this source are hydrocarbons. Spills can occur at industrial sites; along city streets, highway and railroad rights of way; and at airports. Spills and leaks of radioactive substances have taken place at facilities of the Energy Research and Development Administration.

A typical instance of a serious case of ground-water contamination from a spill occurred in the Northeast in 1957 when 30,000 gal. (114 cu m) of jet fuel were spilled on the ground at an Air Force base. The crystalline rock aquifer was so badly contaminated that the original wells supplying the base could not be used for 15 years after the spill took place. In the Northwest, the Department of Ecology of the State of Washington recorded, during the first six months of 1973, nearly 500 complaints of spills, many of which affected ground-water quality.

The accidental spill is an unavoidable hazard inherent in the storing and transportation of fluids. It is in the handling of spills after they have taken place that better protection of ground-water resources can be achieved. In the past, for example, liquids spilled on highways have been simply flushed from the road to adjacent soils at the expense of contamination of a shallow aquifer in order to have a minimal effect on traffic flow. Because time appears to be the most important factor in minimizing ground-water contamination from spills, some state and Federal agencies have developed procedures for reporting spills and leaks to the proper authorities so that effective action can be taken quickly.

Contaminants escaping from leaky and ruptured buried pipes and from storage tanks are another common problem that can affect ground-water quality. Again, the principal contaminants reported are hydrocarbons, which have leaked from gasoline service station and home fuel-oil storage tanks, industrial production facilities, and petroleum product transmission lines.

Details on the number of cases of ground-water contamination due to leakage from buried tanks and pipelines that occur each year are generally not available. However, some statistics are revealing. A review of 203 legal proceedings, involving ground-water contamination in the United States and England, indicated that leaks from refineries, oil storage depots, pipelines, gas mains and gasoline service stations were the source of contamination in 40 cases. 12) Leaks from gasoline and home fuel-oil storage tanks were also responsible for the vast majority of cases of ground-water contamination in the central and lower Susquehanna Basin of Pennsylvania as shown in Table 73. The Pennsylvania Department of Environmental Resources estimates that 2,600 new or replacement storage tanks are buried each year within that state. Failure of the tank is normally the reason for replacement, and the product originally contained has been lost to the ground. In Colorado, the State Oil and Gas Inspection Office records gasoline leaks on a monthly basis. As much as 37,000 gal. (140 cu m) from various sources leaked into the subsurface during one 30-day period. 11)

A well-documented case occurred in southern California in 1968. Thousands of gallons of gasoline were found to have contaminated a broad area underlying the City of Glendale. About 30 wells for observation, containment, and removal were drilled in the problem area, and the clean-up operation involved installation of special facilities for separating gasoline from the contaminated water pumped from the wells.

Transportation pipelines are used for a wide number of materials such as oil, gas, ammonia, coal, and sulfur. Their heaviest use is for the transportation of petroleum products, and natural gas. In 1972, 99 companies operated 174,000 mi (280,000 km) of petroleum pipelines, of which 43,000 mi (69,000 km) were gathering lines and 131,000 mi (211,000 km) were trunk lines. In the same year a total of 5.1 billion bbl (811 million cu m) of crude oil and 3.4 billion bbl (540 million cu m) of oil products moved through trunk lines. 13)

Because interstate pipelines are a major means of transportation, they are regulated by Federal agencies. As leaks of petroleum products can produce a fire or explosion hazard, these regulated pipelines are required to report leaks and spills of more than 50 bbl (8 cu m).

A summary of accidents and volume of liquids lost in interstate pipelines during 1971 is given in Table 74. More than 300 pipeline accidents took place that year, and the total volume of hydrocarbons lost was approximately 250,000 bbl

Table 74. SUMMARY OF INTERSTATE PIPELINE ACCIDENTS FOR 1971. [14)]

Commodity	No. of Accidents	Percent of Total	Loss (Barrels)	Percent of Total
Crude Oil	172	55.9	115,760	47.2
Gasoline	51	16.6	42,001	17.1
L.P.G.	39	12.7	39,887	16.3
Fuel Oil	21	6.8	13,724	5.6
Diesel Fuel	5	1.6	6,953	2.8
Condensate	5	1.6	3,658	1.5
Jet Fuel	4	1.3	2,236	0.9
Natural Gasoline	4	1.3	8,743	3.6
Anhydrous Ammonia	3	1.0	9,810	4.0
Kerosene	2	0.6	700	0.3
Alkylate	2	0.6	1,585	0.7
Total:	308	100.0	245,057	100.0

Note: Barrels multiplied by 0.16 equals cubic meters.

(40,000 cu m). The most frequent cause of pipeline failure was external corrosion (102 cases), followed by damage from excavating machines (67 cases).[14]

Hydrocarbon spills and leaks pose very difficult environmental problems. Small spills may be absorbed or adsorbed in the unsaturated zone, but in large spills a substantial quantity of fluid will percolate down to the water table. Depending on the density and miscibility of the fluid it will tend to float or mix with the ground water. Removal by pumping is difficult as the fluids frequently react differently (two-fluid systems) and thus, interfere with recovery. Low-density, low-miscibility fluids, such as refined petroleum products, have been removed by skimming from the top of the water table in a few places.

One of the most serious consequences of pipeline and tank leakage into the soil is that oils and petroleum products in even trace quantities will render potable water objectionable because of taste and odor. In sufficiently high concentrations, the vapors of lighter fractions of petroleum products, liquified petroleum gas, and natural gas can seep into basements, excavations, tunnels, and other underground structures. These vapors mix with the air in the cavity and constitute a severe explosion or fire hazard in the presence of open flame or sparks.

Chemicals such as ammonia and other agricultural or industrial chemicals can have toxic properties. For example, ammonia will add to the nitrification of ground water, while acids lower the pH of ground water which, in turn, will accelerate the solution of soil solids and heavy metals.

MINE DRAINAGE

Ground-water contamination associated with extensive mining operations is prevalent in the northeast, northwest, and to some extent, in the southwest regions. Most mining operations encounter ground water, and drainage of highly mineralized water from mine workings can cause ground-water contamination.

Dewatering of mines to allow work to proceed below the water table causes water levels to fall and may result in air contact and oxidation of exposed sulfide-bearing minerals. The most common ore--sulfide mineral association is that of coal and pyrite (iron disulfide), but other associations exist, and such ores as galena (lead ore) and sphalerite (zinc ore) are themselves sulfides. Sulfide minerals in the ground-water environment are normally stable. However, when exposed to air, they oxidize to a form that is easily leached by water to form sulfuric acid and a soluble sulfate compound of the principal metal ion.

The oxidation of sulfide minerals does not by itself cause ground-water contamination. However, percolating surface water from streams or from rainfall entering the mine, leaches the minerals and may transport them downward to the water table. After a mine is abandoned, and dewatering operations are suspended, the local water table rises up through the oxidized minerals, accelerating the leaching. For this reason, abandoned mines, including strip mines, are a greater source of ground-water contamination than are operating mines.

In a study of ground-water quality in Appalachia, high iron and sulfate concentrations and low pH in ground water were traced to coal mining operations.[15] Even with cessation of mining it was estimated that decades would be required before the ground water would become usable again. In northwestern Pennsylvania, acid mine drainage moved downward from strip mines into underlying aquifers through abandoned oil and gas wells and rock fractures, increasing the iron and sulfate content of the ground water.

Thousands of active and abandoned metal mines in the western United States contribute to the acid drainage problem. In Montana, over 100 lead, silver, and copper mines discharge acid water. In Washington, drainage from abandoned gold mines is believed to be the source of high manganese in individual well-water supplies, and in Idaho, cases of cattle poisoning were reported to have been caused by arsenic leachates from abandoned mines. Radium concentration in uranium mine drainage can be raised from 50 to 200 times above background as a result of oxidation and ore leaching. High concentrations of Ra-226 have been found in ground waters of the uranium mining district in Shirley Basin, Wyoming.

There is considerable concern about the impact on ground-water quality from large-scale coal strip mining planned in Montana and Wyoming. Although western coal is lower in sulfur content (as pyrite) than eastern coal beds, acid water could be produced. This acid drainage could contaminate important water-bearing sandstones associated with these coal beds.

Measures to correct drainage of poor quality water from abandoned mines typically are prohibitive in cost. These may include sealing of mine openings to prevent drainage or to reduce entrance of precipitation, flooding with water to eliminate air contact with acid forming minerals, and chemical treatment of drainage water.

SALT-WATER INTRUSION

Intrusion of salty water into fresh-water aquifers in coastal areas is one form of ground-water contamination that has been widely recognized for many years. Salt water occurs naturally in water-table and artesian aquifers in coastal areas and pumping from wells can induce the mineralized water to intrude into fresh-water zones.

The large number of individual and widely publicized cases of salt-water intrusion (see Table 73) has led to the development of strict controls over diversion of ground water in the coastal plain states of the northeast. These controls on pumpage have been most effective in eliminating salt-water intrusion as a critical problem in this region.

Contamination of wells with sea water does not appear to be a major problem in the northwest. However, in the Gulf Coast area, salt-water encroachment has affected a number of important and heavily pumped ground-water areas including Baton Rouge and Lake Charles in Louisiana and Houston, Galveston-Texas City, and Matagorda-Lavoca Bay in Texas.

California has had serious problems of salt-water encroachment in many of its coastal basins. Various agencies in the state have established programs to reverse the movement of intruding saline water, the most well known of which involves the placement of a series of "barriers." The barriers, such as those in the Los Angeles area, are established by injecting non-saline water at a line of wells whose axis roughly parallels the ocean shore. Some of these barriers have been successful in reversing the hydraulic gradient in the affected aquifer so that flow is toward the sea instead of toward fresh-water supply wells.

An even more critical problem than salt-water intrusion in coastal areas, is that which can occur inland. More than two-thirds of the conterminous United States is underlain by water containing more than 1,000 ppm of dissolved solids and many inland fresh-water aquifers are hydraulically connected with saline ground water.[16)] In most cases, the heavier mineralized water underlies the fresh water. Where wells are too deep or where excessive pumping modifies the hydraulic gradient, saline water may be drawn into zones formerly containing fresh water.

Unlike coastal intrusion, potential problems associated with inland saline ground-water bodies have not been studied in detail. Regulatory controls over diversion of ground water and well construction have not been developed to the degree that they have in coastal areas.

WATER WELLS

Water wells under certain conditions can be sources of ground-water contamination. Typical examples are where a casing has been corroded or ruptured, where a well screen or an open bore hole interconnects two separate aquifers, or where the surface casing has not been adequately sealed in soil or rock. Water wells can serve as a means for transmission of contaminants from one aquifer to another or from the land surface to an aquifer.

In some of the south-central states, improperly constructed and abandoned water wells are considered by many public agency officials to be the most significant cause of ground-water contamination. Problems are especially prevalent in cavernous limestones, such as those in the Edwards Plateau of Texas and the Ozark Plateau of Arkansas. Unplugged abandoned water wells tapping artesian brine aquifers have resulted in reported cases of contamination of shallow ground-water supplies in a number of counties in Texas. In Florida, abandoned wells that flow salty water are considered a major ground-water contamination problem.

In the northeast, salt-water intrusion in coastal areas has been aggravated at numerous locations by the presence of corroded well casings, which allow salt water to enter fresh-water aquifers either from an underlying or an overlying saline-water aquifer or from an adjacent salty surface-water body. A classic example occurred in Baltimore, Maryland, where highly acidic industrial wastes in the water-table aquifer corroded the casings of more than 1,000 abandoned wells. Saline-water intrusion caused by pumping in the same shallow aquifer affected the deeper fresh-water artesian aquifer because the leaky abandoned wells acted as conduits, allowing poor quality water to migrate into the deeper artesian aquifer.

A few states have adopted regulations and codes governing well construction and the plugging of abandoned wells. However, it is difficult to enforce these regulations because records showing where operating water wells have been drilled over the past 50 years are incomplete. Licensing of well-drilling contractors in many states has been moderately effective in improving well construction practices. However, enforcement of construction standards is difficult considering the more than 500,000 new water wells drilled each year.

OIL AND GAS WELLS

Contamination of ground water can occur from poorly constructed, old, and/or damaged oil and gas wells. Unsealed or uncapped abandoned test and production wells also can provide convenient pathways and allow upward movement of contaminants in a similar fashion as described in the previous section for water wells. Cases of contamination of ground water from these sources are reported in practically all oil and gas producing states. For example, highly mineralized water under artesian pressure has moved upward through uncapped and leaking abandoned oil and gas wells in New York and Kentucky, contaminating fresh ground water.[17,18] In Kentucky, potable ground water was changed to a sodium chloride type with chloride content as high as 51,000 ppm (as compared to less than 60 ppm before oil production).[19]

An unplugged oil test well in Glynn County, Georgia, which penetrated shallow fresh-water aquifers, has allowed upward migration of salt water, resulting in a chloride content of up to 7,780 ppm in previously potable ground water. The high chloride water extended 1.5 mi (2.4 km) along the hydraulic gradient, and additional pumping in the area may hasten contamination of nearby water wells.[20]

Leaking oil and gas wells often compound ground-water contamination problems from waste disposal practices. In Limestone County, Texas, 600 abandoned oil and gas wells that were improperly plugged added to ground-water degradation, which had already occurred due to brine disposal in surface pits.[21] Similar conditions are reported from other oil-field areas in Texas.[22,23]

One of the most serious cases of ground-water contamination both in terms of area and thickness of aquifer affected has occurred in West Virginia. More than 50,000 deep oil and gas wells have been drilled, almost all of which penetrate artesian salt water. Many of these wells have never been plugged, and it was common practice in the past to salvage the casing before abandoning the well. An estimated 25,000 of these uncased and unplugged deep wells have altered the hydrologic system and allowed salt water to flow up to and into fresh-water aquifers. Throughout perhaps two-thirds of the state, at least a 200-ft (60-m) thickness of fresh-water aquifer has been contaminated. Locally, where salt water has been injected to increase oil and gas recovery, the rise in pressure has driven the salt water even further upward, contaminating a 400-ft (122-m) thickness of fresh-water aquifer.[24]

Oil and gas producing states, aware of the potential problem of ground-water contamination, now have regulations regarding minimum casing length, and cementing and abandonment procedures. However, strict enforcement of these regulations is necessary to insure better well construction and plugging of abandoned wells.

SURFACE WATER INFILTRATION

Much of the ground water pumped in the nation is derived from sand and gravel aquifers in river valleys. Where such aquifers are in hydraulic connection with surface water, replenishment of water withdrawn by wells is partly from river infiltration. If the river water is of poor quality, contaminants enter the aquifer and degrade ground-water quality. Most of the cases of ground-water contamination from surface-water infiltration are a direct consequence of discharge of waste fluids or irrigation return flow to surface-water bodies. In coastal areas, some wells tapping fresh-water aquifers have become saline as a result of infiltration of sea water during tidal inundations.

Contamination of ground water by infiltrating surface water also may occur when chemical reactions take place between the induced water and native water and/or the aquifer material. Mineralization of ground water by this mechanism is quite common in the glaciated regions of the United States where wells in alluvium and glacial outwash commonly derive recharge from adjacent streams. High concentrations of iron and manganese, leached from the sediments, cause discoloration and bad taste, and may encourage bacterial growth in well water, which leads to clogging of well screens and other water system operational problems.

When the surface water is contaminated, this process is accentuated because of the reducing environment created by infiltrated river water moving through iron-rich unconsolidated sediments. In one case, infiltrated water from a polluted tributary of the Hudson River in New York dissolved iron and manganese in the sand and gravel sediments tapped by a high-capacity caisson well. Manganese concentrations in the water from the caisson well rose from less than one ppm when first pumped to more than 14 ppm after several months of operation. Treatment for the high concentration of manganese was considered to be uneconomic. The well, used for municipal supply, was abandoned.

In another case in New York State, a public supply well has occasionally yielded water with a concentration of lead that is three times the maximum limit allowed for potable supplies. The source of the contaminant is concluded to be a river several hundred feet away, which provides a major portion of the recharge to the well. The amount of the contaminant reaching the well depends on the character of industrial discharges to the stream and the river stage.

AGRICULTURAL ACTIVITIES

Agricultural practices responsible for contamination of ground water are, in order of importance: irrigation return flow, application of chemical fertilizers or animal wastes, man-caused changes in vegetation, and use of pesticides.

Irrigation return flow is water diverted for irrigation purposes that finds its way back into an existing or potential water supply. This process concentrates salts by evapotranspiration and can introduce chloride and other substances from irrigated lands into a ground-water reservoir by means of infiltration. Contaminants in irrigation return flows may originate from many sources including the applied water, soils, fertilizers, and pesticides.

Irrigation return flow is considered a major problem which has led to a large number of areally extensive ground-water contamination cases. In the southwestern and south central states, for example, ground-water quality has deteriorated from irrigation return flows in the Rio Grande basin of New Mexico and Texas. Other problem areas include the Pecos River valley in New Mexico and Texas and the Arkansas River valley in Oklahoma and Arkansas. In California, degradation of ground-water quality on a broad scale has been reported in the San Joaquin basin.

In the northwestern states, it is estimated that there are over 2 million acres (809,000 ha) of saline land within the region. A few of the larger areas currently experiencing irrigation return flow problems are: the valleys of the Grand, Platte, and Arkansas Rivers in Colorado; the Yakima Valley in Washington; Larimer County in Wyoming; Rosebud County in Montana; the Snake River valley in Idaho; and the lower Columbia River basin in Washington. One of the most severe and best studied instances of a problem related to irrigation return flow is in the Grand Valley of Colorado, where a high percentage of the irrigated acreage has become marginal because of a high water table and concentrated salts. It has been estimated that approximately 37 percent of the total salt load from the Upper Colorado Basin is associated with irrigation return flows in this area.

Irrigation return flows from agricultural practices are and

will continue to be a major source of ground-water contamination within the foreseeable future. In some areas, the problem could decrease in severity as new techniques are developed for application and management of irrigation waters and more efficient use is made of crop types.

The use of chemical fertilizers has doubled during the past 20 years from 20 million tons (18 million tonnes) in 1950 to 40 million tons (36 million tonnes) in 1970. The use of nitrogen in fertilizer has increased even more rapidly than phosphorus or potassium because it generally stimulates crop yields to a greater degree. In 1950, nitrogen constituted 6.1 percent of all fertilizer used; in 1970 it had risen to 20.4 percent.[13] As a result, nitrogen applications in excess of the amounts removed by the crop are common. Nitrate can migrate with percolating water and enter the ground-water system. High nitrate content is reported in ground water in many agricultural areas, and fertilizers are certainly an important factor in this type of degradation of water quality.

Dryland farming, especially prevalent in the northern Great Plains, appears to be a significant source of contamination on a regional level. In dryland farming, no irrigation takes place, and plants depend entirely on precipitation to obtain their necessary moisture. Prior to settlement of the Great Plains, the natural vegetation consisted of buffalo grass, which consumed relatively large quantities of water and left no surplus for deep percolation. Present crop-fallow farming practices have reduced evapotranspiration losses, and more water moves down below the root system.

Poorly permeable shale and sandstone overlain by glacial till extend over large areas of the Great Plains. Because the shale is relatively impermeable, percolating water mounds up in the overlying till and begins to move downslope. The large supply of natural soluble salts in the subsoil, till, and shale is leached by the moving ground water. Eventually the ground water discharges at the surface as a seep and evaporates, leaving the salt residue behind. The discharge water commonly has a dissolved solids concentration in excess of 25,000 ppm.

Saline seeps are becoming a regional problem in Montana, where 80,000 acres (32,000 ha) of crop land have already been affected with a loss of farm income of $5 million per year.[11]

The term "pesticides" encompasses algicides, herbicides, fungicides, and insecticides and is a general term for that group of chemicals used to control organisms which limit crop growth or proliferation. Large quantities of pesticide chemicals are used in the United States. Pesticide contamination of ground water is less common than nitrate contamination as many pesticides degrade naturally through microbial metabolism, hydrolysis, volatilization or exposure to sunlight. In contrast, such organics as chlorinated hydrocarbons are particularly resistant to decay, and are very stable in soil. DDT (now outlawed), for example, decomposes at a rate of only 5 percent per year. In 1971, production and sales of organic pesticides amounted to 1.1 billion lb (0.5 billion kg).[25]

The most frequently mentioned cases of ground-water contamination from pesticides are those related to spills in the vicinity of a well or an irrigation canal that is infiltrating water to an aquifer. However, a few documented instances of contamination from the application of pesticides have been noted in the northeast. For example, arsenate compounds used for insect control in the blueberry barrens of Maine have been found in shallow ground waters. In another case, water from a sand and gravel well in Massachusetts was contaminated by pesticides containing chlorinated hydrocarbons sprayed on cranberry bogs.

In summary, contamination of ground water from agricultural activities is difficult to control because so many of the problems are in isolated rural areas. Changes in water management and farming practices may be effective in reducing degradation of ground-water quality. Severe problems develop when former farm areas become urbanized, and ground-water use for drinking-water purposes increases. Ground-water quality degradation also can be aggravated by heavy application of fertilizers and pesticides by individual home owners on relatively small lots in the urbanized area.

HIGHWAY DEICING SALTS

The use of large amounts of soluble salts for road maintenance during winter months has led to a significant number of cases of ground-water contamination in the northern latitudes.

There are two principal ways in which road salt can contaminate ground water. Salt-laden runoff from roads can percolate into soils adjacent to highways and eventually reach the water table. Rain falling on uncovered storage piles at highway maintenance garages can dissolve the salt and infiltrate into shallow aquifers. The latter generally is considered to be the more serious problem because of the very high concentrations of chloride entering the ground-water system as a slug of contaminant.

The quantities of deicing salts used in the conterminous United States in the winter of 1966-67 are listed in Table 75. Also shown are quantities of deicing salts applied per single-lane mile in the winter of 1965-66. Pennsylvania used 637,000 tons (578,000 tonnes) of salt for its highways in the winter of 1966-67, and the states of Michigan, Minnesota, and New York each applied over 400,000 tons (360,000 tonnes). Over 20 tons of salt per single-lane mile (11 tonnes per single-lane kilometer) were applied during the 1965-66 winter in Washington, D. C., Massachusetts, Pennsylvania, and Illinois.

Because of the large amounts of salt spread and stored in the northeast, water from many aquifers, especially sand and gravel deposits in the glaciated region, has shown a disturbing rise in chloride and sodium concentration. Complaints of salt contamination of water from individual wells are so common in New England that several states have established annual budgets to allow for replacement of affected wells.

In Maine, 100 randomly selected wells adjacent to major highways were sampled over a three-year period. Natural chloride concentrations in various aquifers in Maine are normally less than 20 ppm; yet the average April concentration for chloride from the selected wells was 171 ppm. Similarly, for a municipal well field in Burlington, Massachusetts, which had been contaminated by saline water, the U. S. Geological Survey calculated a "salt budget" to estimate the contributions of various chloride sources in the basin to the aquifer at the site. Eighty-five percent of the contamination was related to sources of highway deicing salts including a nearby salt pile, and only 15 percent was attributed to other sources such as septic tanks.[10)]

There appears to be no adequate substitute for highway deicing salts. However, general recognition of the potential for contaminating water supplies has resulted in many states embarking on programs to reduce the quantities of salt spread per winter storm. Equipment modification and driver education has been quite successful. In addition, although the practice is restricted by availability of funds, highway departments are enclosing many salt storage piles to protect them against contact with precipitation.

ATMOSPHERIC CONTAMINANTS

Very few studies have been made of the possible effects of

Table 75. USE OF DEICING SALTS IN THE CONTERMINOUS UNITED STATES IN 1965-66 AND 1966-67.

State	Tons of deicing salts applied per single-lane mile per year by the State Highway Departments and Toll Authorities for the period 1965-66 [26]: User	Tons	Reported total tons of sodium chloride and calcium chloride used in the winter of 1966-67 by all users [27]
Region I			
Connecticut	State Highway Dept.	8.98	104,000
Delaware	do	4.48	8,000
District of Columbia	do	37.51	36,000
Maine	do	9.30	100,000
	Toll Authorities	9.45	
Maryland	State Highway Dept.	6.82	133,000
Massachusetts	do	20.70	196,000
	Toll Authorities	23.51	
Michigan	State Highway Dept.	5.91	416,000
Minnesota	All	-	412,000
New Hampshire	State Highway Dept.	11.95	118,000
New Jersey	do	3.33	57,000
	Toll Authorities	8.83	
New York	State Highway Dept.	7.50	477,000
	Toll Authorities	17.04	
Pennsylvania	State Highway Dept.	-	637,000
	Toll Authorities	32.08	
Rhode Island	All	-	48,000
Virginia	State Highway Dept.	3.57	99,000
	Toll Authorities	3.60	
Vermont	State Highway Dept.	18.22	90,000
West Virginia	do	2.80	64,000
	Toll Authorities	3.84	
Wisconsin	State Highway Dept.	4.60	228,000
Region II			
Illinois	State Highway Dept.	3.42	259,000
	Toll Authorities	24.21	
Indiana	State Highway Dept.	7.69	243,000
Iowa	do	2.73	56,000
Kansas	do	0.62	27,000
	Toll Authorities	1.65	

Table 75 (continued). USE OF DEICING SALTS IN THE CONTERMINOUS UNITED STATES IN 1965-66 AND 1966-67.

State	Tons of deiciing salts applied per single-lane mile per year by the State Highway Departments and Toll Authorities for the period 1965-66 — User	Tons	Reported total tons of sodium chloride and calcium chloride used in the winter of 1966-67 by all users
Region II (Cont'd)			
Kentucky	State Highway Dept.	-	61,000
Missouri	do	0.39	37,000
Nebraska	All	-	10,000
North Dakota	State Highway Dept.	0.08	3,000
Ohio	All	-	523,000
South Dakota	do	-	3,000
Region III			
Alabama	State Highway Dept.	5.00	-
Arkansas	All	-	1,000
Florida	do	-	-
Georgia	State Highway Dept.	0.06	-
Louisiana	All	-	-
Mississippi	State Highway Dept.	0.30	-
North Carolina	do	2.40	19,000
Oklahoma	All	-	7,000
South Carolina	do	-	-
Tennessee	do	-	-
Texas	Toll Authorities	1.33	3,000
Region IV			
Arizona	All	-	-
California	State Highway Dept.	3.33	11,000
Colorado	do	1.93	7,000
Idaho	do	0.18	1,000
Montana	do	0.52	4,000
Nevada	All	-	4,000
New Mexico	State Highway Dept.	1.50	7,000
Oregon	do	0.04	1,000
Utah	do	2.46	28,000
Washington	do	0.15	2,000
Wyoming	do	0.06	1,000

Note: Tons/mile multiplied by 0.56 equals tonnes/km.

airborne contaminants on ground water, and only a few cases of ground-water contamination have been reported from this source.

One of the more obvious sources of airborne contaminants is that of pesticides sprayed by crop-dusting airplanes. However, such contamination can not readily be distinguished from direct application to crops at the surface and will not be considered here. Annual emissions of air pollution constituents in the United States are given in Table 76. By far the largest volume is from operation of motor vehicles. Typical concentrations of particulate contaminants in the atmosphere in urban and nonurban environments are shown in Table 77. Such contaminants are washed down to the land surface by precipitation and can travel downward to the water table. Concentration of heavy metals and acid-forming gases in the atmosphere have generated concern. In a study conducted in several Delaware watersheds, it was discovered that the cadmium concentration in rainfall frequently exceeded the EPA maximum drinking water concentration of 10 ppb. [28)]

Analyses of precipitation in the montane regions of New England have revealed the presence of lead, cadmium, and mercury. Lead concentrations ranged from approximately 4 ppb to 67.7 ppb. With the EPA maximum allowable lead concentration in drinking water set at 50 ppb, at least one sample from the research area exceeded the EPA limit. Cadmium concentrations ranged from about 0.1 to 2.3 ppb, and mercury, from about 0.025 to 0.3 ppb. These values for cadmium and mercury are below the EPA limits for these heavy metals. [29)]

In arid or semiarid regions, very little precipitation percolates to ground water and any atmospheric pollution that may occur above these areas may not affect ground-water quality. Alternatively, in humid regions having permeable soils, the effect of air pollution on ground-water quality may be more in evidence.

One case of airborne contamination has been suspected as the source of ground-water contamination in Michigan. Chromium-laden dust discharged through ventilators on an industrial plant roof settled to the ground, where it accumulated. Rainfall washed the chromium down to the water table where it migrated through the aquifer to a well field in response to pumping. [30)]

Table 76. ANNUAL EMISSIONS OF AIR POLLUTION CONSTITUENTS IN THE UNITED STATES. (In millions of tons). [14)]

	Carbon Monoxide	Sulfur Oxides	Nitrogen Oxides	Hydro-carbons	Particulate Matter
Motor Vehicles	66	1	6	12	1
Industry	2	9	2	4	6
Power Plants	1	12	3	1	3
Space Heating	2	3	1	1	1
Refuse Disposal	1	1	1	1	1

Note: Tons multiplied by 0.9078 equals tonnes.

Table 77. CONCENTRATIONS OF SELECTED PARTICULATE CONTAMINANTS IN THE ATMOSPHERE IN THE UNITED STATES FROM 1957 TO 1961. (Micrograms per cubic meter.) [14]

	Urban		Nonurban	
	Mean	Maximum	Mean	Maximum
Suspended particulates	104	1,706	27	461
Benzene-soluble organics	7.6	123.9	1.5	23.55
Nitrates	1.7	24.8	-	-
Sulfates	9.6	94.0	-	-
Antimony	(a)	0.230	-	-
Bismuth	(a)	0.032	-	-
Cadmium	(a)	0.170	-	-
Chromium	0.020	0.998	-	-
Cobalt	(a)	0.003	-	-
Copper	0.04	2.50	-	-
Iron	1.5	45.0	-	-
Lead	0.6	6.3	-	-
Manganese	0.04	2.60	-	-
Molybdenum	(a)	0.34	-	-
Nickel	0.028	0.830	-	-
Tin	0.03	1.00	-	-
Titanium	0.03	1.14	-	-
Vanadium	(a)	1.200	-	-
Zinc	0.01	8.40	-	-
Radioactivity	(b)4.6	(b)5,435.0		

(a) Less than minimum detectable quantity
(b) Picocuries per cubic meter

REFERENCES CITED

1. Rima, D. R., and R. C. Vorhis. 1972. Ground-water contamination in the northeastern states: a preliminary inventory. U. S. Geological Survey, Water Resources Division.

2. Vorhis, R. C., D. R. Rima, and L. F. Emmett. 1973. Ground-water contamination-southeastern states: a preliminary inventory. U. S. Geological Survey, Water Resources Division.

3. Kume, J., and others. 1973. Ground-water contamination-central states: a preliminary inventory. U. S. Geological Survey, Water Resources Division.

4. Bader, J. S. 1973. Ground-water contamination-western states: a preliminary inventory. U. S. Geological Survey, Water Resources Division.

5. Emmett, L. F. 1973. Ground-water contamination-western Great Lakes states: a preliminary inventory. U. S. Geological Survey, Water Resources Division.

6. Geraghty & Miller, Inc. 1975. Comprehensive water quality management plan - central and lower Susquehanna River Basin - draft report on ground-water conditions. Prepared for Gannett Fleming Corddry and Carpenter, Inc.

7. Geraghty & Miller, Inc. 1975. Unpublished information in company files.

8. Fuhriman, D. K., and J. R. Barton. 1971. Ground water pollution in Arizona, California, Nevada and Utah. U. S. Environmental Protection Agency, Water Pollution Control Research Series 16060ERW, December.

9. Scalf, M. R., J. W. Keeley, and C. J. LaFevers. 1973. Ground water pollution in the south central states. U. S. Environmental Protection Agency, Environmental Protection Technology Series, EPA-R2-73-268.

10. Miller, D. W., F. A. DeLuca, and T. L. Tessier. 1974. Ground-water contamination in the northeast states. U. S. Environmental Protection Agency, Environmental Protection Technology Series, EPA-660/2-74-056.

11. van der Leeden, F., L. A. Cerrillo, and D. W. Miller. 1975. Ground-water pollution problems in the northwestern United States. U. S. Environmental Protection Agency, Ecological Research Series, EPA-660/3-75-018.

12. Davis, P. N. 1974. Ground water pollution: case law theories for relief. Missouri Law Review, Vol. 39, No. 2. pp. 117-163.

13. U. S. Department of Commerce. 1974. Statistical abstract of the United States 1974. Social and Economic Statistics Administration, Bureau of the Census.

14. Meyer, C. F., ed. 1973. Polluting groundwater: some causes, effects, controls and monitoring. U. S. Environmental Protection Agency, Environmental Monitoring Series, EPA-600/4-73-001b.

15. Emrich, G. H., and G. L. Merritt. 1969. Effects of mine drainage on ground water. Ground Water, Vol. 7, No. 3. pp. 27-32.

16. Geraghty, J. J., and others. 1973. Water atlas of the United States. Water Information Center, Port Washington, New York.

17. Crain, J. 1969. Ground-water pollution from natural gas and oil production in New York. New York State Water Resources Commission Report of Investigation No. RI-5. 15 pp.

18. Krieger, R. A., and G. E. Hendrickson. 1960. Effects of Greensburg oil field brines on the streams, wells, and springs of the Upper Green River basin, Kentucky. Kentucky Geological Survey Report of Investigations 2, Series X. 36 pp.

19. Hopkins, H. T. 1963. The effect of oil field brines on the potable ground water in the Upper Big Pitman Creek basin, Kentucky. Kentucky Geological Survey Report of Investigations 4, Series X. 36 pp.

20. Wait, R. L., and M. J. McCollum. 1963. Contamination of fresh water aquifers through an unplugged oil test well in Glynn County, Georgia. Georgia Geological Survey Mineral Newsletter, Vol. 16, No. 3-4. pp. 74-80.

21. Burnitt, S. C., and others. 1962. Reconnaissance survey of salt water disposal in the Mexia, Negro Creek, and Cedar Creek oil fields, Limestone County, Texas. Texas Water Commission Memo Report 62-02. 27 pp.

22. Burnitt, S. C. 1963. Reconnaissance of soil damage and ground-water quality, Fisher County, Texas. Texas Water Commission Memo Report 63-02. 49 pp.

23. Fink, B. E. 1965. Investigation of ground- and surface-water contamination near Harrold, Willbarger County, Texas. Texas Water Commission Report LD-0365. 23 pp.

24. Vorhis, R. C. 1975. Personal communication.

25. Hansberry, R. 1966. Industry's concern with pesticide residues. Pesticides and their effects on soils and water. Soil Science Society of America.

26. Hanes, R. E., L. W. Zelazny, and R. E. Blaser. 1970. Effects of deicing salts on water quality and biota literature review and recommended research. National Cooperative Highway Research Program Report 91, Highway Research Board, National Academy of Sciences.

27. Edison Water Quality Laboratory. 1970. Environmental impact of highway deicing: interim report. U. S. Environmental Protection Agency, Water Quality Office, 11040 QCG.

28. Biggs, R. B., and others. 1973. Trace metals in several Delaware watersheds. Water Resources Center, University of Delaware, Newark, Delaware. 47 pp.

29. Schlesinger, W. H., W. A. Reiners, and D. S. Knopman. 1974. Heavy metal concentrations and deposition in bulk precipitation in montane ecosystems of New Hampshire. U. S. A. Environmental Pollution 6:39-47.

30. Deutsch, Morris. 1963. Ground-water contamination and legal controls in Michigan. U. S. Geological Survey Water-Supply Paper 1691. 79 pp.

SECTION XVI

EXISTING FEDERAL LEGISLATION

SUMMARY

Conscious effort and legislation toward comprehensive water pollution control began with the Water Pollution Control Act of 1948. This Act was primarily concerned with abatement of stream pollution, and it directed the Surgeon General to "prepare or adopt comprehensive programs for eliminating or reducing the pollution of interstate waters and tributaries thereof and improving the sanitary conditions of surface and underground waters."

Several other pieces of Federal legislation since 1948 provide further legal methods to protect ground water from contamination.

1. Section 208 of the Federal Water Pollution Control Act Amendments of 1972 (PL 92-500) establishes a planning function which provides for areawide and statewide waste treatment management. This planning must specifically include a process to identify and control pollution from surface and underground mine runoff, the disposal of residual waste, and the disposal of pollutants on land or in subsurface excavations. EPA's role, as set forth by Section 304(e) is to provide guidance and information, but EPA has no implementation authority.

   Section 402 of PL 92-500 establishes the National Pollutant Discharge Elimination System (NPDES) which is a program for issuing permits for point source discharges of pollutants. Section 402 also requires states to control the discharge of pollutants into wells. However, Section 502 excludes from the definition of pollutants "water, gas, or other material which is injected into a well to facilitate production of oil or gas, or water derived in association with oil or gas production and disposed of in a well, if the well used either to facilitate production or for disposal purposes is approved by authority of the state in which the well is located, and if such state determines that such injection or disposal will not result in the degradation of ground or surface water resources." This exclusion therefore removes wells used in association with oil and gas production from regulations under Section 402.

2. The Solid Waste Disposal Act of 1965, as amended in 1970,

contains no specific reference to ground water. However, guidelines developed under the Act provide for ground-water protection resulting from polluting activities and surface drainage and also for site development to minimize the impact on ground water. These guidelines are only mandatory for Federal agencies, but they serve as recommended practices for non-Federal agencies.

3. The National Environmental Policy Act (NEPA) of 1969 (PL 91-190) requires that all Federal agencies prepare environmental impact statements on major Federal or Federally regulated actions significantly affecting the quality of the environment. EPA has promulgated regulation for implementation of NEPA which lists ground-water protection as a significant parameter in determining the need for an EIS.

4. The discharge of radioactive wastes have been regulated from the beginning. However, there have been many significant problems, and an Interagency Working Group consisting of representatives of the Nuclear Regulatory Commission (NRC), the Energy Research and Development Agency (ERDA) and EPA has been formed to evaluate radioactive waste management and disposal.

5. The Safe Drinking Water Act of 1974 (PL 93-523) requires the regulation of underground injection which may endanger underground drinking water sources. The provisions of the Act will produce a Federal/State cooperative effort which is based on Federally set minimum standards and regulations administered by the states. The practices to be covered under the Act include "deep" and "shallow" waste disposal wells, oil-field brine disposal wells and secondary recovery wells, and engineering wells.

   Section 1424(e) of the Act (the Gonzalez Amendment) provides that if EPA determines an area has an aquifer which is the sole or principal drinking water source and which, if contaminated, will cause a significant hazard to health, EPA may delay or stop commitment of any Federal financial assistance to projects which may result in contamination of the aquifer.

CONSTITUTIONAL AUTHORITIES

The Constitution of the United States is the supreme law of the land (Art. VI, Cl. 2). Under it, the Federal government is limited to those powers expressly delegated or reasonably implied; all other powers "not prohibited by it to the states, are reserved to the states respectively, or to the people" (Amendment 10). The Federal authority over water is found in the following clauses of the Constitution: General Welfare Power, Commerce Clause, War Power, and the General Power to Enact Laws. Other relevant clauses include the clauses on Interstate Compacts, International Treaties and the Proprietary Power. In addition, the due process and "taking" clauses play a significant role.

The Commerce power has been interpreted to extend over the means of interstate transportation. The River and Harbor Act of 1826 is based on the concept that under the Commerce Clause the power of Congress comprehends navigation within the limits of every state of the Union. All navigable waters have been treated as public property, as have non-navigable reaches of navigable waters and their non-navigable tributaries. The Property Clause gives Congress unlimited power over the use of the public domain, including the power to dispose of land and water thereon together or separately. The Reclamation Act of 1902 was based on this power which has generally been concerned with Federal lands and the establishment of their water rights as appurtenant to land ownership.

The General Welfare Clause and the Taxing Power have been limited only by the requirement that they be exercised for the common benefit as distinguished from mere local purpose. The power of the Federal government to establish large scale projects affecting and using water is clear. Some water resource development projects have been based on the War Power (e.g., Wilson Dam, TVA). However, the War Power is an emergency power more suited to crises in development than crises in contamination. The Treaty Power has been used both with the Indians and with Mexico and Canada to protect international waterways such as the St. Lawrence, Columbia, Colorado, and Rio Grande Rivers.

In 1911, Congress gave blanket consent for interstate compacts for conserving the forests and water supply of the United States. The most comprehensive interstate compact is the Delaware River Basin Compact (1961) which is concerned with conservation, utilization, development, management, and control of water resources "in, on, under, or above ground, including related uses of land, which are subject to beneficial use, ownership or control," throughout the basin. This is the only interstate compact wherein ground water is mentioned specifically.

DESCRIPTION OF FEDERAL LAWS APPLICABLE TO GROUND-WATER PROTECTION

The Federal laws discussed below are limited to those for which EPA has whole or partial responsibility. Federal laws administered by other agencies may contain provisions to protect ground water.

Refuse Act

Early Federal legislation concerning water contamination included the Refuse Act comprising Sections 9 through 20 of the River and Harbor Act of 1899, which is still in effect today. The statute was originally intended merely to prevent obstructions to navigation based on the Commerce Power of Congress: it prohibits the discharge of refuse of any kind into navigable waters or their tributaries from any vessel, from the shore, from a wharf, manufacturing establishment or mill of any kind, but does not apply to fluid wastes from streets and sanitary sewers. Permits could be issued for deposit of material where anchorage and navigation would not be injured.

In recent years the statute has been used increasingly against pollution of navigable waters. Case law has extended the concept of "refuse" beyond waste products to include valuable products such as oil and gasoline. Since 1971, the Act has become the basis for a permit system administered by the Department of the Army in coordination with EPA and in at least one case in the northeast, has been used successfully to litigate in an action involving ground-water contamination. In this instance, action was brought by the Federal government against the operator of a landfill because of a non-permitted discharge of leachate to a navigable river.

Water Pollution Control Act of 1948

Conscious effort and legislation toward comprehensive water pollution control began with the Water Pollution Control Act of 1948 (62 Stat. 1155, as amended, 33 U.S.C. 466). Cautious in the face of constitutional limitations on Federal power, this Act declared: "The policy of Congress to recognize, preserve, and protect the primary responsibilities and rights of the states in controlling water pollution....and to provide Federal technical services to state and interstate agencies and to industries and financial aid to state and interstate agencies and to municipalities, in the formulation and execution of their stream pollution abatement programs." The Act was concerned primarily with abatement of stream pollution and it directed the Surgeon General to "prepare or adopt comprehensive programs for eliminating or reducing the pollution of interstate waters and tributaries thereof and improving the sanitary condition of surface and underground waters."

Public Law 92-500

The objective of the Federal Water Pollution Control Act, as amended, 1972 (Public Law 92-500) is to restore and maintain the chemical, physical, and biological integrity of the nation's waters, evidently depending upon the Commerce Power for its prime authority: (Title I, Section 101) "It is the national goal that the discharge of pollutants into the navigable waters be eliminated by 1985;" and the national policy is to "develop technology necessary to eliminate the discharge of pollutants into the navigable waters, waters of the contiguous zone, and the oceans." The general provisions of Title V include (in Section 502) definitions of pollutants, navigable waters, territorial seas, contiguous zone, and oceans. All these have been recognized as public waters.

Section 208 of PL 92-500 -

The most effective potential means for controlling ground-water contamination in PL 92-500 is found in Section 208, which provides for statewide and areawide planning for pollution control. The plans developed pursuant to Section 208 are the basis for the establishment of a continuing planning process and are the key vehicles for the control of nonpoint sources of water pollution. In addition to providing for identification of needed waste treatment works, their regulation, and nonpoint source control, Section 208 (b)(2)(G) requires a plan prepared under that Section to include "a process to (i) identify, if appropriate, mine related sources of pollution including new, current, and abandoned surface and underground mine runoff" and to set forth control procedures. Under sub-section (J) the plan must include "a process to control the disposition of all residual waste generated in such area which could affect water quality." The most powerful means to control ground-water contamination in Section 208 is in sub-section (K) which requires the plan to include "a process to control the disposal of pollutants on land or in subsurface excavations within such area to protect ground- and surface-water quality."

While Section 208 is part of a Federal regulatory scheme, the primary responsibility for preparing plans and implementing programs is in the hands of the state and local agencies. EPA has no effective power to force action to protect ground water. EPA can withhold approval of a plan that does not provide adequately for ground-water protection. However, it does not have authority to act if the ground-water provisions of the plan are not implemented.

Section 304(e) of PL 92-500 -

EPA's basic function in relation to Section 208 is set forth in Section 304(e) of PL 92-500. There are three sub-sections under sub-section (e) which refer to guidelines which EPA must issue on nonpoint sources of pollutants. EPA must develop and issue guidelines for identifying and evaluating the nature and extent of nonpoint sources and processes, procedures and methods to control pollution resulting from: (D) the disposal of pollutants in wells or in subsurface excavations; (E) salt-water intrusion resulting from reductions of fresh-water flow from any cause, including extraction of ground water; (F) changes in the movement, flow, and circulation of any navigable waters or ground waters.

These three sub-sections cover a very broad range of ground-water contamination sources including wells, impoundments, landfills, pipelines, septic systems, sea-water intrusion, and overpumping. This section of PL 92-500 is not enforceable as law, however. The guidelines developed can serve only as information which can be used or not used at the discretion of the Federal, state or local governments. The information is not regulatory in nature but educational. As noted above, Section 208 is the basic vehicle in PL 92-500 for controlling nonpoint-source pollution and ground-water contamination. The basic approach taken by EPA to solve these pollution problems is the use of Best Management Practices (BMP's) to prevent or minimize the pollution rather than treatment after the pollution has occurred. Information has been developed by EPA pursuant to Section 304(e) to indicate the BMP's currently recognized as effective for control of different types of pollution.

Section 402 of PL 92-500 -

The National Pollutant Discharge Elimination System (NPDES) is developed under Section 402 of PL 92-500 to issue permits for the discharge of any point source pollutant or combination of pollutants to navigable waters. The section goes on to state that individual states may issue NPDES permits if the program is authorized by EPA. Currently, 27 states have received such approval. One requirement for state approval is that the states must issue permits which control disposal of pollutants into wells.

Wells under this section of law have been interpreted by EPA to mean only "deep" waste injection wells, of which there are less than 400 in the United States. EPA has further interpreted that permits will only be given to cover these wells if there is an associated surface-water discharge of pollutants. Many state programs are not placing these same restrictions on the definition of wells; however, most states are covering only a very small number of wells.

One reason for this narrow coverage is the definition of pollutant which is included in Section 502 of PL 92-500. It states that pollutant does not mean water, gas or other material which is injected into a well to facilitate production of oil or gas, or water derived in association with oil or gas production and disposed in a well if the well is approved by authority of the state in which it is located.

Solid Waste Disposal Act

The Solid Waste Disposal Act, passed in 1965, was the first Federal legislation establishing a Solid Waste Program. It was amended in 1970 to add resource recovery provisions, and was extended without major revision since then. It has currently expired, and the program is now functioning under a general appropriation to EPA.

The Act contained no direct reference to ground water. The findings and purposes show an awareness that improper management of wastes creates hazards to public health including pollution of water resources. Section 212 of the act contained provisions for submittal of a report to Congress on hazardous waste disposal. However, no legislation was forthcoming as a result of that report. Section 207 provided planning grants (no longer funded), which included such factors as water pollution control among funded activities, and Section 209 provided that EPA would recommend guidelines and model codes consistent with water quality standards. In practice, none of the 207 planning grants fully addressed ground-water protection.

The EPA guidelines (42CFR:460-464, 1971) state that provisions are to be made to insure that no pollution of surface or ground water is created and surface drainage is to be diverted to control infiltration at the site. Additionally, the guidelines suggest that the hydrology of the site be evaluated in order to design site development as to minimize the impact on ground-water resources (Section 241-202-(C)). These guidelines are mandatory for Federal agencies which operate or contract for the operation of disposal services, and they serve as recommended practices for non-Federal agencies (state and sub-state). The Act did not contain regulatory provisions, and its demonstration grant program did not address water protection.

National Environmental Policy Act

The National Environmental Policy Act (NEPA) 1969 (PL 91-190), implemented by Executive Order 11514 of March 5, 1970 and the Council on Environmental Quality's Guidelines of August 1, 1973 requires that all agencies of the Federal government prepare detailed environmental impact statements (EIS) on proposals for legislation and other major Federal actions significantly affecting the quality of the human environment. Although NEPA does not create a regulatory process, it does provide for the development of a decision-making document which is central to requiring Federal agencies to consider the environmental impact of their actions. It is, therefore, one way in which the ground-water impact of a Federal action may be raised.

EPA has promulgated regulations for implementing NEPA for non-regulatory projects which the agency has funded, as well as for New Source NPDES permits. In both these regulations ground water is listed as a significant parameter in determining the need for an EIS. In the regulations for non-regulatory EPA funded projects (April 14, 1975) Section 6.200, the criteria for determining when to prepare an EIS, states, "....particular attention should be given to....significant changes in surface- or ground-water quality or quantity...."

The regulation for EIS preparation for New Source NPDES permits (October 9, 1975) lists as one of the criteria to be used in determining the need for an EIS, "....the new source may directly or through induced development have a significant adverse effect upon....surface- or ground-water quality or quantity."

The Safe Drinking Water Act of 1974 (PL 93-523)

The Safe Drinking Water Act of 1974 (PL 93-523) is to assure the provision of safe drinking water to all Americans served by public water supply systems. Two mechanisms have been developed to meet this goal. It is required that all public water systems meet minimum water quality standards. These standards include bacteria as measured by *Escherichia coli* fecal indicator; organic pesticides; inorganics including iron, manganese, arsenic, lead, mercury, chromium, and cadmium; and radioactive materials including radium and strontium. Secondly, Part C of the Act develops a program to protect underground drinking water sources. The Safe Drinking Water Act is to be a Federal/state cooperative effort which is based on Federally set minimum standards and regulations administered by the states.

Part C sets the basic guidance under which EPA must develop minimum state requirements. These requirements are that programs shall (Section 1421(b)(1):

1. prohibit any underground injection which is not authorized by permit or rule issued by the state.

2. require that the applicant for a permit must satisfy the state that underground drinking water sources are not endangered.

3. include inspection, monitoring, record keeping and reporting requirements.

4. apply to underground injection by Federal agencies and by other persons on Federal land.

5. not interfere with or impede (A) underground injection of brine or other fluids brought to the surface in connection with oil and gas production, or (B) underground injection for secondary or tertiary recovery of oil unless such requirements are essential to assure that underground sources of drinking water are not endangered.

Three other important provisions are made which affect ground-water protection. First, underground injection is defined as "the subsurface emplacement of fluids by well injection" (Section 1421(d)). Second, endangerment means the presence of a contaminant which may prevent a public system from complying with any national primary drinking water standard or otherwise adversely affect the public health (Section 1421(d)). And thirdly, the Administrator shall determine each state which needs an underground injection program to protect drinking water sources. If the state, so listed, does not obtain primary enforcement authority, then EPA will administer the programs (Section 1422(a)).

The Act and the supporting Committee Report (report no. 93-1185) produce a dilemma on the intent of the Act when defining underground injection. As stated in Section 1421, the Act appears to cover only "wells" whose specific function is to dispose of waste fluid underground. Under this narrow interpretation, less than 400 industrial and municipal "deep" injection wells and about 10,000 brine disposal wells are included. As can be seen from Sections XI and XIII of this report, these two sources of ground-water contamination are not of prime importance.

The House Report (page 31) states "underground injection" is intended to be broad enough to cover any contaminant which may be put below ground level and which flows or moves, whether the contaminant is in semi-solid, liquid, sludge or any other form or state. This definition is not limited to the injection of waste or to injection for disposal purposes; it is intended also to cover the injection of brines and the injection of contaminants for extraction or other purposes.

The only exemptions are septic tanks or other individual residential waste disposal systems. This language appears to indicate that the intent of the Act is to regulate <u>all</u> ground-water contamination sources including landfills, municipal surface impoundments and industrial surface impoundments, as well as other activities such as sea-water intrusion control, geothermal development, LP gas and natural gas storage (covered in Section XV as non-waste disposal sources of contamination).

Currently, EPA is taking the position that the regulation developed under Part C will cover only "deep" and "shallow" waste disposal wells, oil-field brine disposal wells and secondary recovery wells, and engineering wells (including barrier wells, subsidence control wells, geothermal wells, solution mining wells and gas storage wells), and is deferring the inclusion of other practices such as industrial surface impoundments pending further study. This coverage has been chosen after a legal evaluation of the legislative history, the Act and the House Report.

The handling of the endangerment issue in the Act and supportive report produces a problem. Strictly interpreted the Act takes a total non-degradation position. In essence nothing can be put underground (by whatever activities are eventually covered) because there is no possible way to insure that a drinking water source (which the report defines as 10,000 ppm total dissolved solids) will never be endangered. This is a result of the long residence times of contaminants in ground water and the unknown future development of ground water as a drinking water source.

Finally, the requirements for listing states which need a program to protect underground sources may produce difficulties for ground-water protection. Under the interpretation of "underground injection" chosen by EPA there are several states which have only a small number of underground injection facilities. EPA might decide that these states do not require an underground injection control program under the Safe Drinking Water Act, thereby possibly eliminating them from future coverage required by changed circumstances.

The Safe Drinking Water Act gives EPA a tool to regulate activities which may contaminate ground water, but at this time it is difficult to determine which activities will be regulated and what the impact will be on ground-water quality.

Section 1424(e) of the Act, known as the Gonzalez Amendment, provides another method for protection of ground-water quality. If EPA determines upon its own initiative or by petition that an area has an aquifer which is the sole or principal drinking water source for the area and which, if contaminated, will cause a significant health hazard, EPA may delay or stop commitment of any Federal funds for projects which may result in contamination of the sole-source aquifer. The Federal projects which are covered are not limited to underground injection but can include ground-water development or other activities in recharge zones or any other activity which may contaminate ground water. Unlike the rest of Part C, which gives the states primary enforcement authority, this section is to be administered by EPA only.

Areas of potential difficulty with this section are: (1) Federal/state relationship, (2) only Federal projects are covered and major ground-water contamination sources are not Federal projects, and (3) the definition of "Federal financial assistance," i.e., does it include VA or FHA loans to individual homeowners?

Radioactive Materials

Regulatory control of radioactive materials is exercised by several Federal agencies under general radiation protection guidance and standards provided by EPA for such agencies. In some cases, these agencies exercise regulatory control directly; in others, they delegate some control to states under Federally authorized agreements.

EPA, under the authority transferred to it by Reorganization Plan No. 3 (from the Atomic Energy Act of 1954, as amended), currently is preparing generally applicable environmental radiation protection standards. Excluding mining and radioactive waste disposal facilities, these standards will limit the radioactive exposures from all Nuclear Regulatory Commission (NRC) licensees and Energy Research and Development Administration (ERDA) operated facilities (via all pathways including ground water) to the public outside the boundaries of such facilities.

Authority for licensing and regulation of the use of (a) nuclear "source material," (b) "special nuclear material," and (c) radioactive "by-product material" is provided for in the Atomic Energy Act of 1954, as amended; and in the Energy Reorganization Act of 1974. Under these authorities, NRC has been exercising Federal jurisdiction for licensing and regulation of all commercial nuclear energy facilities, and also has been assigned jurisdiction for licensing and regulation of certain Federal facilities, including facilities used primarily for the receipt and storage of high-level radioactive wastes resulting from activities licensed under the Atomic Energy Act of 1954, as amended. In implementing these authorities, NRC's objective is to protect the health and safety of the public. In carrying out this objective, NRC strives to limit the planned releases of radioactivity, in air emissions and in water effluents from individual activities, to levels that cumulatively will not exceed radiation exposures to the public, published by EPA in its proposed generally applicable environmental radiation protection standards. These standards will apply to areas outside the boundaries of the individual activities. Federal responsibility for unplanned releases from these activities belongs directly to NRC.

Except for a few cases licensed by NRC, ERDA regulates the radioactivity levels of planned releases in air emissions and in water effluents from ERDA owned and operated and from ERDA contractor operated facilities using "source, special and by-product materials." Here again, the objective is to protect the health and safety of the public, and to control planned releases so as to meet the EPA environmental radiation protection standards outside the boundaries of such facilities. Federal responsibility for unplanned releases from such facilities belongs directly to ERDA.

In general, close coordination and cooperation is developing between NRC, ERDA, and EPA in Federal control of radioactive releases and in protection of the public from radioactive contamination.

Thus, Federal authorities appear adequate for protection of public health from radioactive pollution released from "source, special and by-product materials" and reaching the public via any pathway including ground water. On the other hand, as illustrated by specific examples discussed in other sections of this report, hindsight indicates that implementation of these authorities has not been fully effective, particularly in areas of potential water contamination from management of radioactive wastes.

For limitation of releases of radioactive contaminants to ground water, other than those emanating from "source, special and by-product materials," the scope of Federally established authority may turn out to be inadequate. Except for related regulation provided in certain provisions of the Safe Drinking Water Act (PL 93-523) and for regulation of some surface-water releases under the NPDES, no known Federal authority exists for explicit limitation of radioactivity specifically for ground water. Radioactive contaminants in ground water can be present in releases from the mining, milling, processing, and utilization of ores, coal, and minerals containing naturally radioactive impurities.

Thus, potential control deficiency arises because the Atomic Energy Act of 1954, as amended, provides for regulation of nuclear "source materials" (defined by the Commission to be containing over 1/20 of one percent of either uranium or thorium or any combination thereof). Hence, this authority does not provide control of uranium and thorium which often are found in nature as impurities (less than 1/20 of one percent) in ores, coal, and minerals; and it does not provide for control of naturally occurring radium and radon, nor of radioactive materials produced by accelerators.

While in the ground, the doses of radioactivity from naturally radioactive materials are shielded by earth cover and do not cause a threat to the public. When mined and brought to the earth's surface, however, unless properly processed and managed, such radioactivity may prove dangerous to public health. Examples of such naturally radioactive materials include fluorspar, bauxite, coal, phosphates, and copper and titanium ores. Currently, the tremendous increase in the use of phosphates for fertilizers, and the expected increase in the use of coal for energy, is causing EPA to investigate the potential radiological impacts from such accelerated use.

In addition, EPA is investigating the potential impacts of medical radionuclides given to patients and then excreted in local sewage systems. A study of the radioactivity found at one sewage plant indicated it consisted principally of Iodine 131, and the amount (while not enough to constitute a hazard to health) was greater than the amount of Iodine 131 normally released in liquid effluents from a large nuclear power plant. As far as known, no Federal limits have been placed on the amounts of such radioactivity which can be discharged into sewage systems.

CONCLUSIONS

The following observations may be drawn regarding the adequacy of existing Federal legislative programs within the ambit of EPA's authority for protection of ground-water contamination.

1. PL 92-500 primarily focuses on navigable waters rather than ground water or "public waters," meaning both navigable and ground waters.

2. Section 208 of PL 92-500, at least on paper, provides strong language for ground-water protection. However, no mechanism exists for EPA to insure the implementation of ground-water protection plans developed pursuant to the 208 process.

3. The coverage under NPDES (Section 402 of PL 92-500) is extremely limited as regards ground-water protection both because of the language in Section 402 and the narrow definition of "pollutant" in Section 502.

4. Under PL 93-523, the Safe Drinking Water Act, several definitional problems have led to interpretations which limit the coverage of the Act. Some of the major polluting activities may not be subject to the requirements of PL 93-523.

5. The Solid Waste Disposal Act has not proved an effective vehicle to protect ground water. It has currently expired.

6. NEPA, through the EIS process, is a significant vehicle for reviewing and evaluating the ground-water impact of Federal actions.

7. Radioactive wastes present a significant potential contamination threat to ground water. Federal authorities appear adequate for protection of public health; however, implementation has not always been adequate. Limitation by Federal legislation of releases of radioactive contaminants to ground water from other than those emanating from "source, special and by-product materials" may be inadequate. No known limits exist on medical radionuclides given to patients and then excreted into local sewage systems.

8. The Federal power over sub-surface waters is less clear than over surface waters and its exercise may raise Federal-state constitutional questions.

SECTION XVII

STATE AND LOCAL ALTERNATIVES FOR GROUND-WATER QUALITY PROTECTION

SUMMARY

There are a number of requirements that are basic to all state and local ground-water protection programs. Similar to Federal activities, control over ground-water quality has been given a low priority when compared to surface water. This has been due to deficiencies in existing legislation, the lack of funds available for proper staffing, and the diverse interests and priorities of existing agencies.

For maximum effectiveness, rules and regulations for protecting ground water should be designed to: (1) prevent and control unwanted contamination and degradation of both ground- and surface-water quality; (2) provide data necessary to evaluate the nature and areal extent of ground-water contamination and the number of sources of contamination; (3) provide a basis for correcting or mitigating existing cases of ground-water contamination; and (4) provide a regulatory framework within which aquifers can be used for waste treatment and storage.

There are two approaches to the problem of protecting ground-water resources. One is to look at the underground water itself as the resource to be managed and to concentrate on limiting waste discharges and preventing causes of contamination. This first method is the one used to control air and surface-water pollution. It is also the most popular basis for ground-water control regulations.

A second approach is to take into account the ability of aquifers to treat and store wastes and to consider these characteristics as the prime resource to be managed or controlled. The second method is not in common usage, but may become more popular as ground-water technology becomes more sophisticated. An example of this approach could be based upon the possibility of states and local governments acquiring ownership of aquifer pore space (storage capacity) through eminent domain proceedings or other methods.

INTRODUCTION

There are a number of alternatives that are available for implementing state and local ground-water protection programs.

In the preceding sections on waste disposal practices, typical examples of existing state legislation have been outlined. However, these examples and the alternatives described below, no matter how well conceived, are ineffective if implementation is inadequate. As in the Federal situation, control over ground-water quality has been given a low priority when compared to surface water. This has been due to a lack of appreciation of potential and real ground-water problems, which has resulted in deficiencies in existing legislation together with insufficient funding for proper staffing of environmental protection agencies. Implementation of existing legislation also has been hampered by the relatively low level of prevention and abatement technology and the diverse interests and priorities of government, industry, and the public.

The alternatives discussed in this section are not presented as a recommended model law. They are included only to illustrate the availability and variety of controls that can be exerted by state and local agencies to protect ground-water quality.

OUTLINE OF STATE WATER RIGHTS LAW

Water is one of man's "inalienable rights;" a water right is also real property and is protected by the Federal and state constitutions that prohibit the deprivation of property without due process of law or its taking for public use without just compensation. But water rights vary from state to state. In the humid regions (primarily the states east of the Mississippi), ground water has been considered appurtenant to land ownership. The English common law rule of riparian rights is that the water belongs to the people who own the land, and not the state; the landowner owns the ground water as an ingredient of his soil. Several states follow the old English doctrine (no longer accepted in England) that the landowner's water right is absolute and independent of the rights of all others. Other states in humid regions have placed some restriction on the rights of landowners, as exemplified by the "American" rule of reasonable use. In New Hampshire a landowner does not have the right to waste water unnecessarily or to export it from the area. Under the "California" doctrine of correlative rights, the owners of land overlying a ground-water reservoir have correlative rights in a common supply, and each is limited to reasonable and beneficial use of the water. A surplus may be appropriated and exported. Where a surplus has been appropriated to the point where it has become a deficit, the landowners who have continued to pump to meet their needs have rights by mutual prescription; the infringers, after five years, have prescriptive rights; the landowners who have not pumped lose their rights.

In arid regions (primarily west of the Mississippi), rights to water are generally gained by appropriation. Such rights are entitled to protection, except as against the true owner. In states where appropriation is an accepted means of obtaining water rights, beneficial use is the basis, the measure, and the limit of a water right. The appropriative right, in contrast to the landowner right, is to a specific quantity or rate of flow and is lost by non-use. Nevertheless, it is a property right demanding constitutional protection.

In the southwest, some water rights have originated with grants of land and the rights appurtenant thereto by foreign sovereigns. In states which follow the appropriation theory, the historic role of the state has been to prescribe conditions under which rights may be acquired to use water, to record the rights, to adjudicate claims, and to allocate water in accordance with the established rights. Non-consumptive use and contamination are usually kept separate in the public mind and in public agencies. Thus, a water right for non-consumptive use can be granted without specifying any responsibility for the resulting contamination. Each state has adopted its own system of water laws, subject to certain paramount Federal powers. The general rule is that large bodies of water are common property and that water flowing in a natural stream is not the subject of private ownership.

BASIC POLICY ISSUES

The promulgation of reasonable rules and regulations for protection and management of potable ground-water supplies first requires identification of the activities to be regulated, and a basic understanding of how ground water occurs. Uses of aquifers for purposes other than drinking water also must be defined and considered in order to promulgate sufficient rather than overprotective measures. The goal of regulatory controls should be to deflect contaminants away from water-supply aquifers to either resource recovery operations or to other more suitable aquifers.

SCOPE OF REQUIRED REGULATORY AUTHORITY

The major physical interrelationships between surface water and ground water strongly suggest that regulation of ground-water quality be under the same administrative and organizational system as surface-water quality. The following kinds of authority in basic legislation are listed here as examples of rules and regulations:

1. Authority to require permits and licenses for specific activities: (a) permits to discharge wastes to or into underground waters, (b) permits for appropriating and using ground water (diversions, allocations), (c) permits to construct or abandon wells, (d) permits or licenses to construct waste disposal facilities which discharge to the land and underground waters, (e) licensing of well drillers, (f) licensing of operators of waste disposal facilities, and (g) licensing of waste haulers.

2. Authority to promulgate and change rules and regulations such as: (a) construction standards (wells and disposal facilities), (b) ground-water quality standards and criteria, (c) uses for ground water and uses for aquifers based on the geology and hydrology of an area, and (d) exemptions from the requirements of regulation.

3. Authority to establish charges for use of ground water, and utilization of the waste attenuation and storage capacity of aquifers.

4. Authority to establish permit and license fees.

5. Authority to require monitoring of potential sources of contamination.

6. Authority to enforce laws and regulations and enforce penalties for violations.

7. Authority to set requirements for public hearings for permits.

8. Authority to form an entity capable of owning and/or operating waste disposal systems designed to discharge to the land and ground waters.

EFFECTIVE RULES AND REGULATIONS

For maximum effectiveness, rules and regulations for controlling ground-water contamination should be designed to: (1) prevent and control unwanted contamination and degradation of ground- and surface-water quality, (2) provide data necessary to evaluate the nature and areal extent of ground-water contamination and the causes of contamination, (3) provide a basis for correcting or mitigating existing cases of ground-water contamination, and (4) provide a regulatory framework within which aquifers can be used for waste treatment and storage.

There are two approaches to the problem of controlling

ground-water contamination. One is to look at the underground water itself as the resource to be managed and to concentrate on limiting waste discharges and preventing causes of contamination. This first method is the one used to control air and surface-water pollution. It is also the most popular basis for ground-water control regulations.

A second approach is to take into account the ability of aquifers to treat and store wastes and to consider these characteristics as the prime resource to be managed or controlled. The second method is not in common usage, but may become more popular as ground-water technology becomes more sophisticated. An example of this approach could be based upon the possibility of states and local governments acquiring ownership of aquifer pore space (storage capacity) through eminent domain proceedings or other methods. In the Annotated Code of Maryland there is a section called the "Prince Georges County Underground Storage Act" which illustrates this concept as applied to storage of natural gas. 1) The act states, in part:

> "For the privilege of using geological strata beneath the surface of the earth in Prince Georges County for underground storage of gas ...a gas storage company shall pay the county an underground storage fee."

PRIMARY REGULATORY CONTROL MECHANISMS

When faced with the multiplicity of ground-water contamination causes and sources, the question becomes "to permit or not to permit." To deal with this problem, these sources and causes have been divided into four categories (Table 78). The first two categories concern discharges of contaminants that are wastes or waste waters, the third category concerns discharges of contaminants that are not wastes, and the fourth category consists of those causes of ground-water quality degradation not related to discharges. Some of the sources or causes of ground-water contamination could fall under more than one category, for example, some lagoons may be designed to discharge to land and ground waters.

As a general rule, all Category I (Table 78) causes will require a discharge or injection control permit for each project. Exceptions can be made, for example, on the basis of existing permit systems. A regulatory agency can also decide for political or economic reasons, to simply exempt the discharge activity from permit requirements if the impact on aquifers and underground waters is not considered significant. Category II causes will require approval of construc-

Table 78. CLASSIFICATION OF SOURCES AND CAUSES OF GROUND-WATER POLLUTION USED IN DETERMINING LEVEL AND KIND OF REGULATORY CONTROL.

WASTES		NON-WASTES	
CATEGORY I	CATEGORY II	CATEGORY III	CATEGORY IV
Systems, facilities or activities designed to discharge waste or waste waters (residuals) to the land and ground waters	Systems, facilities or activities which may discharge wastes or waste waters to the land and ground waters	Systems, facilities or activities which may discharge or cause a discharge of contaminants that are not wastes to the land and ground waters	Causes of ground water pollution which are not discharges
LAND APPLICATION OF WASTE WATER - sprary irrigation, infiltration-percolation basins, overland flow	SURFACE IMPOUNDMENTS - waste holding ponds, lagoons and pits	BURIED PRODUCT STORAGE TANKS AND PIPELINES	SALT-WATER INTRUSION - sea water encroachment, upward coning of saline ground water
SUB-SURFACE SOIL ABSORPTION SYSTEMS - (septic systems)	LANDFILLS AND OTHER EXCAVATIONS - landfills for industrial wastes, sanitary landfills for municipal solid wastes, landfills for municipal water and waste water treatment plant sludges, other excavations (e.g., mass burial of livestock)	STOCKPILES - highway deicing salt stockpiles, ore stockpiles	RIVER INFILTRATION
WASTE DISPOSAL WELLS AND BRINE INJECTION WELLS	ANIMAL FEEDLOTS	APPLICATION OF HIGHWAY DEICING SALTS	IMPROPERLY CONSTRUCTED OR ABANDONED WELLS
DRAINAGE WELLS AND SUMPS	LEAKY SANITARY SEWER LINES	PRODUCT STORAGE PONDS	FARMING PRACTICES - (e.g., dry land farming)
RECHARGE WELLS	ACID MINE DRAINAGE	AGRICULTURAL ACTIVITIES - fertilizers and pesticides, irrigation return flows	
	MINE SPOIL PILES AND TAILINGS	ACCIDENTAL SPILLS	

tion standards (again, exceptions can be made). A permit for Category II could be required for activities which posed an exceptional threat of ground-water contamination such as lagoons and landfills. Category III causes will require facility construction standards and/or guidelines and manuals (e.g., tons/mile limits on highway deicing salts, corrosion-proof buried storage tanks, covered stockpiles). For Category IV causes, other types of regulatory controls will be needed in addition to facility construction standards, guidelines and manuals (e.g., controls on ground-water withdrawals, limits on discharges of contaminants to streams, and constraints on land use).

SECONDARY REGULATORY CONTROL TECHNIQUES

Although issuance or denial of permits and approval for activities which may contaminate ground water can be handled on a case-by-case basis, consistency alone will require that standards be established upon which to base these decisions. In addition, manpower constraints will necessitate concentrating control efforts on specific causes of contamination in specific areas. No state, for example, has sufficient trained personnel to deal with the thousands of sources of ground-water contamination categorized in Table 78. However, there must be sufficient qualified people to make the hard choices of what to do first and where to do it.

The purpose of ground-water quality standards is to provide a yardstick for determining the amount of change to permit in native ground-water quality. It does not mean that native ground-water quality must be made to meet these standards and, unlike surface waters, it does not necessarily mean that ground waters already contaminated will be or can be improved in quality to meet these standards.

Ground-water quality standards are basically applicable to water supply uses for aquifers. These, in turn, are governed by native ground-water quality, aquifer storage capacity and the ability of the aquifer to yield water to wells. Existing state standards recognize at least three classes of native ground-water quality: (1) high quality, meeting or exceeding (with minor undesirable, naturally occurring elements) Federal drinking water standards; (2) intermediate quality suitable for industrial and agricultural use and for possible use as a potable water supply after treatment; and (3) low quality (saline water).[1,2,3] These standards should also explicitly or implicitly recognize differences between the saturated and unsaturated portions of aquifers, ground water as a source of recharge to surface waters, differences in aquifer transmissive characteristics, and the designation of aquifers or sections thereof for waste treatment and storage.

Water-quality standards are by no means an answer to all of the multiple problems of ground-water contamination. In fact, promulgating such standards may raise more questions than are answered. For example, there are many aquifers where the native ground-water quality is superior to Federal drinking water standards or criteria, and many aquifers where the native quality is inferior to these standards but which are the only water supplies reasonably available. Thus, there is the problem of degrading ground waters to drinking water standards and protecting usable ground water inferior to those standards.

Suggested Guidelines for Ground-Water Quality Standards

1. All reasonable supply uses for ground water should be included in the standards (water for agriculture and industrial process water). To protect these uses, a general limiting level of ground-water quality needs to be established. A total dissolved solids concentration of 5,000 ppm would be a satisfactory limit.

2. An upper limit for protection of drinking water supplies also needs to be established. A total dissolved solids concentration of 2,500 ppm is a conservative value. The 1962 Drinking Water Standards [4)] included a limit for total dissolved solids of 500 ppm, if other less mineralized sources were available. The 1972 EPA report on water-quality criteria [5)] "recognized that a considerable number of supplies with dissolved solids in excess of 500 ppm are used without any obvious ill effects," and therefore, did not recommend a dissolved solids limit. Because of the variability in native ground-water quality, a limit of 500 ppm for potable ground-water supplies is suggested for areas where present concentrations are below this value.

3. Parameters for which maximum contaminant levels have been established by either Federal criteria or regulations should represent upper limits for potable ground-water quality standards.

4. Aquifer productivity characteristics, in addition to native ground-water quality, should be considered in determining uses for ground water. In general, aquifers containing ground water with a relatively low total dissolved solids concentration should be protected over a wider range of transmissivities than aquifers containing

ground water with a higher dissolved solids concentration (Figure 79). Use of aquifer transmissive characteristics to delineate classifications is not simple, especially in light of the differences between fractured rock, carbonate rock, and unconsolidated aquifers. Furthermore, one could conceivably allow degradation of a Class I aquifer to Federal drinking water standards while placing more severe restrictions on a Class II aquifer on the grounds that any change in the quality of water from the latter could make it completely unusable as a drinking-water source.

5. The interrelationships between ground water and surface water should be recognized in the standards. For example, augmenting streamflow with ground water is considered a "use" in North Carolina. 2)

6. The ability of geologic formations (saturated and unsaturated) to treat wastes should be recognized.

7. Other, indirect, water-supply uses of aquifers and ground water, such as fresh-water storage in saline aquifers and low-flow augmentation of streams, should be recognized and protected.

8. Standards should be designed to meet individual state needs.

GROUND-WATER EFFLUENT LIMITS AND WASTE LOAD ALLOCATIONS

Effluent limits are a mechanism for placing specific numerical limits (quantity and quality, or loading) on multiple sources of wastes entering the saturated or unsaturated portion of an aquifer. For surface-water contamination control, effluent limits are placed on point-source discharges. For ground-water contamination control, effluent limits will apply to a much broader range of activities. Examples include limits on nitrogen loading for land application of municipal waste waters; limits on types of materials disposed of in sanitary landfills; limits on specific contaminants injected into an aquifer by means of a disposal well; effluent limits on waste treatment facilities which discharge to a stream that recharges an aquifer; and a pollutant discharge limitation requirement for surface impoundments.

Waste-load allocations for streams are a way of achieving surface-water quality standards. Their purpose is to establish the actions necessary to maintain or improve existing surface-water quality and provide a basis for effluent limitations. Allowable contaminant loads are allocated to indi-

Class I Aquifer means an aquifer having a transmissivity greater than 100 gal/day/ft and a permeability greater than 10 gal/day/sq ft. In addition, the total dissolved solids concentration for natural water in each aquifer shall be less than 500 ppm. The regulatory agency may designate an aquifer as Class I where it serves as the sole water supply for any persons regardless of limitations on aquifer transmissivity or permeability. This aquifer type is designated in order to protect those aquifers used, or capable of being used, for a fresh drinking water supply.

Class II Aquifer means an aquifer having a transmissivity greater than 1,000 gal/day/ft, a permeability greater than 100 gal/day/sq ft and natural water with a total dissolved solids concentration of between 500 and 2,500 ppm. This aquifer class is designated to protect those aquifers of poorer natural quality than Class I which are used, or may reasonably be expected to be used, as a source of drinking water. The regulatory agency may designate an aquifer as Class II where it serves as the sole water supply for any persons, and a reasonable alternative aquifer or surface water source is not available, regardless of limitations on aquifer transmissivity or permeability.

Class III Aquifer means an aquifer having a transmissivity greater than 10,000 gal/day/ft, a permeability greater than 100 gal/day/sq ft and natural water with a total dissolved solids concentration of between 2,500 and 5,000 ppm. This aquifer class is designated in order to protect brackish water aquifers for use for industrial water supply, agricultural water or a potential fresh-water supply after desalinization. Where feasible, the regulatory agency will maintain the quality of waters in Class III aquifers so as to be usable for these purposes.

Class IV Aquifer means all aquifers other than Class I, Class II, or Class III aquifers. This aquifer type is designated in order to provide for the use of some aquifers for waste storage and treatment.

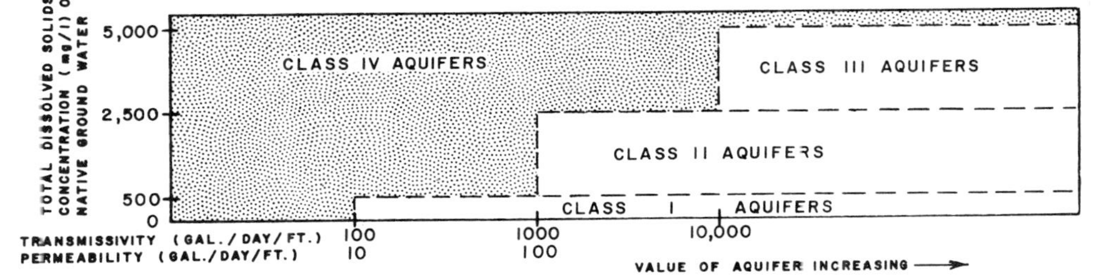

Figure 79. Guidelines for ground-water quality standards.

vidual waste treatment facilities that contribute contaminants to a given segment of a stream. In general, the application of the waste-load allocation concept to underground waters will be for the purpose of maintaining or limiting degradation of existing ground-water quality. A waste-load allocation will generally apply to an entire aquifer or basin while an effluent limit is site specific. The waste-load allocation concept means that there is a finite allocable waste handling capacity, assignable by government, to an aquifer. It is one method of dealing with the difficult technical problem of determining the "waste assimilative capacity" of the land and aquifers.

Although soils and sediments have a capacity to attenuate wastes, there are many cases of ground-water contamination (e.g., landfills) where the assimilative capacity of the aquifer system has been overwhelmed. Even though it is not yet possible to precisely determine aquifer assimilative capacity, estimates and ranges can be established (e.g., cation exchange capacity of soil). In addition, the percentage removal of contaminants can be estimated for treatment techniques such as land application of waste water and conventional waste treatment processes; therefore, waste-disposal techniques can be adjusted to compensate for uncertainties in determining allocable waste capacities for aquifers.

Determination of Waste-Load Allocations

Simple examples of a determination of a waste-load allocation for an aquifer, using nitrogen as the contaminant to be allocated, is shown in Figures 80 and 81. The first case (Figure 80) shows a method for obtaining a crude estimate of the maximum number of septic systems permissible per square mile without causing an increase in nitrate concentration of ground water exceeding drinking water standards (10 ppm as N). This example could be for a hypothetical water-table aquifer for which only a bare minimum of basic data is available. It is assumed that the nitrate concentration in the native ground water is zero, that dilution by precipitation is the only mechanism acting to alter the concentration of nitrogen, and that all nitrogen in the septic system effluent is converted to nitrate.

The second example (Figure 81) shows the use of a more sophisticated analysis for Massachusetts [6]) based on stream low-flow characteristics. An increase in the total dissolved solids concentration of stream base flow, which is actually dependent upon the quality of ground-water discharge in the basin, is shown as a function of the number of septic systems per square mile. By knowing the percentage

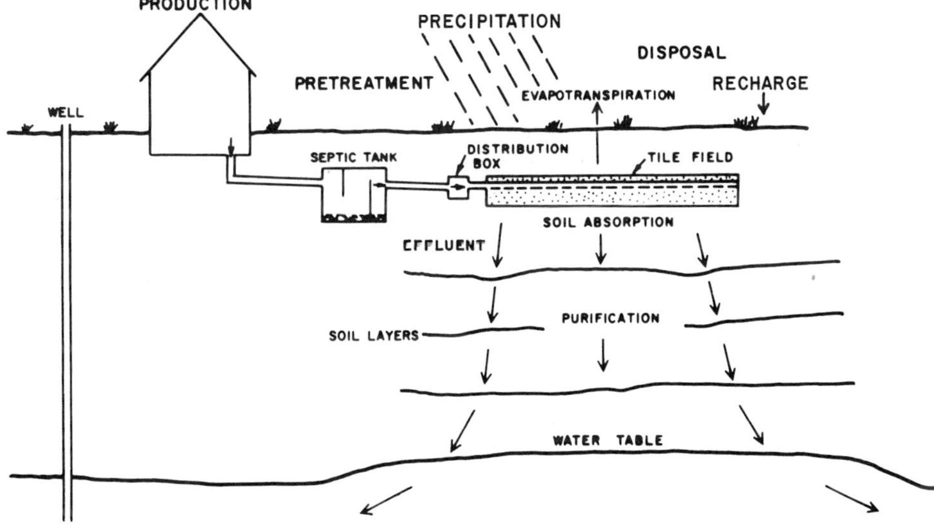

R = Recharge from precipitation (700 gpd/acre)
NS = Average nitrogen concentration in septic system effluent (30 ppm)
NG = Allowable increase of nitrogen in ground water (10 ppm)
NR = Nitrogen concentration in recharge (0.1 ppm)
Q = Average flow from septic system (200–350 gpd/system)

$$NG \propto \frac{(NS \times Q) + (NR \times R)}{Q + R}$$

Thus, maximum number of septic systems is 1 to 1.75 per acre for a maximum increase of nitrogen in ground water of 10 ppm.

Figure 80. Example of waste-load allocation based on natural recharge.

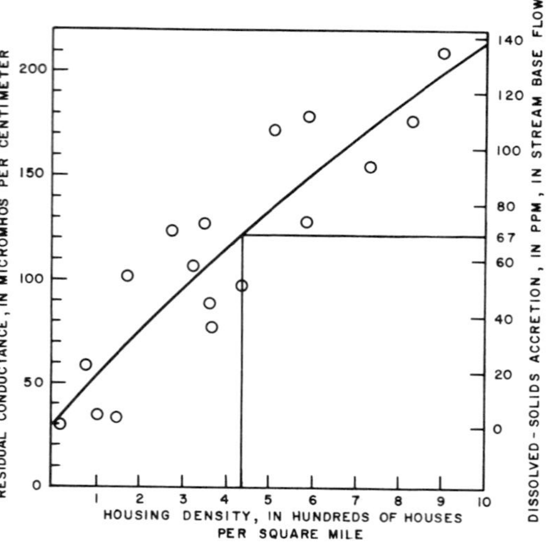

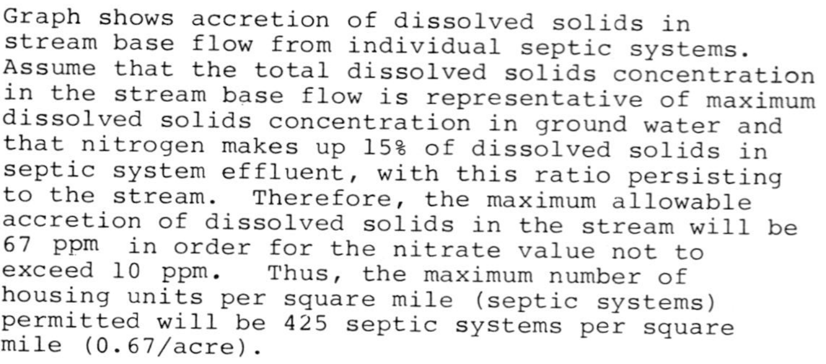

Graph shows accretion of dissolved solids in stream base flow from individual septic systems. Assume that the total dissolved solids concentration in the stream base flow is representative of maximum dissolved solids concentration in ground water and that nitrogen makes up 15% of dissolved solids in septic system effluent, with this ratio persisting to the stream. Therefore, the maximum allowable accretion of dissolved solids in the stream will be 67 ppm in order for the nitrate value not to exceed 10 ppm. Thus, the maximum number of housing units per square mile (septic systems) permitted will be 425 septic systems per square mile (0.67/acre).

Figure 81. Example of waste-load allocation based on ground-water discharge.

of nitrogen making up the total dissolved solids, one can estimate the maximum number of septic systems allowable per square mile without causing the ground water to exceed the nitrogen limits for drinking water.

These examples indicate that, for these or similar areas, the waste-load allocation in terms of the number of septic systems per square mile should be roughly 640 systems or one system/acre (2.5 systems/ha). Waste-load allocation calculations for most aquifer systems will be crude, at best. More sophisticated analyses by computer modeling may be possible.

Hazardous Wastes

In general, hazardous wastes are largely industrial, about 10 million tons/yr (9 million tonnes/yr). About 90 percent by weight are generated in the form of liquid streams, of which approximately 40 percent are inorganic and 60 percent organic materials. [7] Furthermore, disposal of hazardous wastes on the land will increase as a result of air and surface-water pollution controls. Therefore, whether or not a waste that is to be disposed of to an aquifer is hazardous could determine the type of regulatory control applied (permit or rule, such as construction standards).

Caution must be used in applying regulatory control philosophies to hazardous waste disposal practices that may affect ground water. Even after treatment processes are applied to wastes, there will remain a residue that will require disposal on either land or in the ocean. As ocean dumping becomes less desirable, land and eventually ground waters will become the ultimate sinks.

When regulatory constraints for protecting underground waters increase disposal costs, it is likely that reuse of industrial waste residuals will take place in-plant. Thus, systems such as a waste exchange similar to those operating in several European nations, may become economically feasible. [8] By contrast, recovery of residuals from municipal wastes will likely be on the disposal end through waste application on marginal or prime agricultural land. The main ground-water contamination issue in land application of municipal wastes will be nitrate. EPA National Interim Primary Drinking Water Standards establish a mandatory maximum level for nitrate of 10 ppm as N. [9] Under these circumstances, nitrate discharged to underground waters may have to be classified as a hazardous waste.

GROUND-WATER QUALITY MANAGEMENT AREAS

Focusing limited resources on specific problem areas rather than trying to control all causes of ground-water contamination for all areas of a state, county, or water management district, for example, requires the use of available data on geology, hydrology, use of ground water, and causes of ground-water contamination. Based on such information, a region can be divided into ground-water quality management areas. This classification can be used to provide a sound technical basis for determining where underground injection control or discharge permits are or are not required (e.g., recharge areas, and limestone aquifers) or for prohibiting disposal of hazardous wastes in certain areas. The following are some of the criteria that may be used to establish management areas to facilitate decision-making:

1. Unconfined (water-table) vs. confined (artesian) aquifers

   Water-table aquifers are generally subject to greater water-quality degradation from all of the sources of ground-water contamination than confined aquifer systems.

2. Aquifer recharge areas vs. discharge areas

   Recharge areas are obvious places where contaminants may enter ground waters. For water-table aquifers, the recharge area may be the total land surface area of the aquifer. Therefore, a large portion of a state could be classified as a potential recharge area. However, the concept may be applicable as a criterion for ground-water quality management when applied to inclined confined aquifer systems of coastal plains or discrete basins with water-table aquifers, because the boundaries of the recharge area can be more easily defined.

   It should also be noted that many aquifers are brim-full of water and are discharging to surface streams. Under these conditions, the region involved can not be classified as a recharge area. However, subsequent development of the aquifer can lower ground-water levels creating a new recharge area, which would require protection. Such a change in conditions can be determined by a continuing program of monitoring pumpage and water levels.

3. Surface-water drainage basins and structural basins

   Drainage basins could be used for unconfined aquifer systems but would not be readily applicable to confined aquifers. In many cases, structural basins may form suitable management boundaries for an aquifer system. Boundaries could be based on the degree of restriction to ground-water flow in and out of the basin.

4. Consolidated vs. unconsolidated aquifers

Movement of ground water through consolidated rock aquifers is generally controlled by rock fracture patterns while movement through unconsolidated aquifers is controlled by horizontal and vertical permeability differences of gravels, sands, silts and clays. Consequently, the mechanics of ground-water contamination will normally be considerably different.

5. Carbonate vs. non-carbonate consolidated aquifers

Carbonate rock aquifers often have high transmissivities because of solution openings. This characteristic results in their being more susceptible to contamination and allowing contaminants to move relatively long distances in short periods of time. Certain other aquifers, such as basalt aquifers, may be considered in the same way as carbonate rock.

6. Unsaturated zones of aquifers

The suitability of the unsaturated zone for treatment of wastes varies considerably. The thickness and type of material in the unsaturated zone could be used as a basis for delineating management areas.

7. Pollution causing activities

Management areas can be established on the basis of coal and other mining activities, or concentrations of particular sources of contamination, such as large population centers using individual septic systems.

8. Use of aquifers for public water supplies

Ground-water quality management areas can be designated on the basis of present use of underground water for public water supply and/or future use. For example, areas planned for high densities of domestic wells or aquifers which have been traditionally used to supply community water systems would receive special attention.

An example of the use of management areas is shown in Figure 82 where six areas are designated. Areas 1 through 4 are

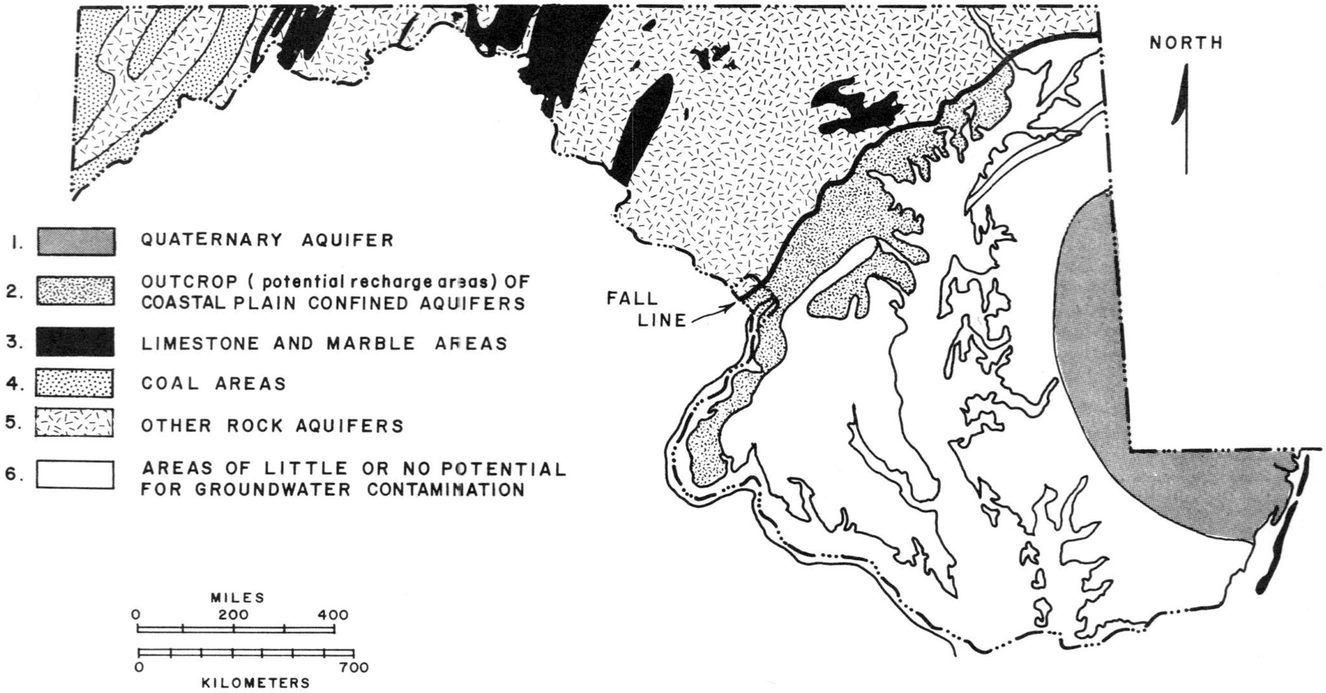

NORTH

1. QUATERNARY AQUIFER
2. OUTCROP (potential recharge areas) OF COASTAL PLAIN CONFINED AQUIFERS
3. LIMESTONE AND MARBLE AREAS
4. COAL AREAS
5. OTHER ROCK AQUIFERS
6. AREAS OF LITTLE OR NO POTENTIAL FOR GROUNDWATER CONTAMINATION

FALL LINE

MILES
0 200 400

0 700
KILOMETERS

Figure 82. Example of ground-water quality management areas for Maryland.

deemed critical for the protection of underground waters from contamination, and controls on sources of ground-water contamination are more restrictive than elsewhere.

STATE ENVIRONMENTAL POLICY ACTS

Many states have environmental policy acts which may be useful as a regulatory tool for controlling ground-water contamination. These usually call for an environmental impact statement (EIS) for actions taken by state agencies. In some states, the requirements extend to local government agencies and possibly to private activities which require state permits. Rules and regulations for evaluating these impact statements are generally broad enough to include review for impact on ground water.

CONTROLS ON WATER APPROPRIATIONS (WITHDRAWALS) AND WELL CONSTRUCTION

Regulatory controls on water appropriations (water rights) and well construction (permits) can be applied to problems involving ground-water quality for some causes of contamination which are not discharges (Table 78, Category IV) or for existing cases of contamination.

The problem of what to do in a case of existing ground-water contamination has always been difficult. The contamination usually remains after the source has been eliminated, and removal is not often feasible. Thus, there has been a tendency to say that "little or nothing can be done about ground-water contamination after it has occurred." There is, however, one rather obvious action: restrict or prohibit the use of water from an aquifer that has been contaminated.

Within the legal framework under which each state must operate, withdrawal of ground water could be limited in aquifers affected by contamination. It would be the task of the proper public agency to determine "critical zones" around each known significant case of ground-water contamination. In each zone, ground-water diversion would be restricted from the standpoint of either the quantity that can be pumped or the purpose for which it can be used. Wells and other monitoring techniques would aid in determining when and how to modify the areal extent of such a zone over a period of time. Figure 83 and Table 79 illustrate how this system might be utilized.

In the theoretical case illustrated by the diagram, it is assumed that no ground-water pumpage presently exists within the three zones and that the mechanism causing ground-water

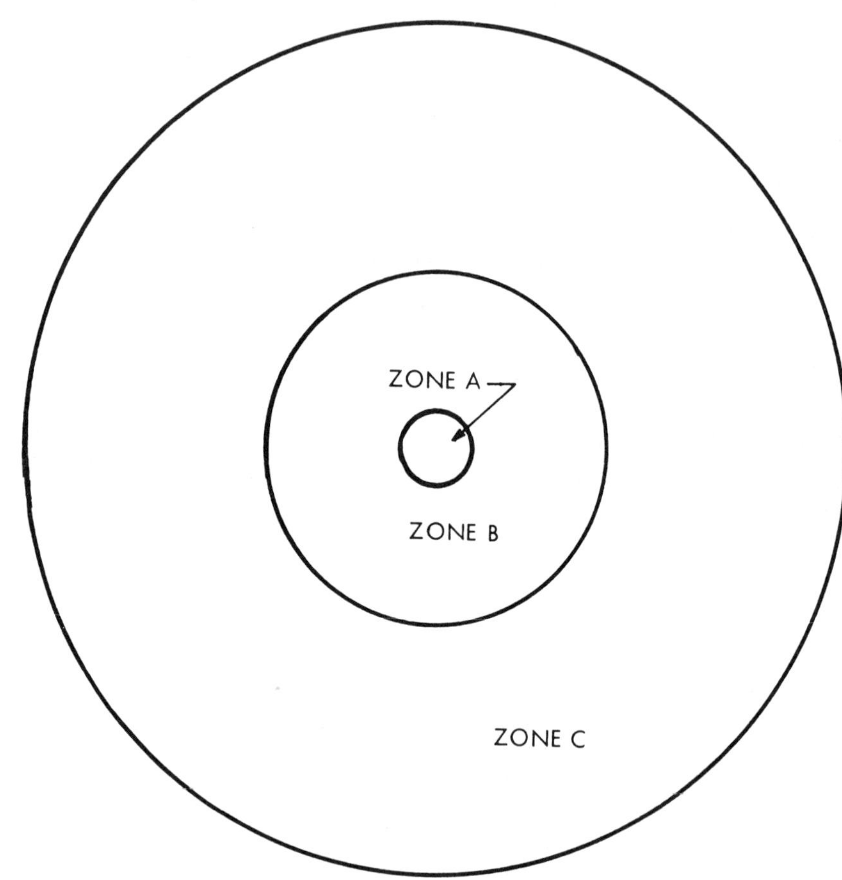

Note: See Table 79 for explanation.

Figure 83. Map of theoretical critical zones. 10)

Table 79. RESTRICTIONS ON GROUND-WATER USE IN THE CRITICAL ZONES SHOWN ON FIGURE 83. [10)]

Zone	Description	Restrictions on use of water-table aquifer	Restrictions on use of shallowest artesian aquifer
A	Area in which water-table aquifer already contains contaminant or ground-water quality is threatened because of proximity to contaminated area.	1. No ground-water pumpage permitted except where poor quality water can be used safely for special purposes or the contaminant can be successfully removed by treatment. 2. Ground-water quality monitored.	1. Pumpage regulated so that head is maintained above water table; otherwise pumpage not permitted. 2. Well construction strictly regulated to guard against inter-aquifer exchange of contaminated water. 3. Well-water quality periodically monitored.
B	Area in which natural process such as adsorption, dispersion, and ion exchange will have reduced the concentration of the contaminant significantly but not to a level acceptable for potable water supplies.	1. Ground-water pumpage limited to prevent significant increase in rate of travel of contaminated water. 2. Ground-water use for potable water supplies not permitted unless contaminant can be successfully removed by treatment. 3. Ground-water quality strictly monitored.	1. Pumpage regulated so that head is maintained above the water table in Zone A but can be lower than water table within this zone. 2. Well construction regulated.
C	Area in which natural processes will have reduced the concentration of the contaminant to a level acceptable for potable water supplies.	1. Ground-water pumpage limited to prevent significant increase in rate of travel of contaminated water. 2. Ground-water quality monitored.	1. Proposed ground-water users warned that pumpage may be restricted in the future if significant ground-water contamination spreads to Zone B.

contamination has been eliminated. Also, it is assumed that the water-table and underlying artesian aquifers are separated by a uniform layer of low permeability, which is quite effective in retarding the vertical movement of ground water and also in removing a substantial portion of the contaminant through such natural processes as adsorption and ion exchange. Taking these assumptions into account, regulations could be established for controlling ground-water use, as outlined, in order to minimize the effects of the problem. Of course, consideration of each new ground-water diversion proposed in any of the zones would require an evaluation of the effects of prior approvals for ground-water pumpage.

Permits for drilling water wells, and water-well construction and abandonment standards, can be effective control techniques for certain ground-water contamination problems. There are many instances reported of individual water supply wells which produce contaminated ground water as shown by the presence of coliform bacteria. Many of these instances are not really cases of contaminated ground water but improper or ineffective sealing (grouting) of the well casing from surface contaminants and/or inadequate disinfection procedures. In addition, enforcement of standards for construction and abandonment of oil and gas wells and holes drilled for soil sampling, geophysical surveys, etc., can help minimize ground-water contamination problems.

LICENSING OF WELL DRILLING CONTRACTORS, WASTE HAULERS AND OPERATORS OF WASTE DISPOSAL FACILITIES

Licensing or certifying persons responsible for certain activities can be a valuable method for controlling ground-water contamination. Enforcement of well-construction standards is a significant problem because of the difficulty of inspecting what amounts to a hole in the ground. After-the-fact inspection, common in other trades (e.g., electrical, plumbing), is seldom completely satisfactory for water wells. However, licensed professional drilling contractors can help minimize this problem through self-regulation.

The concept of licensing or certifying operators of waste treatment and disposal facilities was developed when it was recognized that a sewage treatment plant is no better than the people who operate and maintain it. The same concept is applicable to facilities which discharge wastes to aquifers. Proper operation of these facilities is critical for the prevention of ground-water contamination. An example of this problem is the growing acceptance of land application of municipal waste waters and the need for proper operation of a rather complex process related to soil, vegetation, and climatic conditions.

Licensing of haulers of hazardous industrial wastes appears to be an important technique for controlling the disposal of these substances, and providing an inventory of their nature and quantity. In general, there are far fewer waste haulers than waste generators. This can help make the administration of a hazardous waste management program simpler than would otherwise be possible. The key factor for ground-water contamination control is to make certain that hazardous wastes are only disposed of and treated at approved facilities.

PUBLIC NOTIFICATION AND INVOLVEMENT

Public participation and awareness is an essential part of a ground-water protection program. There normally will be requirements for public hearings on applications for permits to discharge or inject wastes to aquifers. Conventional notification techniques such as newspaper advertisements and mailing lists are likely to be ineffective in the case of ground water. Generally, the people most likely to be affected by discharge of wastes to ground waters will be property owners near the proposed facility. Therefore, a simple requirement (in addition to advertisement, etc.) is for permit applicants to notify, by certified mail, adjacent property owners. This will insure attendance by the public at hearings on applications to discharge or inject wastes to underground waters.

Another illustration of the need for public notification is a state regulatory agency's restricting ground-water withdrawals in the area around a source of contamination. Notification by the state agency to the property owners of restrictions (and possible decline in property value) on the use of their land will foster increased public awareness about ground-water contamination.

CHARGES

The "polluter pays" principle is a common concept for control of surface-water contamination. Its application to underground waters presents some interesting challenges and applications.

Effluent Charges

As applied to ground water, effluent charges could be used to: (1) require minimum waste treatment levels before discharge to aquifers, (2) allocate the waste treatment capacity of aquifers among competing applicants for such use, (3) help fund efforts to mitigate existing ground-water contamination, and (4) encourage compliance with any effluent limits or water-quality standards established for ground waters. [11)]

Resource Utilization Charges

A resource utilization charge is similar, in some respects, to an effluent charge except that it has a much broader application and can be much more useful. [12)] The effluent charge concept focuses on the contaminant before it is discharged to an aquifer while the resource utilization charge concept is concerned with how the aquifer handles the contaminant.

This concept is applicable to aquifers and would be based on (1) costs of administering the aquifers (planning, monitoring, allocating, inspecting, and enforcing); (2) "rent" on the scarcity value of the natural capacity of aquifers to treat and store wastes; and (3) controlling, where it exists, overuse of aquifers for waste storage and treatment by establishing a price for such use.

Licensing and Permit Fees

These charges generally are used to cover only the administrative costs of the control program. In some cases they may be based on the potential that the activity regulated may become a problem. Considerable moneys could be generated by these charges because of the potentially large number of activities encompassed by a ground-water contamination control program (permits to drill wells, discharge or injection control permits).

Bonds

Bonds can be used as incentives to comply with the law or to help alleviate concern about uncertainties which are inherent in the control of ground-water contamination.

Disincentives

"A disincentive is a monetary charge levied by government on conduct which is not illegal but which does impose social costs, for the principal purpose of discouraging the conduct." [13)] A disincentive in the form of a pollution tax may be as effective for controlling ground-water contamination as ground-water quality standards and effluent limits. [14)]

Problems Associated with Levying Charges

A basic problem in assessing effluent charges for ground-water contamination is the selection of the parameters to be used. In the case of surface water, biochemical oxygen demand, chemical oxygen demand, and suspended solids are parameters usually chosen as the basis for such charges. These have little meaning for control of disposal facilities designed to discharge wastes and waste waters to aquifers (Table 78, Category I) because they are largely removed during passage through both the unsaturated and saturated zones.

For this reason, effluent charges, and to a lesser degree resource utilization charges and disincentives, must be based on persistent dissolved substances in the waste. Examples are total dissolved solids, nitrate, chloride, and heavy metals.

EXEMPTIONS

From a practicable standpoint, it will not be possible to completely control all the sources and causes of ground-water contamination listed in Table 78. It would be foolish to even attempt to issue underground injection control or discharge permits for the hundreds of thousands of septic systems installed annually. Yet, these systems are a major cause of ground-water contamination. The answer to this problem is a qualified exemption, whereby permit requirements are waived for a specific kind or class of waste discharges to ground waters (e.g., septic systems for individual homes, selected agricultural practices) while still requiring that these systems comply with minimum construction and operational guidelines.

Another example would be to waive permit requirements for that class of discharges involving artificial recharge of water-supply aquifers with renovated waste water. Ground-water quality standards could be the only regulatory tool applied, based on the reasonable assumption that the persons discharging to the aquifer would be the same ones using it for water supply.

Exemptions may be applied where there are existing categorical permit programs such as those for solid-waste disposal. In some cases, stiffening the ground-water protection aspects of an existing program could obviate the need for a separate ground-water discharge permit.

IMPLEMENTATION OF A GROUND-WATER CONTAMINATION CONTROL PROGRAM

Establishment of methods to ensure compliance with a program designed to control ground-water contamination will be facilitated by the use of the following techniques to assure maximum application of laws, regulations and permit requirements, etc.

1. Control of surface-water contamination has resulted in an increased generation of industrial and municipal waste sludges. Conditions placed on National Pollutant Discharge Elimination System permits require reports on pollutants removed from point-source waste streams (quantity, quality, disposal area, etc.). In this way, an inventory can be obtained of waste sludges (residuals) which could end up on the land and, eventually, in ground waters.

2. Where law allows, a regulatory agency can impose conditions requiring proper disposal of waste waters (discharge permits as a prerequisite to issuance of ground-water appropriation permits).

3. If lending institutions are notified of rules and regulations requiring control of discharges to ground waters, it will then be difficult for certain projects to obtain financing without first having obtained the required permits or approval.

4. Federal OMB Circulars A-95 and A-98 or State Clearinghouse Review contain procedures for approval of projects which receive state and/or Federal funds. There is a provision for environmental control agencies to comment on such projects.

5. Many waste-disposal facilities are designed to discharge to ground water (e.g., spray irrigation) as a result of requirements necessary to meet stream standards. "Need and Adequacy Statements" (Section 8 of PL 92-500) are required before Small Business Administration loans for environmental control applications can be processed. A state ground-water contamination control agency can be designated as the party responsible for determining the need and adequacy of projects which discharge to ground waters.

6. Local agencies are often the first to be contacted by potential applicants for waste-disposal facilities. Agreement can be reached with local agencies not to is-

sue local permits (e.g., building permits) until necessary ground-water contamination control permits have been obtained.

7. Septic tank cleaners, product storage tank cleaners, and waste haulers may be a source of waste discharges to aquifers. They should be checked on a periodic basis in order to ensure that licensing and reporting requirements are being fulfilled.

MONITORING GROUND-WATER QUALITY

The problem of monitoring ground-water quality deserves its own special place in any discussion about controlling contamination. This problem, which is manageable for control of surface-water pollution, becomes highly complex for ground water.

There are four basic reasons for monitoring ground-water quality. In order of increasing technical complexity, they are:

1. To control quality of drinking water supplies (monitoring requirements of the Safe Drinking Water Act).

2. To establish existing values for various ground-water quality parameters in order to evaluate changes in quality and long-term trends.

3. To investigate and evaluate potential ground-water contamination causes and sources.

4. To regulate and evaluate known (permitted) discharges to ground water (compliance monitoring).

Sampling public water supply wells (raw water) for compliance with the Safe Drinking Water Act should be straightforward, provided that it is coordinated with baseline monitoring networks. The main problem is to make sure that all the parameters necessary to interpret changes in ground-water quality are measured in addition to those of public health significance (e.g., all major anions and cations and field pH).

Establishing baseline values for ground-water quality will likely necessitate the drilling of wells designed specifically for this purpose. Because such a program can be very expensive, any network of monitoring wells should be supported with data from the routine sampling of existing wells. Regulation of drinking water supplies already calls for periodic sampling to determine the safety of water supply systems. These data can be reviewed to help determine long-term trends in ground-water quality. The best approach for investigating and evaluating existing or potential ground-water contamination sources is to inventory and examine the actual facilities that may be causing problems and then decide which ones would require the installation of monitoring wells, so that water samples can be collected and analyzed.

The function of observation wells for a new facility which has received a permit or approval, is for compliance monitoring only. Because this is not a case of existing contamination, the purpose of the observation well should not be for detecting ground-water contamination. A monitoring well cannot be a substitute for good facility design. Failure to recognize this distinction will result in monitoring wells becoming a pacifier for ground-water regulatory officials (Figure 84).

In other words, monitoring wells can serve as a safety device to detect failure of properly designed waste holding lagoons; to determine the performance of waste disposal facilities properly designed to use the land and aquifers for waste treatment; and to guard against not recognizing failure of a waste disposal facility properly designed to use aquifer storage space for wastes.

When this concept is used, selection of the location for and type of monitoring wells to comply with permits becomes straightforward. However, detailed location, depth, etc., will often require a sophisticated geohydrologic analysis. For a surface impoundment holding industrial wastes, the permit or approval may contain a specific limitation on discharge of contaminants to the aquifer. A monitoring well would be installed as close as possible to the impoundment. A suction lysimeter (device for obtaining a water sample from the unsaturated zone) would be installed directly beneath the impoundment. Other devices, such as underdrains and alarm-triggering liquid-level sensors for the impoundment, could also be used.

For a spray irrigation system for municipal wastes, the permit would specify an effluent limit (e.g., secondary treatment), a waste load allocation (e.g., nitrogen limit), and perhaps a maximum permitted rise in the elevation of the water table. In addition, a ground-water quality standard of drinking water could be specified. Monitoring wells for measuring compliance with quality standards would be located near the outer boundary of the spray area to measure system performance and changes in water levels. Furthermore, the

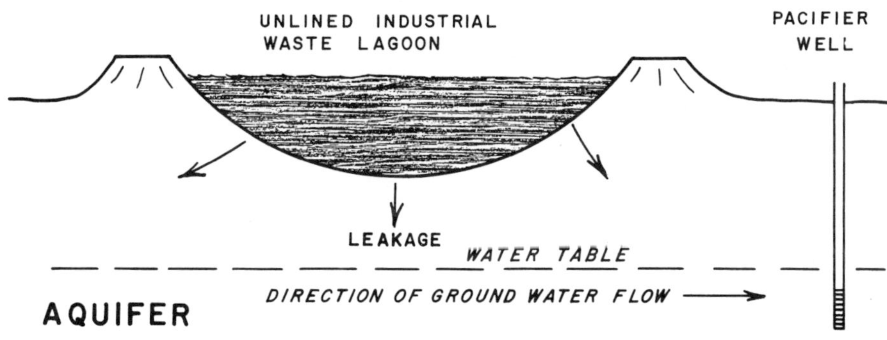

INCORRECT USE OF MONITORING WELL

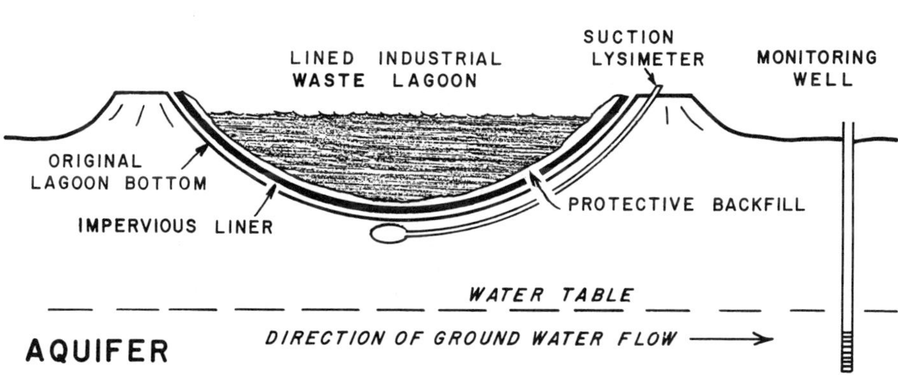

CORRECT USE OF MONITORING WELL

Figure 84. Correct and incorrect use of compliance monitoring wells.

effluent would be monitored before application to the land, and other operational features would be regulated (e.g., harvesting practices).

It should be noted that a major operational problem with monitoring wells is evaluating the data that will come pouring into the regulatory agency. Personnel must be available to analyze these data to determine if any changes are taking place.

RELATIONSHIP BETWEEN GROUND-WATER AND SURFACE-WATER CONTAMINATION CONTROL

Ground-water regulatory controls should not be limited to protecting underground waters but also should be used to protect surface waters and encourage environmentally sound waste disposal practices. One example would be the placing of constraints on a discharge to an aquifer from land application of municipal waste waters. The problems of runoff and nitrogen-rich ground-water underflow into a stream may be more important than changes in the ground water itself.

Ground water, surface water, generation of waste residuals, and waste disposal and reuse are all interrelated. A case in point would be a municipal sewage treatment plant in an industrial city that has separate sewers and storm drains. Limits on the municipal discharges are established to improve surface-water quality. This then results in an increase in sludge production, which now has to be disposed of. However, there are high levels of heavy metals (from industrial waste waters) which make the sludge unacceptable for application on agricultural land used for crops. Thus, since the sludge has to be disposed of somewhere on the land, ground waters (and possibly surface waters) will likely be contaminated by nitrate and heavy metals.

The answer to this problem is to require pretreatment of the industrial waste waters before they are discharged to the sewage treatment plant. The sludge, now relatively free of heavy metals, can be applied on agricultural land for growing crops. Harvesting the crops will remove nitrogen (recover the resource) and protect ground waters. The industrial sludges will now be concentrated and many will be suitable for recycling. The remainder, of course, must be treated and disposed of in a secure land disposal area.

TYPICAL PROBLEMS AND RISKS IN IMPLEMENTING A GROUND-WATER CONTAMINATION CONTROL PROGRAM

Decisions made and policies established for controlling

ground-water contamination can sometimes have undesirable side effects. The following are some typical examples:

1. A policy decision could be made to protect only those aquifers which supply or could be expected to supply a public water system. On the surface, this sounds reasonable and might reduce the size of the protection program to within manageable limits. However, there is then the problem of how to define a "public water supply aquifer." If this definition is based on aquifer productivity as determined by the definition of a public water system in the Safe Drinking Water Act (15 service connections or regularly serving 25 individuals), then one would protect all aquifers capable of producing about 2 gpm (11 cu m/day) -- or nearly every square mile in the United States.

2. A decision is made not to protect non-carbonate consolidated aquifers, on the reasonable basis of assumed "low" productivity, in order to limit the size of the control program. Subsequently a sophisticated well location technique, such as fracture-trace analysis, is developed which changes a "low" productivity aquifer into a moderately productive one.

3. A decision is made to control only industrial or hazardous waste disposal. Major ground-water contamination problems are then caused by uncontrolled or improper land application of municipal waste waters and sludges.

4. In order to protect ground-water quality, a limit is established on the number of individual septic systems permitted per square mile in the recharge area of a particular aquifer. Problems develop (low yield domestic wells, contamination, etc.) and it turns out that the building density is too low to economically justify public water and/or sewer service.

Obviously, a ground-water contamination control program will involve risks that are inherent in all decisions made with inadequate data. There are, however, some actions that can be taken to help mitigate these risks. These include: (1) time limits on permits that are short enough to correct mistakes but not so short as to create an administrative nightmare of renewals (minimum 2 years, maximum 5 years); (2) technically competent geohydrologic project evaluations as a condition precedent to issuing any permit or approval; (3) requirements for projects to be staged in order to fully evaluate whether or not a disposal system will work; and (4) public ownership and/or operation of certain waste disposal facilities (e.g., perpetual care of hazardous waste disposal sites).

SITING OF WASTE DISPOSAL FACILITIES

One problem and risk in implementing a ground-water protection program is the selection and approval of sites for carrying out such waste disposal activities as landfilling, impounding industrial effluent, and spreading sludge. Throughout this report, proper siting of waste facilities has been emphasized as essential to minimizing ground-water contamination. Unfortunately, there is no way to scientifically guarantee that ground-water contamination will not occur at a specific site. One principal reason for this is the large number of variables that must be considered when predicting long-term effects. Twenty such variables are listed in Table 80. In most cases, determining the favorability of such variables is not possible because the technology is nonexistent or the cost of obtaining the required data is prohibitive.

Nevertheless, proper geologic and hydrologic investigation can go far to minimize the risk of poor decisions on site permitting. Of equal importance is adequate source control, including such factors as timing the discharge, regulating the quantity and quality of discharge, pretreating the waste, and containing the discharge.

In addition to technical considerations, site selection is also subject to social, economic and political pressures. For example, the best natural location for a sanitary landfill in a particular region may be unacceptable to the local citizenry because of potential odor or traffic problems. The public would also be concerned over a decline in property values adjacent to the landfill. Finally, accepting wastes from another community, if the landfill is part of a regional plan, can have adverse political repercussions for those in local government.

With such technical and socioeconomic difficulties to overcome and with more and more liquid and solid wastes going to the land, the availability of favorable sites for disposal, spreading, or treatment of wastes may be the prime factor determining how successful environmental controls will be in the future. Regional land-use planning is one promising method for helping to assure this availability, and Section 208 of PL 92-500 appears to be a suitable vehicle for beginning to address this high-priority item.

Table 80. THE PRINCIPAL VARIABLES IN LAND DISPOSAL. [15]

Element	Variables
Waste	1. Composition 2. Type of treatment
Application	1. Method 2. Loading rate
Soil	1. Texture [a] 2. pH 3. Organic matter content 4. Cation exchange capacity 5. Percent base saturation 6. Depth 7. Slope
Climate	1. Temperature regime 2. Precipitation regime
Vegetation	1. Uptake 2. Management
Ground Water	1. Depth to zone of saturation 2. Nature of zone of aeration 3. Natural quality of ground water 4. Physical nature of aquifer 5. Chemical nature of aquifer

a) This includes infiltration rate, permeability, and available moisture capacity, each of which could well be considered as a separate variable.

ECONOMIC IMPACTS

Although it is not possible to determine the full cost of the application of rules and regulations for controlling ground-water contamination, some rough estimates can be made.

Impact on State Government

The cost to states of implementing a control program will basically be in the areas of permits, monitoring, and enforcement. Estimates of the magnitude of this cost can be made by a comparison with surface-water contamination control programs.

1. Compliance monitoring and follow-through on permits. This activity for surface waters involves setting effluent limits on specific parameters by means of permits. Enforcement personnel oversee the collection of water samples and the measurement of waste flows. The results are analyzed and evaluated. Presumably, corrective action is initiated if permit limits are exceeded. For control of ground-water contamination, the procedure could be similar but only when applied to the waste flow before it is discharged to either the saturated or unsaturated zones of an aquifer.

   Major additional costs will be incurred for ground-water compliance monitoring where the land and aquifers are used for waste treatment and storage as well as for water supply. Decisions to take action are based on the trend, with time, of specific quality parameters and not on whether or not a specific numerical value is exceeded. Naturally, this will require trained personnel. For example, a yearly trend of increases in chloride and hardness from landfill observation wells would serve as the basis for corrective action even though ground-water quality standards had not been exceeded. In the case of salt-water intrusion, increasing chloride concentrations with time would serve as a warning, requiring action by a public agency, even though drinking-water standards for chloride had not been exceeded in water from the monitoring well.

2. In the early days of surface-water contamination control, it was always possible to "walk streams" and find point-source discharges. Samples could be taken and it was relatively simple to prove that there was an actual discharge of pollutants. The problem of delineating existing sources of ground-water contamination will be difficult and expensive to solve. Evaluation of a simple

leaky lagoon can take several weeks and cost thousands of dollars. Then, there is the problem of making the leap from the demonstration by a ground-water hydrologist that there is contamination to assembly of sufficient legal proof that the contaminant which enters the ground at point "A" is the same, or basically the same, as the contaminant which appears at a well some distance away from the source.

3. Monitoring costs (sampling stations, chemical analyses) for ground-water contamination control may be considerably greater than for surface-water monitoring. Most monitoring wells will probably be shallow, less than 50 ft (15 m) deep, reflecting the vulnerability of water-table aquifers to contamination. An order-of-magnitude estimate for drilling costs (4-in. or 10-cm well) would be $10.00/ft ($33/m); however, set-up charges could substantially distort this figure. In addition, there will be the cost of a geohydrologic study to determine where and how deep to drill.

4. The sampling frequency for ground-water quality monitoring will be much less than for surface water (quarterly samples would be more than adequate for most cases). This may be offset by the need for a more complete analysis of the samples. Chemical quality data can be a useful tool for evaluating the dynamics of ground-water movement, and the concentrations of major cations and anions need to be determined in addition to specific parameters applicable to quality standards.

   Ultimate program costs can be estimated by assigning man-day requirements for permitting, monitoring, and enforcement activities for each source or cause of ground-water contamination. For example, it has been estimated that for holding ponds and lagoons, these activities will require 1.6 man-years for each 25 systems, and that 4.8 man-years per million people in a state will be needed for landfills and other excavations.[16)] Obviously, it will be necessary to determine how many lagoons, waste-piles, etc., there are in each state.

The first efforts of a program will involve setting of priorities, determining the number and significance of causes of ground-water contamination, and assessing the environmental damage of existing pollution. It is assumed that the funds required for this effort will be in addition to existing state moneys for environmental control. Another alternative would be to reallocate some of the resources presently being used to eliminate contaminants from the air and surface waters to the control of the ultimate disposal and reuse of these same contaminants. Funds should follow the contaminant path.

Impact on Industry

The economic impacts on industry of controlling ground-water contamination will be the cost of obtaining and maintaining permits and approvals, and increased costs for proper storage and disposal of wastes and other materials which may contaminate ground water.

Permit costs will be largely for supplying data sufficient to show that the proposed project will not pose an unnecessary risk of ground-water contamination. (An excellent example of these data requirements can be found in "data modules" required by, and available from the Pennsylvania Department of Environmental Resources.) 17) Other permit related costs will be installation of observation wells and periodic submission of data (water levels, chemical analyses).

Improved facility design and construction will be a major cost to industry for control of ground-water contamination. For example, the Maryland Department of Transportation investigated the cost and feasibility of environmentally sound coverings for highway deicing chemicals. 18) It was concluded that a Domar building, originally designed for the storage of wheat, had the flexibility in its basic dome shape to adapt easily to the size and natural angle of repose of any bulk, stockpiled material. Complete costs of properly covering 85 salt storage sites with Domar buildings was about $5.3 million (1973 installed prices for basic 82-ft (25-m) diameter buildings with a storage capacity of 1,500 tons (1,360 tonnes) each).

Waste disposal and storage facilities, such as landfills and lagoons, will have to be lined in many cases. For example, actual 1973 costs for securing fluoride wastes in subsurface solid-waste disposal cells for an east coast aluminum company were estimated at $7.00/cu yd ($9.15/cu m). This included a 30-mil thick neoprene liner at $0.50/sq ft ($5.40/sq m) and site preparation. A neoprene cover would increase the price by nearly $2.00/cu yd ($2.60/cu m). These high costs forced the industry to actively pursue the concept of recycling the fluoride wastes. At a western mining operation, construction and material costs for trenching to intercept flow of potentially contaminated ground water and the lining of lagoons for holding mine-waste effluent exceeded $3 million.

It should be noted that liner costs for a facility such as a landfill represent only part of the cost of controlling contamination. The problem of what to do with the landfill leachate still remains as there usually will be a definable surface discharge. This contaminant discharge will then have to be treated before release to a stream or back to the aquifer system (e.g., spray irrigation).

Impact on the Public

Direct cost to the public of protecting ground-water quality will be reflected in such items as improved well construction (grouting of annular space), increase in individual home lot size where septic systems are used, and increase in costs of solid-waste and sewage disposal services. Some of these costs may be more apparent than others. For example, the cost of cement grouting the annular space of a typical domestic well might have a significant impact on the individual consumer. However, the one-time capital cost of proper design and construction for a sanitary landfill will be relatively small at the consumer end because the major costs for solid-waste disposal are in transportation. Of course, increased costs incurred by industry to correct problems and by public agencies to oversee protection programs will be passed on to the public in the form of higher priced products and higher taxes. There also will be the environmental benefits of protecting ground-water and surface-water quality and encouraging resource recovery of waste residuals.

REFERENCES CITED

1. Maryland Water Resources Administration. 1973. Rules and regulations .08.05.04.04, groundwater quality standards.

2. North Carolina Department of Natural and Economic Resources. 1975. Groundwater quality management and monitoring program for North Carolina. Circular No. 16.

3. van der Leeden, Frits. 1973. Ground-water pollution features of Federal and state statutes and regulations. U. S. Environmental Protection Agency Report EPA-600/4-73-001a.

4. U. S. Public Health Service. 1962. Drinking water standards. Public Health Service, Publication No. 956.

5. U. S. Environmental Protection Agency. 1973. Water quality criteria.

6. Morrill, G. B., III, and L. G. Toler. 1973. Effect of septic-tank wastes on quality of water, Ipswich and Shawsheen River basins, Massachusetts. U. S. Geological Survey, Journal of Research 1(1):117.

7. U. S. Environmental Protection Agency. 1974. Report to Congress, disposal of hazardous wastes.

8. U. S. Environmental Protection Agency. 1975. Industrial waste management. Report No. SW-156. Pages 69-77.

9. U. S. Environmental Protection Agency. 1975. National interim primary drinking water regulations. Federal Register 40(248):Part IV. December 24.

10. Miller, D. W., F. A. DeLuca, and T. L. Tessier. 1974. Ground-water contamination in the northeast states. U. S. Environmental Protection Agency EPA-6602-74-056.

11. U. S. Environmental Protection Agency. 1972. Development of a state effluent charge system.

12. Bower, B. T., and R. H. Forste. 1974. A proposal for resource utilization charges in the State of Maryland. Maryland Environmental Service, Annapolis, Maryland.

13. U. S. Environmental Protection Agency. 1974. Economic disincentives for pollution control: legal, political and administrative dimensions.

14. Wenders, J. T. 1975. Methods of pollution control and the rate of change in pollution abatement technology. Water Resources Research 11(3):393-396.

15. Wilson, G. R. 1975. Impact of land disposal of sludges on ground water. Pages 193-199 in Information Transfer, Inc. Proceedings of the 1975 national conference on municipal sludge management and disposal. Rockville, Maryland.

16. U. S. Environmental Protection Agency. In press. A manual of laws, regulations, and institutions for control of ground-water pollution.

17. Pennsylvania Department of Environmental Resources. 1972. Spray irrigation manual. Publication No. 31.

18. Maryland Department of Transportation, State Highway Administration. 1973. Stockpiling deicing chemicals with environmental safeguards.

SECTION XVIII

ACKNOWLEDGEMENTS

Project Manager and Editor: Mr. David W. Miller
Geraghty & Miller, Inc.

Project Officers: Mr. Stephen C. James
Office of Solid Waste Management Programs, EPA

Mr. William E. Thompson
Office of Water Supply, EPA

Chairman, Working Group: Mr. George A. Garland
Office of Solid Waste Management Programs, EPA

The following have contributed to one or more chapters of this report.

Geraghty & Miller, Inc., Port Washington, New York

Mr. David W. Miller
Mr. George R. Wilson
Mr. Frits van der Leeden
Mr. William J. Seevers
Dr. Olin C. Braids
Mr. Douglas R. MacCallum
Mr. Thomas L. Tessier
Mr. Paul H. Roux
Mr. James D. Miller

Dr. David K. Todd, Consulting Engineer, Berkeley, California
Dr. Edward Clark, Consulting Engineer, South Miami, Florida
Mr. Harry E. LeGrand, Consulting Hydrogeologist, Raleigh, North Carolina
Dr. Robert D. Varrin, Consulting Hydrologist, Newark, Delaware
Mr. Arnold Schiffman, Chief of Ground-Water Services, Maryland Water Resources Administration, Annapolis, Maryland

Eugene A. Hickok & Associates, Wayzata, Minnesota

Mr. Eugene A. Hickok
Mr. Norman C. Wenck
Mr. Michael A. Zagar

Resources Development Associates, Los Altos, California

Mr. Robert W. Campbell
Dr. Raymond E. Borton
Mr. Bruce W. McCall
Dr. Harold E. Thomas

National Water Well Association, Worthington, Ohio

Dr. Jay H. Lehr
Mr. Laurence E. Sturtz
Mr. James R. Hanson
Dr. Wayne A. Pettyjohn
Mr. Truman Bennett

U. S. Environmental Protection Agency, Washington, D. C.
Office of Solid Waste Management Programs

Mr. Alan Corson
Mr. Tim Fields
Mr. George A. Garland
Mr. Toby Goodrich
Mr. Phil Hawk
Mr. Stephen C. James
Mr. Emery Lazar
Mr. Alfred Lindsey
Mr. Murray Newton
Mr. Les Otte
Mr. Kenneth Shuster
Mr. Burnell Vincent

Office of Water Supply

Mr. William Thompson

Special appreciation is accorded to reviewers Mr. Truett De Geare, Mr. Bernard Stoll, Ms. Joyce Corry, Mr. Bruce Weddle, Mr. Dale Mosher, and Mr. Edwin L. Hockman of the U. S. Environmental Protection Agency, and to other Geraghty & Miller, Inc. staff members, particularly Mr. Boris J. Bermes, Ms. Nola P. Gillies, Mr. Robert L. Stollar, Mr. Vincent P. Amy, Mr. Lawrence A. Cerrillo, Mr. Michael F. Wolfert, Mr. Ronald Rioux, Mrs. Elaine LaBella, Mr. David K. Shaw, and Ms. Sandra Michalowski. Especial thanks are extended to Ms. Marie Edmonds for preparation of the manuscript and Mr. Joseph J. Lewandowski, Jr., for supervision of all graphics.

SECTION XIX

APPENDIX A - GLOSSARY

Acidization - The process of forcing acid through a well screen or into the limestone, dolomite, or sandstone making up the wall of a bore hole. The general objective of acidization is to clean incrustations from the well screen or to increase permeability of the aquifer materials surrounding a well by dissolving and removing a part of the rock constituents.

Anion - An atom or radical carrying a negative charge.

Annular Space (Annulus) - The space between casing or well screen and the wall of the drilled hole or between drill pipe and casing.

Aquiclude - A saturated, but poorly permeable bed, formation, or group of formations that impedes ground-water movement and does not yield water freely to a well or spring. However, an aquiclude may transmit appreciable water to or from adjacent aquifers, and where sufficiently thick, may constitute an important ground-water storage unit.

Aquifer - A geologic formation, group of formations, or part of a formation that is capable of yielding a significant amount of water to a well or spring.

Aquitard - Used synonymously with aquiclude.

Artesian - The occurrence of ground water under greater than atmospheric pressure.

Artesian (Confined) Aquifer - An aquifer bounded by aquicludes and containing water under artesian conditions.

Artificial Recharge - The addition of water to the ground-water reservoir by activities of man.

Backwashing - The surging effect or reversal of water flow in a well. Backwashing removes fine-grained material from the formation surrounding the bore hole and, thus, can enhance well yield.

Barrier Well - A pumping well used to intercept a plume of contaminated ground water. Also a recharge well that delivers water to or in the vicinity of a zone of contamination under sufficient head to prevent the further spreading of the contaminant.

Base Flow - The flow of streams composed solely of groundwater discharge.

Biochemical Oxygen Demand (BOD) - A measure of the dissolved oxygen consumed by microbial life while assimilating and oxidizing the organic matter present in water.

Bore Hole - An uncased drilled hole.

Brine - A concentrated solution, especially of chloride salts.

Casing - Steel or plastic pipe or tubing that is welded or screwed together and lowered into a bore hole to prevent entry of loose rock, gas, or liquid or to prevent loss of drilling fluid into porous, cavernous, or fractured strata.

Cation - An atom or radical carrying a positive charge.

Chemical Oxygen Demand (COD) - The amount of oxygen, expressed in parts per million, consumed under specified conditions in the oxidation of organic and oxidizable inorganic matter in waste water, corrected for the influence of chlorides.

Coliform Group - Group of several types of bacteria which are found in the alimentary tract of warm-blooded animals. The bacteria are often used as an indicator of animal and human fecal contamination of water.

Cone of Depression - The depression, approximately conical in shape, that is formed in a water-table or potentiometric surface when water is removed from an aquifer.

Connate Water - Water that was deposited simultaneously with the geologic formation in which it is contained.

Consumptive Use - That part of the water withdrawn that is no longer available because it has been either evaporated, transpired, incorporated into products and crops, or otherwise removed from the immediate water environment.

Contamination - The degradation of natural water quality as a result of man's activities, to the extent that its usefulness is impaired. There is no implication of any specific limits, since the degree of permissible contamination depends upon the intended end use, or uses, of the water.

Curie - The quantity of any radioactive material giving 3.7×10^{10} disintegrations per second. A picocurie is one trillionth of a curie, or a quantity of radioactive material giving 22.2 disintegrations per minute.

Drainage Well - A well that is installed for the purpose of draining swampy land or disposing of storm water, sewage, or other waste water at or near the land surface.

Dry Well - A bore hole or well that does not extend into the zone of saturation.

Effluent - A waste liquid discharge from a manufacturing or treatment process, in its natural state, or partially or completely treated that discharges into the environment.

Eutrophication - The reduction of dissolved oxygen in natural and man-made lakes and estuaries, leading to deterioration of the esthetic and life-supporting qualities.

Evapotranspiration - The combined processes of evaporation and transpiration.

Exfiltration - The leakage of effluent from sewage pipes into the surrounding soils.

Field Capacity - The moisture content of the soil after water has been removed by deep seepage through the force of gravity. It is the moisture retained largely by capillary forces.

Flow Path - The direction of movement of ground water and any contaminants that may be contained therein, as governed principally by the hydraulic gradient.

Fracture - A break in a rock formation due to structural stresses. Fractures may occur as faults, shears, joints, and planes of fracture cleavage.

Ground Water - Water beneath the land surface in the saturated zone that is under atmospheric or artesian pressure. The water that enters wells and issues from springs.

Ground-Water Reservoir - The earth materials and the intervening open spaces that contain ground water.

Hazardous Waste - Any waste or combination of wastes which pose a substantial present or potential hazard to human health or living organisms.

Head - The height above a standard datum of the surface of a column of water that can be supported by the static pressure at a given point.

Heavy Metals - Metallic elements, including the transition series, which include many elements required for plant and animal nutrition in trace concentrations, but which become toxic at higher concentrations. Examples are: mercury, chromium, cadmium, and lead.

Hydraulic Conductivity - The quantity of water that will flow through a unit cross-sectional area of a porous material per unit of time under a hydraulic gradient of 1.00 at a specified temperature.

Hydraulic Fracturing - The fracturing of a rock by pumping fluid under high pressure into a well for the purpose of increasing permeability.

Hydraulic Gradient - The change in static head per unit of distance along a flow path.

Infiltration - The flow of a liquid through pores or small openings.

Injection Well - A well used for injecting fluids into an underground stratum.

Intermittent Stream - A stream which flows only part of the time.

Ion Exchange - Reversible exchange of ions adsorbed on a mineral or synthetic polymer surface with ions in solution in contact with the surface. In the case of clay minerals, polyvalent ions tend to exchange for nonvalent ions.

Iron Bacteria - Bacteria which can oxidize or reduce iron as part of their metabolic process.

Irrigation Return Flow - Irrigation water which is not consumed in evaporation or plant growth, and which returns to a surface stream or ground-water reservoir.

Leachate - The liquid that has percolated through solid waste or other man-emplaced medium from which soluble components have been removed.

Loading Rate - The rate of application of a material to the land surface.

Mined Ground Water - Water removed from storage when pumpage exceeds ground-water recharge.

Mineralization - Increases in concentration of one or more constituents as the natural result of contact of ground water with geologic formations.

Monitoring (Observation) Well - A well used to measure ground-water levels, and in some cases, to obtain water samples for water-quality analysis.

Nonpoint Source - The contaminant enters the receiving water in an intermittent and/or diffuse manner.

Organic - Being, containing, or relating to carbon compounds, especially in which hydrogen is attached to carbon, whether derived from living organisms or not; usually distinguished from inorganic or mineral.

Overburden - All material (loose soil, sand, gravel, etc.) that lies above bedrock. In mining, any material, consolidated or unconsolidated, that overlies an ore body, especially deposits mined from the surface by open cuts.

Oxidation - A chemical reaction in which there is an increase in valence resulting from a loss of electrons; in contrast to reduction.

Percolate - The water moving by gravity or hydrostatic presure through interstices of unsaturated rock or soil.

Percolation - Movement of percolate under gravity or hydrostatic pressure.

Perennial Stream - One which flow continuously. Perennial streams are generally fed in part by ground water.

Permeability - A measure of the capacity of a porous medium to transmit fluid.

Piezometric Surface - The surface defined by the levels to which ground water will rise in tightly cased wells that tap an artesian aquifer.

Plume - A body of contaminated ground water originating from a specific source and influenced by such factors as the local ground-water flow pattern, density of contaminant, and character of the aquifer.

Point Source - Any discernible, confined and discrete conveyance, including but not limited to any pipe, ditch, channel, tunnel, conduit, well, discrete fissure, container, rolling stock, or concentrated animal feeding operation from which contaminants are or may be discharged.

Potentiometric Surface - Used synonymously with piezometric surface.

Public Water Supply - A system in which there is a purveyor and customers; the purveyor may be a private company, a municipality, or other governmental agency.

Recharge - The addition of water to the ground-water system by natural or artificial processes.

Reduction - A chemical reaction in which there is a decrease in valence as a result of gaining of electrons.

Runoff - Direct or overland runoff is that portion of rainfall which is not absorbed by soil, evaporated or transpired by plants, but finds its way into streams as surface flow. That portion which is absorbed by soil and later discharged to surface streams is ground-water runoff.

Salaquifer - An aquifer which contains saline water.

Saline - Containing relatively high concentration of salts.

Salt-Water Intrusion - Movement of salty ground water so that it replaces fresh ground water.

Saturated Zone - The zone in which interconnected interstices are saturated with water under pressure equal to or greater than atmospheric.

Self-Supplied Industrial and Commercial Water Supply - A system from which water is served to consumers free of charge, or from which water is supplied by the operator of the system for his own use.

Sludge - The solid residue resulting from a process or wastewater treatment which also produces a liquid stream (effluent).

Specific Conductance - The ability of a cubic centimetre of water to conduct electricity; varies directly with the amount of ionized minerals in the water.

Storage (Aquifer) - The volume of water held in the interstices of the rock.

Strata - Beds, layers, or zones of rock.

Subsidence - Surface caving or distortion brought about by collapse of deep mine workings or cavernous carbonate formations, or from overpumping of certain types of aquifers.

Surface Resistivity (Electric Resistivity Surveying) - A geophysical prospecting operation in which the relative values of the earth's electrical resistivity are interpreted to define subsurface geologic and hydrologic conditions.

Surface Water - That portion of water that appears on the land surface, i.e., oceans, lakes, rivers.

Toxicity - The ability of a material to produce injury or disease upon exposure, ingestion, inhalation or assimilation by a living organism.

Transmissivity - The rate at which water is transmitted through a unit width of an aquifer under a unit hydraulic gradient.

Unsaturated Zone (Zone of Aeration) - Consists of interstices occupied partially by water and partially by air, and is limited above by the land surface and below by the water table.

Upconing - The upward migration of ground water from underlying strata into an aquifer caused by reduced hydrostatic pressure in the aquifer as a result of pumping.

Water Table - That surface in an unconfined ground-water body at which the pressure is atmospheric. It defines the top of the saturated zone.

Water-Table Aquifer - An aquifer containing water under atmospheric conditions.

Well - An artificial excavation that derives fluid from the interstices of the rocks or soils which it penetrates, except that the term is not applied to ditches or tunnels that lead ground water to the surface by gravity. With respect to the method of construction, wells may be divided into dug wells, bored wells, drilled wells, and driven wells.

Well Capacity - The rate at which a well will yield water.

Withdrawal - The volume of water pumped from a well or wells.

SECTION XIX

APPENDIX B - ABBREVIATIONS

10^3	-	thousand
10^6	-	million
10^{-6}	-	millionth
acre-ft	-	acre-foot
bbl	-	barrels (oil, 42 gal.)
bgd	-	billion gallons per day
BOD	-	biochemical oxygen demand
Btu	-	British thermal unit
Ca	-	calcium
$CaCO_3$	-	calcium carbonate
CaO	-	lime
cap	-	capita
Cd	-	cadmium
CEC	-	cation exchange capacity
Ci	-	curie
Cl	-	chloride
cm	-	centimetre
CN	-	cyanide
COD	-	chemical oxygen demand
Cu	-	copper
cu ft	-	cubic foot
cu m	-	cubic metre
cu yd	-	cubic yard
CWT	-	hundredweight
Fe	-	iron
ft	-	foot
g	-	gram
gal.	-	gallon
gpcd	-	grams per capita per day
gpd	-	gallons per day
gpm	-	gallons per minute
H_2SO_4	-	sulfuric acid
H_3PO_4	-	phosphoric acid
ha	-	hectare
HCl	-	hydrochloric acid
HNO_3	-	nitric acid
HOAc	-	acetic acid
in.	-	inch
K	-	potassium
K_2O	-	potash
kg	-	kilogram
km	-	kilometre
l	-	litre
lb	-	pound
l/s	-	litre per second

m	-	metre
MBAS	-	methylene blue active substances
Mg	-	magnesium
mgd	-	million gallons per day
mg/l	-	milligrams per litre
$Mg(OH)_2$	-	magnesium hydroxide
mi	-	mile
min	-	minute
ml	-	millilitre
Mn	-	manganese
MPN	-	most probable number
N	-	nitrogen
Na	-	sodium
Na_2CO_3	-	soda ash
NH_3	-	ammonia
NH_4	-	ammonium
Ni	-	nickel
NO_3	-	nitrate
ORG-N	-	organic nitrogen
P	-	phosphorus
P_2O_5	-	formula on which phosphorus content in fertilizer is based
Pb	-	lead
PCB	-	polychlorinated biphenyls
pCi	-	picocurie
PO_4	-	phosphate
ppb	-	parts per billion
ppm	-	parts per million
psi	-	pounds per square inch
pvc	-	polyvinyl chloride
s	-	second
SiO_2	-	silica
SO_4	-	sulfate
sq cm	-	square centimetre
sq ft	-	square foot
sq km	-	square kilometre
sq mi	-	square mile
sq yd	-	square yard
SS	-	suspended solids
TDS	-	total dissolved solids
TOC	-	total organic carbon
TS	-	total solids
TSS	-	total suspended solids
yd	-	yard
yr	-	year
Zn	-	zinc

SECTION XIX

APPENDIX C - CONVERSIONS

To convert	Into	Multiply by
acre	hectare	0.4047
acre-feet	cubic feet	43,560.0
acre-feet	cubic metres	1,234
barrels (oil)	cubic metres	0.159
barrels (oil)	gallons (oil)	42.0
btu	kilogram-calories	0.2520
centimetres	metres	0.01
cubic feet	cubic metres	0.0283
cubic yards	cubic metres	0.7646
feet	centimetres	30.48
feet	metres	0.3048
gallons	cubic metres	0.0038
gallons	litres	3.785
hundredweights (short)	pounds	100
hundredweights (short)	tonnes	0.0454
inches	centimetres	2.540
micrograms	grams	1.0×10^{-6}
miles (statute)	kilometres	1.609
milligrams/litre	parts/million	1.0
million gallons	acre-feet	3.06
million gallons/day	cubic feet/second	1.5472
million gallons/day	cubic metres/second	0.0438
mils	centimetres	0.0025
mils	inches	0.001
parts per million	milligrams/litre	1.0
pounds	kilograms	0.4536
pounds/acre	kilograms/hectare	1.121
square feet	square metres	0.0929
square inches	square centimetres	6.452
square miles	square kilometres	2.590
tonnes	kilograms	1,000.0
tons (short)	tonnes	0.9078

SECTION XIX

APPENDIX D - WATER-QUALITY STANDARDS

The levels of mineralization or contamination which can be tolerated in ground water are dependent upon the intended use for the supply. Recommended water-quality standards are available for agricultural, industrial, and public-supply needs. Certain chemical constituents can be tolerated through a wide range of concentrations without adverse effects even in stringent cases requiring excellent water quality, while other constituents can be acceptable only at minute levels or not at all. The water-quality standards for any particular use are varied and in most cases well documented. It is evident that to list each and every guideline is beyond the scope of this report.

The U. S. Environmental Protection Agency has National Interim Primary Drinking Water Standards for certain constituents and is currently in the process of updating the 1962 Public Health Service recommended limits for others. The new standards and the 1962 recommended limits are shown in the two tables following.

U.S PUBLIC HEALTH SERVICE CHEMICAL STANDARDS OF DRINKING WATER, 1962.

Recommended maximum allowable concentrations where other more suitable supplies are, or can be made available:

Substance	Concentration in ppm
Alkyl Benzene Sulfonate (ABS)	0.5
Arsenic (As)	0.01
Chloride (Cl)	250
Copper (Cu)	1
Carbon Chloroform Extract (CCE)	0.2
Cyanide (CN)	0.01
Iron (Fe)	0.3
Manganese (Mn)	0.05
Phenols	0.001
Sulfate (SO_4)	250
Total Dissolved Solids (TDS)	500
Zinc (Zn)	5

U.S. ENVIRONMENTAL PROTECTION AGENCY NATIONAL INTERIM PRIMARY DRINKING WATER STANDARDS, DECEMBER, 1975.

Maximum contaminant level which shall constitute grounds for outright rejection of the supply:

Substance	Concentration in ppm
Arsenic (As)	0.05
Barium (Ba)	1
Cadmium (Cd)	0.010
Total Chromium (Cr)	0.05
Fluoride (F)	1.4 to 2.4 a)
Lead (Pb)	0.05
Mercury (Hg)	0.002
Nitrate (as N)	10
Selenium (Se)	0.01
Silver (Ag)	0.05

a) Varies with annual average of maximum daily air temperature.

SECTION XIX

APPENDIX E - ESTIMATED NUMBER OF FACILITIES, VOLUMES OF WASTE, AND LEAKAGE TO GROUND WATER.

Waste disposal practice	Estimated total number	Estimated total size	Estimated amount of waste	Estimated leakage to ground
Industrial impoundments				
Treatment lagoons	NA	NA	1,700 bgy	100 bgy
All impoundments	50,000	NA	NA	NA
Land disposal of solid wastes				
Municipal	18,500	500,000 acres	135 mty	90 bgy
Industrial	NA	NA	NA	NA
Septic tanks and cesspools				
Domestic	16,600,000	-	800 bgy	800 bgy
Industrial	25,000	-	NA	NA
Municipal waste water				
Sewer systems	12,000	470,000 mi	5,000 bgy	250 bgy
Treatment lagoons	10,000	20,000 acres	300 bgy	18 bgy
Land spreading of sludge				
Municipal	NA	NA	NA	NA
Manufacturing	NA	NA	NA	NA
Petroleum exploration and development				
Wells	60,000	-	260 bgy	260 bgy*
Pits	NA	NA	43 bgy	43 bgy
Mine waste				
Coal				
Waste water	277	-	77 bgy	8 bgy
Solid waste	NA	173,000 acres	100 mty	600 m lbs/y acid
Other	691	-	860 bgy	100 bgy
Disposal and injection wells				
Agricultural, urban run-off, cooling water and sewage disposal wells	40,000	-	NA	NA
Industrial and municipal injection wells	< 400	-	NA	NA
Animal feeding operations				
Cattle	140,000	50,000 acres	83 mty	NA
Other	NA	NA	7 mty	NA

bgy - billion gallons per year
mty - million tons per year
m lbs/y - million pounds per year

\- - not applicable
* - almost all returned to salt-water aquifers
NA - insufficient data for estimate

APPENDIX E (continued) - ESTIMATED NUMBER OF FACILITIES, VOLUMES OF WASTE, AND LEAKAGE TO GROUND WATER.

EXPLANATION

INDUSTRIAL IMPOUNDMENTS

Within this category, available data make necessary the separation of secondary treatment lagoons from other impoundments such as settling ponds, pits and basins. The total number of all impoundments in the United States is estimated at 50,000. The flow to these impoundments is not known. The total flow to treatment lagoons alone is calculated at 1,700 billion gallons per year. Average leakage to ground from treatment lagoons is estimated at 6 percent. Thus, the total leakage of industrial waste water from secondary lagoons is estimated at 100 billion gallons per year.

LAND DISPOSAL OF SOLID WASTES

Municipal

An estimated 18,500 municipal solid waste land disposal sites in the U.S. cover a total area of approximately 500,000 acres (estimate based on 25 acres per site). Approximately 135 million tons of refuse per year is landfilled. The volume of leachate generated can be estimated based on average infiltration of precipitation in water surplus areas and on site size. It is estimated that 70 percent of the land disposal sites in the U.S. are in water surplus areas and that the average infiltration is 10 inches per year. Thus, municipal sites would generate a total of 90 billion gallons of leachate per year, most of which goes into the ground.

Industrial

The number of and typical size of industrial solid waste land disposal sites are unknown. A large portion of industrial solid waste, including that which is considered hazardous, presently goes to municipal solid waste land disposal sites.

SEPTIC TANKS AND CESSPOOLS

Domestic

There were an estimated 16,600,000 septic tanks and cesspools in the U.S. in 1970. Annual flow to a septic tank or cesspool from an average house is 49,275 gallons (45 gpd/person x 3 persons/house x 365 days/year). Thus, the total flow to septic tanks and cesspools in the U.S. is about 800 billion gallons per year, virtually all of which enters the ground.

Industrial

It is estimated that about 25,000 industrial septic tanks are currently in use, based on the number of industrial establishments in the U.S. using water. However, little information is available regarding flow rates to these systems and no estimate can be made as to total leakage to ground.

APPENDIX E (continued) - ESTIMATED NUMBER OF FACILITIES, VOLUMES OF WASTE, AND LEAKAGE TO GROUND WATER.

MUNICIPAL WASTE WATER

Sewer Systems

There are currently about 12,000 sewer systems in the United States using approximately 470,000 miles of pipe. Approximately 144 million persons were served by sewer facilities in 1970. Based on an estimated 100 gpd/person sewerage flow (including infiltration-inflow, combined sewer flow, illegal drain hook-ups and industrial waste flow to sanitary sewer lines), the total sewerage flow in sanitary sewers is estimated at 5,000 billion gallons per year. Based on available information, sewer leakage on the average is probably around 5 percent of the total, with wide variations from system to system. Thus, the total leakage for all sewer lines in the U.S. is estimated at 250 billion gallons per year.

Treatment Lagoons

There are approximately 5,000 municipal treatment plants in the U.S. which use lagoons as a treatment procedure. Assuming each plant has an average of two lagoons, there would be about 10,000 municipal treatment lagoons. Assuming each lagoon is roughly two acres in size, there would be about 20,000 acres of municipal lagoons in the country.

Municipal treatment plants using treatment lagoons receive an inflow of approximately 300 billion gallons per year. Leakage from these lagoons is estimated at 6 percent. Thus, it is estimated that municipal lagoons leak 18 billion gallons per year into the ground.

LAND SPREADING OF SLUDGE

Municipal

The number and average size of sludge spreading operations for municipalities is not known. It has been calculated that about 4 million dry tons of municipal sludge is generated each year. How much of this quantity is land spread and how much goes to solid waste land disposal sites and lagoons are unknown.

Industrial

The manufacturing sludge category includes effluent treatment sludge, stack scrubber residue, fly and bottom ash, slag and numerous other manufacturing residues. The total number of industrial sludge spreading sites and typical sizes are unknown. Most industrial sludge presently goes to solid waste land disposal sites and lagoons.

APPENDIX E (continued) - ESTIMATED NUMBER OF FACILITIES, VOLUMES OF WASTE, AND LEAKAGE TO GROUND WATER.

PETROLEUM EXPLORATION AND DEVELOPMENT

Disposal Wells

An estimated 60,000 brine injection wells are in use in the U.S. The total estimated brine disposal is 260 billion gallons per year. Almost all goes into salt water aquifers. The volume which finds its way into fresh aquifers is unknown.

Pits and Basins

An estimated 13 billion gallons per year of oil field brine is disposed of into pits and basins, most of which enters the ground, usually a fresh water aquifer.

MINE WASTE

Coal

Waste Water -

The volume of waste water discharged by the 277 coal mining establishments reporting water consumption in 1972, including processing water and collected mine drainage, totalled 77 billion gallons. The volume of this waste water which entered the ground is not known, but based on the typical geology of the major coal mining regions and what is known about disposal practices, it is estimated at roughly 10 percent, or 8 billion gallons.

Solid Waste -

Between 1930 and 1971, almost 200,000 acres have been used for the disposal of coal mining wastes. Of this area, only 27,000 have been reclaimed. A study in Illinois found that each acre of unreclaimed coal waste could generate 198 lb of acidity (as $CaCO_3$) per day. Assuming half the total acreage of refuse was still producing acid, about 3.6 million tons of acid would be generated each year. On comparison of the location of coal waste dumps with ground-water aquifer types, it is estimated that approximately 10 percent of this total, 600 million lbs of acid/year, might enter the ground-water system.

Other

There were about 1,300 active mines (excluding coal, clay, sand, and stone mines) in the U.S. in 1972. The total solid waste from these mines include some 1.5 billion tons of waste rock plus a large volume of other waste materials from various processing procedures. Of the total number of mines, 691 reported substantial water use in 1972. Total waste water discharged was about 860 billion gallons for that year. A rough estimate of the portion of this volume which would have entered aquifer systems is 10 percent or about 100 billion gallons.

APPENDIX E (continued) - ESTIMATED NUMBER OF FACILITIES, VOLUMES OF WASTE, AND LEAKAGE TO GROUND WATER.

DISPOSAL AND INJECTION WELLS

An estimated 40,000 agricultural, urban runoff and sewage disposal wells are in current use in the U.S. The volume of waste water injected into the ground cannot be estimated. In addition, there are less than 300 industrial injection wells currently in use. The volume of waste injected through these wells is not known.

ANIMAL FEEDING OPERATIONS

Cattle

There are currently about 140,000 cattle feeding operations covering some 50,000 acres in the United States. The total waste deposited in these feeding operations was estimated at 83 million tons in 1975. There are insufficient data in the literature to allow a reasonable estimate of the volume of contaminated runoff from these feedlots which enters the ground.

Other

Very little data are available on the effects of other types of feeding operations, such as sheep, hogs and chickens, on ground-water quality. It is estimated that the total volume of waste from these three sources is 7 million tons per year.